COMPENSATING FOR WETLAND LOSSES UNDER THE CLEAN WATER ACT

Committee on Mitigating Wetland Losses

Board on Environmental Studies and Toxicology

Water Science and Technology Board

Division on Earth and Life Studies

National Research Council

NATIONAL ACADEMY PRESS
Washington, D.C.

NATIONAL ACADEMY PRESS 2101 Constitution Avenue, N.W. Washington, D.C. 20418

NOTICE: The project that is the subject of this report was approved by the Governing Board of the National Research Council, whose members are drawn from the councils of the National Academy of Sciences, the National Academy of Engineering, and the Institute of Medicine. The members of the committee responsible for the report were chosen for their special competences and with regard for appropriate balance.

This project was supported by Cooperative Agreement No. C X 827828-01-0 between the National Academy of Sciences and the U.S. Environmental Protection Agency. Any opinions, findings, conclusions, or recommendations expressed in this publication are those of the author(s) and do not necessarily reflect the view of the organizations or agencies that provided support for this project.

Library of Congress Cataloging-in-Publication Data

Compensating for wetland losses under the Clean Water Act / Committee on Mitigating Wetland Losses, Board on Environmental Studies and Toxicology, Water Science and Technology Board, Division on Earth and Life Studies, National Research Council.
    p. cm.
Includes bibliographical references (p.   ).
  ISBN 0-309-07432-0 (hardcover)
  1. Wetlands—Law and legislation—United States. 2. Wetland conservation—Government policy—United States. 3. Wetland mitigation banking—United States. I. National Research Council (U.S.). Committee on Mitigating Wetland Losses.
  KF5624 .C66 2001
  346.7304'6918--dc21
                  2001004921

Compensating for Wetland Losses Under the Clean Water Act is available from the National Academy Press, 2101 Constitution Avenue, N.W., Box 285, Washington, DC 20055; (800) 624-6242 or (202) 334-3313 (in the Washington metropolitan area); Internet: http://www.nap.edu

THE NATIONAL ACADEMIES

National Academy of Sciences
National Academy of Engineering
Institute of Medicine
National Research Council

The **National Academy of Sciences** is a private, nonprofit, self-perpetuating society of distinguished scholars engaged in scientific and engineering research, dedicated to the furtherance of science and technology and to their use for the general welfare. Upon the authority of the charter granted to it by the Congress in 1863, the Academy has a mandate that requires it to advise the federal government on scientific and technical matters. Dr. Bruce M. Alberts is president of the National Academy of Sciences.

The **National Academy of Engineering** was established in 1964, under the charter of the National Academy of Sciences, as a parallel organization of outstanding engineers. It is autonomous in its administration and in the selection of its members, sharing with the National Academy of Sciences the responsibility for advising the federal government. The National Academy of Engineering also sponsors engineering programs aimed at meeting national needs, encourages education and research, and recognizes the superior achievements of engineers. Dr. Wm. A. Wulf is president of the National Academy of Engineering.

The **Institute of Medicine** was established in 1970 by the National Academy of Sciences to secure the services of eminent members of appropriate professions in the examination of policy matters pertaining to the health of the public. The Institute acts under the responsibility given to the National Academy of Sciences by its congressional charter to be an adviser to the federal government and, upon its own initiative, to identify issues of medical care, research, and education. Dr. Kenneth I. Shine is president of the Institute of Medicine.

The **National Research Council** was organized by the National Academy of Sciences in 1916 to associate the broad community of science and technology with the Academy's purposes of furthering knowledge and advising the federal government. Functioning in accordance with general policies determined by the Academy, the Council has become the principal operating agency of both the National Academy of Sciences and the National Academy of Engineering in providing services to the government, the public, and the scientific and engineering communities. The Council is administered jointly by both Academies and the Institute of Medicine. Dr. Bruce M. Alberts and Dr. Wm. A. Wulf are chairman and vice chairman, respectively, of the National Research Council.

COMMITTEE ON MITIGATING WETLAND LOSSES

Staff

SUZANNE VAN DRUNICK, Study Director
RUTH CROSSGROVE, Editor
BARBARA O'HARE, Editor
MIRSADA KARALIC-LONCAREVIC, Information Specialist
LEAH PROBST, Senior Project Assistant
JENNIFER SAUNDERS, Project Assistant
JAMIE YOUNG, Research Associate

OTHER REPORTS OF THE BOARD ON ENVIRONMENTAL STUDIES AND TOXICOLOGY

A Risk-Management Strategy for PCB-Contaminated Sediments (2001)
Toxicological Effects of Methylmercury (2000)
Strengthening Science at the U.S. Environmental Protection Agency: Research-Management and Peer-Review Practices (2000)
Scientific Frontiers in Developmental Toxicology and Risk Assessment (2000)
Modeling Mobile-Source Emissions (2000)
Toxicological Risks of Selected Flame-Retardant Chemicals (2000)
Copper in Drinking Water (2000)
Ecological Indicators for the Nation (2000)
Waste Incineration and Public Health (1999)
Hormonally Active Agents in the Environment (1999)
Research Priorities for Airborne Particulate Matter: I. Immediate Priorities and a Long-Range Research Portfolio (1998); II. Evaluating Research Progress and Updating the Portfolio (1999); III. Early Research Progress (2001)
Ozone-Forming Potential of Reformulated Gasoline (1999)
Risk-Based Waste Classification in California (1999)
Arsenic in Drinking Water (1999)
Brucellosis in the Greater Yellowstone Area (1998)
The National Research Council's Committee on Toxicology: The First 50 Years (1997)
Toxicologic Assessment of the Army's Zinc Cadmium Sulfide Dispersion Tests (1997)
Carcinogens and Anticarcinogens in the Human Diet (1996)
Upstream: Salmon and Society in the Pacific Northwest (1996)
Science and the Endangered Species Act (1995)
Wetlands: Characteristics and Boundaries (1995)
Biologic Markers (5 reports, 1989-1995)
Review of EPA's Environmental Monitoring and Assessment Program (3 reports, 1994-1995)
Science and Judgment in Risk Assessment (1994)
Ranking Hazardous Waste Sites for Remedial Action (1994)
Pesticides in the Diets of Infants and Children (1993)
Issues in Risk Assessment (1993)
Setting Priorities for Land Conservation (1993)
Protecting Visibility in National Parks and Wilderness Areas (1993)
Dolphins and the Tuna Industry (1992)
Hazardous Materials on the Public Lands (1992)

Science and the National Parks (1992)

Animals as Sentinels of Environmental Health Hazards (1991)

Assessment of the U.S. Outer Continental Shelf Environmental Studies
    Program, Volumes I-IV (1991-1993)

Human Exposure Assessment for Airborne Pollutants (1991)

Monitoring Human Tissues for Toxic Substances (1991)

Rethinking the Ozone Problem in Urban and Regional Air Pollution
    (1991)

Decline of the Sea Turtles (1990)

Copies of these reports may be ordered from
the National Academy Press
(800) 624-6242
(202) 334-3313
www.nap.edu

Acknowledgments

Many individuals assisted the committee and National Research Council (NRC) staff in their task to create this report. We are especially grateful for the outstanding assistance provided by Lisa Morales, U.S. Environmental Protection Agency. We are also appreciative of the generous support provided by John Goodin, U.S. Environmental Protection Agency; John Studt, U.S. Army Corps of Engineers; Robert Brumbaugh, U.S. Army Corps of Engineers; Benjamin Tuggle, U.S. Fish and Wildlife Service; Thomas Bigford, National Marine Fisheries Service; Kathryn Conant, National Marine Fisheries Service; and Susan Marie Stedman, National Marine Fisheries Service.

Field trips held in conjunction with committee meetings helped the committee better understand the complexities of mitigating wetland losses. We would like to express our appreciation to the following people, who assisted the committee and NRC staff during these field trips:

Washington, DC
Michael Bean, Environmental Defense Fund
George Beston, Maryland Department of the Environment
Denise Clearwater, Maryland Department of the Environment
Timothy Searchinger, Environmental Defense Fund
Julie Sibbing, National Audubon Society

Orlando, Florida
William Ainslie, U.S. Environmental Protection Agency
William Barnard, Greater Orlando Aviation Authority
Constance Bersok, Florida Department of Environmental Protection

Stephen Brooker, U.S. Army Corps of Engineers
Edwin Edmundson, South Florida Water Management District
Mark Evans, U.S. Army Corps of Engineers
Monica Folk, The Nature Conservancy
Paul Gray, Audubon of Florida
Kathleen S. Hale, Environmental Management and Design
Michael Norland, National Park Service
Robert Robbins, South Florida Water Management District
William Streever, U.S. Army Corps of Engineers
Edward Swakon, EAS Engineering, Inc.
Jora Young, The Nature Conservancy

Northbrook, Illinois
Sue Elston, U.S. Environmental Protection Agency
Jeanette Gallihugh, U.S. Fish and Wildlife Service
Donald Hey, Wetlands Research, Inc.
Kerry Leigh, Christopher B. Burke Engineering, Ltd.
Michael Miller, Illinois State Geologic Survey
James Minor, Illinois State Geologic Survey
Charles Paine, Max McGraw Wildlife Foundation
James Robb, Indiana Department of Environmental Management
Joseph Roth, Corlands
John Ryan, Land and Water Resources, Inc.
David Siebert, Wisconsin Department of Natural Resources

Irvine, California
Rich Ambrose, University of California, Los Angeles
Gerhard Bombe, Orange County Parks and Recreation Department
Anthony Bomkamp, Glenn Lukos & Associates
Peter Bowler, University of California, Irvine
William Bretz, University of California Natural Reserve System, San
    Joaquin Marsh Reserve
Richard Broming, Rancho Mission Viejo
Mary Kentula, U.S. Environmental Protection Agency and National
    Health and Environmental Effects Research Laboratory
Victor Leipzig, Huntington Beach Wetlands Conservancy
Molly Martindale, U.S. Army Corps of Engineers
Thomas Mulroy, Science Applications International Corporation
Eric Stein, PCR Environmental, Inc.
Mark Sudol, U.S. Army Corps of Engineers, Los Angeles District
Sat Tamaribuchi, The Irvine Company
Sherry Teresa, Center for Natural Lands Management
Kenneth Thompson, Irvine Ranch Water District
William Tippets, California Department of Fish and Game, South
    Coast Region

Acknowledgment of Review Participants

This report has been reviewed in draft form by individuals chosen for their diverse perspectives and technical expertise, in accordance with procedures approved by the NRC's Report Review Committee. The purpose of this independent review is to provide candid and critical comments that will assist the institution in making its published report as sound as possible and to ensure that the report meets institutional standards for objectivity, evidence, and responsiveness to the study charge. The review comments and draft manuscript remain confidential to protect the integrity of the deliberative process. We wish to thank the following individuals for their review of this report:

Donald Hey, Wetlands Research, Inc.
Thomas Kelsch, National Fish and Wildlife Foundation
Joseph S. Larson, Environmental Institute
Jay A. Leitch, North Dakota State University
Patrick O'Brien, Chevron Research and Technology Company
Rutherford Platt, University of Massachusetts-Amherst
Timothy Searchinger, Environmental Defense Fund
Donald Siegel, Syracuse University
Margaret Strand, Oppenheimer, Wolff & Donnelly, LLP

Although the reviewers listed above have provided many constructive comments and suggestions, they were not asked to endorse the conclusions or recommendations, nor did they see the final draft of the report before its release. The review of this report was overseen by John Cairns, Jr., Virginia Polytechnic Institute and State University, and William M. Lewis, Jr., University of Colorado. Appointed by the National Research Council, they were responsible for making certain that an independent examination of this report was carried out in accordance with institutional procedures and that all review comments were carefully considered. Responsibility for the final content of this report rests entirely with the authoring committee and the institution.

Preface

The U.S. Army Corps of Engineers (Corps) and the U.S. Environmental Protection Agency (EPA) share responsibility for regulating the mitigation (lessening of impacts) of damages to wetlands. In response to a request from EPA, the National Research Council (NRC) formed the Committee on Mitigating Wetland Losses to evaluate mitigation practice as a way to restore and maintain the quality of the nation's waters, particularly as regulated under Section 404 of the Clean Water Act.

The committee reviewed the available literature on replacement of wetland functions, considered both restoration and creation efforts, visited several mitigation sites around the United States, and then evaluated both the ecological performance of mitigation projects and the institutions under which mitigation projects are conducted (permittee-responsible mitigation banks and in-lieu fee programs). At a series of five meetings, the committee worked in a truly interdisciplinary and collaborative manner to develop the conclusions and recommendations presented in this report.

The committee is grateful for the briefings and the assistance provided by the staff of EPA, the Corps, the U.S. Fish and Wildlife Service, and the National Marine Fisheries Service.

The committee is also grateful for the excellent and untiring support provided by the NRC staff, who organized the meetings and field trips and kept us on track in addressing the major tasks, as well as the fine details in report preparation. Dr. Suzanne van Drunick, our outstanding project director, kept the process on track and made sure that the report was coherent. We all benefited greatly from the help of Jennifer Saunders, who followed Leah Probst as project assistant. Ruth Crossgrove, Mirsada Karalic-Loncarevic, and Barbara O'Hare helped with the many details that made the report ready for publication. Dr. David Policansky initiated the project, and we thank him for providing stimulating discussions. Dr. James Reisa's suggestions improved the Executive Summary.

The committee members were exemplary in their dedication to this complicated task; without their expertise, hard work, and timely responses, completion of the project would not have been possible.

Joy B. Zedler
Chair, Committee on Mitigating Wetland Losses

Contents

EXECUTIVE SUMMARY   1

1   INTRODUCTION   11
    Important Terms, 13
    No Net Loss and the Section 404 Program, 16
    The Committee's Task, 20

2   OUTCOMES OF WETLAND RESTORATION
        AND CREATION   22
    Introduction, 22
    Five Wetland Functions, 27
    Factors That Contribute to the Performance of
        Mitigation Sites, 35
    Recommendations, 45

3   WATERSHED SETTING   46
    Watershed Organization and Landscape Function, 46
    Wetland Function and Position in the Watershed, 47
    Watershed-Scale Patterns of Wetland Losses, 57
    A Watershed Template for Wetland Restoration
        and Conservation, 58
    Conclusions, 59
    Recommendations, 59

4   WETLAND PERMITTING: HISTORY AND OVERVIEW   60
    Evolution of Compensatory Mitigation Requirements in the
        CWA Section 404 Program, 60
    General Mitigation Requirements, 61
    General Corps Mitigation Requirements, 63
    CWA Section 404 Mitigation Requirements, 64
    Mitigation Banking, 67
    In-Lieu Fees, 69
    The Clean Water Act and the Goal of No Net Loss, 70
    Section 404 Permit Process, 73
    Inspection and Enforcement, 80

5   COMPENSATORY MITIGATION MECHANISMS UNDER
        SECTION 404   82
    Location of the Compensatory Mitigation Action, 83
    Legal Responsibility for the Mitigation, 86
    Relationship of Mitigation Actions to Permitted
        Activities (Timing), 88
    The MBRT Process, 91
    Stewardship Requirements, 91
    A Taxonomy, 92
    Recommendation, 93

6   MITIGATION COMPLIANCE   94
    Mitigation Planning, 95
    Mitigation Design Standards, 97
    Project Implementation, 101
    Compliance with Permit Conditions, 103
    Mitigation Ratios, 108
    Monitoring of Mitigation Projects, 110
    Monitoring Duration, 112
    The Compliance Record, 113
    Conclusions, 121
    Recommendations, 122

7   TECHNICAL APPROACHES TOWARD ACHIEVING NO
        NET LOSS   123
    Operational Guidelines for Creating or Restoring Wetlands
        That Are Ecologically Self-Sustaining, 123
    Wetland Functional Assessment, 128
    The Floristic Approach, 129

Habitat Evaluation Procedures and the Hydrogeomorphic
    Approach, 131
HGM as a Functional Assessment Procedure, 132
Recommendations, 136

8   INSTITUTIONAL REFORMS FOR ENHANCING
    COMPENSATORY MITIGATION   138
Introduction, 138
A Watershed-Based Approach to Compensatory
    Mitigation, 140
Improvements in Permittee-Responsible Mitigation, 149
Expectations for the Regulatory Agency, 154
Third-Party Mitigation, 160
Support for Increased State Responsibilities, 165
Recommendations, 166

REFERENCES   169

APPENDIXES

A   Survey of Studies: Comparison of Mitigation and
    Natural Wetlands   189

B   Case Studies   199
Everglades National Park, 199
Coyote Creek Mitigation Site, 201
North Carolina Wetland Restoration Program, 208

C   Analyses of Soil, Plant, and Animal Communities
    for Mitigation Sites Compared With Reference Sites   211

D   California Department of Fish and Game, South Coast
    Region; Guidelines for Wetland Mitigation   217

E   Examples of Performance Standards for Wetland Creation
    and Restoration in Section 404 Permits and an Approach
    to Developing Performance Standards   219

F   Memorandum for Commanders, Major Subordinate
    Commands, and District Commands, April 8, 1999   234

G Army Corps of Engineers Standard Operating Procedures
for the Regulatory Program 239

H Selected Attributes of 40 Common Wetland Functional
Assessment Procedures 285

I Function, Factors, and Values Considered in Section 404
Permit Reviews 292

J Biographical Sketches of Committee Members 294

GLOSSARY 299

INDEX 305

Tables and Figures

FIGURES

FIGURE 1-1   Area of wetland impacts permitted, mitigation required by the permit, and the anticipated gain in wetland area as a result of permits issued by the U.S. Army Corps of Engineers regulatory program from 1993 to 2000, 19

FIGURE 2-1   Percent plant cover on created or restored coastal wetlands on the Atlantic and Gulf of Mexico (GOM) coasts, 41

FIGURE 2-2   Long-term data for salt marshes constructed in San Diego Bay, 43

FIGURE 3-1   Comparison between observed and DRAINMOD (hydrological model) simulated water-table depths for a wetland restoration site in Craven County, N.C., 1996, 55

FIGURE 4-1   Mitigation sequencing, 66
FIGURE 4-2   Section 404 of the CWA permit process flow chart, 75
FIGURE 4-3   Approach to the nationwide permit process, 77
FIGURE 4-4   U.S. Army Corps of Engineers enforcement chart for inspection and noncompliance, 81

FIGURE 6-1   Water-table position and duration of root zone saturation
              for wetland site that satisfies the jurisdictional hydrology
              criteria (5% of growing season) as compared with wetland
              site that satisfies the criteria (12% of the growing season),
              105
FIGURE 6-2   Year-to-year variations in water-table depth and duration
              of root zone saturation for a wetland site that satisfies ju-
              risdictional hydrology criteria at least 5% of the growing
              season, 106
FIGURE 6-3   Year-to-year variation of the longest period that wetland
              hydrological criteria satisfied. Results obtained from long-
              term simulation modeling using DRAINMOD, 107

FIGURE B-1   Conceptual model of factors facilitating the invasion of
              Schinus terebinthifolius, 200

TABLES

TABLE 1-1   Wetland Losses Due to Agricultural and Nonagricultural
              Causes, 17
TABLE 1-2   Percent Loss by Cause and Acres Lost, 18

TABLE 2-1   Summary of Results from Study of a Created Salt Marsh
              Constructed as a Mitigation Site in North Carolina (1991),
              42
TABLE 2-2   Time Toward Equivalency for Soil, Plant, and Animal
              Components in Wetland Restoration Projects Compared
              with That of Natural Reference Wetlands, 42

TABLE 4-1   Listing of Current Nationwide Permits, 78

TABLE 5-1   Taxonomy of Compensatory Mitigation Mechanisms, 84

TABLE 6-1   Required Mitigation as Restoration, Creation, and Enhance-
              ment for Permits Issued under Permitting Programs, 96
TABLE 6-2   Review of Corps Permits Issued Nationwide, 98
TABLE 6-3   Mitigation Initiated for Permits Requiring Mitigation, 102
TABLE 6-4   Parameters Measured in 110 Compensatory Wetland Miti-
              gation Projects in California from 1988 to 1995, 107
TABLE 6-5   Mitigation Ratios Required and the Actual Ratios Met,
              Based on Post-Construction Evaluation (assumes complete
              compliance in meeting permit conditions), 109

TABLE 6-6   Mitigation Ratios (Area Basis) and Achievement Rates (%) for Different Wetland Types in Southern California, 110

TABLE 6-7   Frequency of Monitoring for Permits That Required Mitigation, 111

TABLE 6-8   Permit Requirements and Compliance for Five Replacement Wetlands Investigated in Ohio, 114

TABLE 6-9   Index of Functional Equivalency for Four Constructed Salt Marshes in Relationship to Natural Sites in Paradise Creek, Southern California, 115

TABLE 6-10   Ecological Parameters in Paired Replacement and Reference Wetlands in Massachusetts, 116

TABLE 6-11   Comparison of the Percentage of Permits Meeting Their Requirements and Percentage of Those Permits Meeting Various Tests of Ecological Functionality or Viability, 117

TABLE 6-12   Compliance (Based on Permit Number) for When the Mitigation Plan Was Fully Implemented, 118

TABLE 6-13   Compliance (Area Basis) for Mitigation That Was Attempted Based on Field Inspection or Monitoring Reports, 119

TABLE 6-14   Ranking of Compliance for 30 Sites in San Francisco Bay That Were Issued Section 404 Permits, 120

TABLE 6-15   Results from an Analysis of Compliance for 17 Mitigation Projects with Field Investigation in Western Washington, 120

TABLE 6-16   Summary of Data from Previous Tables on Wetland Permit Implementation, Compliance, Ecological Success, and Monitoring Frequency, 121

TABLE A-1   Survey of Studies: Comparison of Mitigated and Natural Wetlands, 190

TABLE C-1   Analysis of Soil, Plant, and Animal Communities for Mitigation Sites Compared With Reference Sites, 212

TABLE E-1   Summary of Performance Standards from Selected Section 404 Permits Requiring Compensatory Mitigation, 222

TABLE H-1   Selected Attributes of 40 Common Wetland Functional Assessment Procedures, 286

COMPENSATING FOR WETLAND LOSSES UNDER THE CLEAN WATER ACT

Executive Summary

Wetlands are complex ecosystems that, depending on their type and on circumstances within a watershed, can improve water quality, provide natural flood control, diminish droughts, recharge groundwater aquifers, and stabilize shorelines. They often support a wide variety of plants and animals, including rare and endangered species, migratory birds, and the young of commercially valuable fishes. Their beauty and diversity contribute recreational value.

The current high regard for wetlands, however, contrasts with earlier practices of draining and filling prior to the mid-1970s. Some past federal policies encouraged wetland conversion to promote agricultural, commercial, and residential development; mosquito control; and other activities that benefited society. By the 1980s the wetland area in the contiguous United States had decreased to approximately 53% of what it had been in the 1780s.

In recent years, concern about the loss of wetlands in the United States has led to federal efforts to protect wetlands on both public and private lands. Provisions in the Clean Water Act especially, the Food Security Act, several court rulings, and government policies, regulations, and directives regulate discharge of pollutants to wetlands and the filling of wetlands.

A principal objective of the Clean Water Act is "to restore and maintain the chemical, physical, and biological integrity of the Nation's waters." The U.S. Army Corps of Engineers and the U.S. Environmental Protection Agency define the "waters of the United States" to include most wetlands. This interpretation recognizes that some wetlands improve water quality through nutrient cycling and sediment trapping and retention; it is based on the judgment that some goals of the Clean Water Act cannot be achieved if wetlands are not protected. Indeed, in 1989, President Bush stated that "no net loss" of wetlands was a goal of his administration, and that was reflected in interagency agreements soon afterward.

The Clean Water Act prohibits the discharge of materials, such as soil or sand, into waters of the United States, unless authorized by a permit issued under Section 404 of that act. The Corps of Engineers, or a state program approved by the Environmental Protection Agency, has authority to issue such permits and to decide whether to attach conditions to them. To achieve no net loss of wetlands within the Section 404 program, a permittee is first expected to avoid deliberate discharge of materials into wetlands and then to minimize discharge that cannot be avoided. When damages are unavoidable, the Corps of Engineers can require the permittee to provide "compensatory mitigation" as a condition of issuing a permit.

Compensatory mitigation specifically refers to restoration, creation, enhancement, and in exceptional cases, preservation of other wetlands as compensation for impacts to natural wetlands. The permit recipient, either on a permit-by-permit basis or within a single-user mitigation bank, carries out "permittee-responsible" mitigation. In third-party mitigation (i.e., commercial mitigation bank, in-lieu fee program, cash donation, or revolving fund program), another party accepts a payment from the permittee and assumes the permittee's mitigation obligation. Most compensatory mitigation has been done by permit recipients, rather than by third parties.

The Committee on Mitigating Wetland Losses, which prepared this report, was established by the National Research Council to evaluate how well and under what conditions compensatory mitigation required under Section 404 is contributing toward satisfying the overall objective of restoring and maintaining the quality of the nation's waters. The committee reviewed examples of wetland restoration and creation projects in Florida, Illinois, and southern California that were required as a condition of Section 404 permits; received briefings from outside experts; and conducted an extensive review of the scientific literature on wetlands, government data and reports, and information provided by a wide variety of experts and organizations.

THE COMMITTEE'S PRINCIPAL FINDINGS

Conclusion 1: The goal of no net loss of wetlands is not being met for wetland functions by the mitigation program, despite progress in the last 20 years.

A recent study by the U.S. Fish and Wildlife Service suggests that the rate of loss of wetland area has slowed over the past decade. From 1986 to 1997, the estimated annual rate of wetland loss (58,545 acres per year) was about 23% that of the previous decade. Wetland losses due to agriculture declined precipitously, and there were significant reductions in losses due to urban and rural development. The decrease in wetland loss due to development may be attributable to the 404 permit process; however, the available data are not sufficient for drawing a firm conclusion.

The Corps of Engineers keeps data on the areas of permitted fill and areas of compensatory mitigation required as a condition for permits. From 1993 to 2000, approximately 24,000 acres of wetlands were permitted to be filled, and 42,000 acres were required as compensatory mitigation on an annual basis. Thus, 1.8 acres were supposed to be mitigated (i.e., gained) for every 1 acre permitted (i.e., lost). If the mitigation conditions specified in permits were actually being met, this ratio suggests that the 404 permit program could be described as resulting in a net gain in jurisdictional wetland area and function in the United States. The committee, however, found that the data available from the Corps were not adequate for determining the status of the required compensation wetlands. In addition, the data do not report the wetland functions that were lost due to the permitted fill. Further, the literature on compensatory mitigation suggests that required mitigation projects often are not undertaken or fail to meet permit conditions. Therefore, the committee is not convinced that the goal of no net loss for permitted wetlands is being met for wetland functions. The magnitude of the shortfall is not precisely known and cannot be determined from current data.

Recommendations

- The wetland area and functions lost and regained over time should be tracked in a national database. This database could include the Corps of Engineers' Regulatory Analysis and Management System database.
- The Corps of Engineers should expand and improve quality assurance measures for data entry in the Regulatory Analysis and Management System database.
- The Corps of Engineers, in cooperation with states, should encourage the establishment of watershed organizations responsible for tracking, monitoring, and managing wetlands in public ownership or under easement.

Conclusion 2: A watershed approach would improve permit decision making.

Wetland functions must be understood within a watershed framework in order to secure the purposes of the Clean Water Act. The federal guidelines for permit decision making express a strong preference for compensation as near the permitted impact site as possible and for the same wetland type and functions. The committee concluded that such a preference for on-site and in-kind mitigation should not be automatic, but should follow from an analytically based assessment of the wetland needs in the watershed and the potential for the compensatory wetland to persist over time.

On-site compensation is typically constrained by hydrological conditions that are likely to have been or are being modified by the developments requiring mitigation. Hydrological conditions, including variability in water levels and water flow rates, are the primary driving force influencing wetland development, structure, functioning, and persistence. Proper placement within the landscape of compensatory wetlands to establish hydrological equivalence is necessary for wetland sustainability. The ability to achieve desired outcomes within a specific location is also a function of the degree of degradation of the hydrological conditions, soils, vegetation, and fauna at the site. The more degraded the local site and the more degraded the watershed, the less likely it will support a high-quality project. Thus, opportunities for in-kind compensation need to be sought within a larger landscape context.

Even with a suitable position in the landscape, the ability to establish desired wetland functions will depend on the particular function, the restoration or creation approach used, and the degree of degradation at the compensation site. Landscape position, hydrological variability, species richness, biological dynamics, and hydrological regime all are important factors that affect wetland restoration and mitigation of loss. Some wetland types—in particular, fens and bogs—cannot be effectively restored with present knowledge. Mitigation efforts that do not include a proper assessment of such factors are unlikely to contribute to the goals of the Clean Water Act.

Recommendations

- Avoidance is strongly recommended for wetlands that are difficult or impossible to restore, such as fens or bogs.
- Site selection for wetland conservation and mitigation should be conducted on a watershed scale in order to maintain wetland diversity, connectivity, and appropriate proportions of upland and wetland systems needed to enhance the long-term stability of the wetland and riparian systems. Regional watershed evaluation would greatly enhance the protection of wetlands and/or the creation of wetland corridors that mimic natural distributions of wetlands in the landscape.
- All mitigation wetlands should become self-sustaining. Proper

placement in the landscape to establish hydrogeological equivalence is inherent to wetland sustainability.

• The biological dynamics should be evaluated in terms of the populations present in reference models for the region and the ecological requirements of those species.

• The science and technology of wetland restoration and creation need to be based on a broader range of studies involving sites that differ in degree of degradation, restoration efforts, and regional variations. Predictability and effectiveness of outcomes should then improve.

• Hydrological variability should be incorporated into wetland mitigation design and evaluation. Except for some open-water wetlands, static water levels are not normal. Because of climatic variability, it should be recognized that many wetland types do not satisfy jurisdictional criteria every year. Hydrological functionality should be based on comparisons to reference sites during the same time period.

• Riparian wetlands should receive special attention and protection, because their value for stream water quality and overall stream health cannot be duplicated in any other landscape position.

A mitigation site needs to have the ability to become self-sustaining. This means that the hydrological processes that define a wetland in the ecosystem need to be present and expected to persist in perpetuity. To aid regulators and mitigators in designing projects that will become ecologically self-sustaining, the committee offers 10 operational guidelines.

Operational Guidelines for Creating or Restoring Self-Sustaining Wetlands

1. Consider the hydrogeomorphic and ecological landscape and climate.
2. Adopt a dynamic landscape perspective.
3. Restore or develop naturally variable hydrological conditions.
4. Whenever possible, choose wetland restoration over creation.
5. Avoid over-engineered structures in the wetland's design.
6. Pay particular attention to appropriate planting elevation, depth, soil type, and seasonal timing.
7. Provide appropriately heterogeneous topography.
8. Pay attention to subsurface conditions, including soil and sediment geochemistry and physics, groundwater quantity and quality, and infaunal communities.
9. Consider complications associated with wetland creation or restoration in seriously degraded or disturbed sites.
10. Conduct early monitoring as part of adaptive management.

Conclusion 3: Performance expectations in Section 404 permits have often been unclear, and compliance has often not been assured nor attained.

The attainment of no net loss of wetlands through both permittee and third-party mitigation requires that performance requirements for individual compensation sites be clearly stated and that the stated requirements will be met by the parties responsible for the mitigation. Some mitigation sites studied by the committee have met the criteria for permit compliance and are, or show promise of, developing into functional wetlands. However, in many cases, even though permit conditions may have been satisfied, required compensation actions were poorly designed or carelessly implemented. In other cases, the location of the mitigation site within the watershed could not provide the necessary hydrological conditions and hence the desired plant and animal communities, including buffers and uplands, necessary to achieve the desired wetland functions.

At some sites, compliance criteria were being met, but the hydrological variability that is a defining feature of a wetland had not been established. Concern that sites might not meet hydrological criteria used to define wetlands in the permitting process often encouraged construction of permanently flooded open-water wetlands. In some situations, seasonally and intermittently flooded or saturated wetlands would have better served the needs of the watershed. Compliance criteria sometimes specified plant species that the site conditions could not support or required plantings that were unnecessary or inappropriate. Monitoring is seldom required for more than 5 years, and the description of ecosystem functions in many monitoring reports is superficial. Legal and financial mechanisms for assuring long-term protection of sites are often absent, especially for permittee-responsible mitigation.

Long-term management is especially important, because wetland restoration and creation sites seldom achieve functional equivalency with reference sites or comply with permit requirements within 5 years. Up to 20 years may be needed for some wetland restoration or creation sites to achieve functional goals. The amount of time needed to become fully functional depends on the type of wetland, its degree of degradation, conditions in the surrounding watershed, and uncertainties in the application of scientific understanding. Once wetlands become fully functional, long-term stewardship, including monitoring or periodic assessment, is critical to achieving the goals of the Clean Water Act. "Long-term stewardship" implies a time frame typically accorded to other publicly valued natural assets, such as parks. This time frame emphasizes the importance of developing mitigation wetlands that are self-sustaining, so that the long-term costs are not unmanageable. The committee recommends three general goals to ensure compliance of sites that contribute to the watershed. The committee made nine specific recommendations to achieve these goals.

General Goals

• Individual compensatory mitigation sites should be designed and constructed to maximize the likelihood that they will make an ongoing ecological contribution to the watershed; this contribution should be specified in advance.

• Compensatory mitigation should be in place concurrent with, and preferably before, permitted activity.

• To ensure the replacement of lost wetland functions, there should be effective legal and financial assurances for long-term site sustainability and monitoring of all compensatory wetland projects.

Specific Recommendations

• Impact sites should be evaluated using the same functional assessment tools as used for the mitigation site.

• Mitigation projects should be planned with and measured by a broader set of wetland functions than are currently employed.

• Mitigation goals must be clear, and those goals carefully specified in terms of measurable performance standards, in order to improve mitigation effectiveness. Performance standards in permits should reflect mitigation goals and be written in such a way that ecological viability can be measured and the impacted functions replaced.

• Because a particular floristic assemblage might not provide all the functions lost, both restoration of community structure (e.g., plant cover and composition) and restoration of wetland functions should be considered in setting goals and assessing outcomes. Relationships between structure and function should be better known.

• The Corps of Engineers and other responsible regulatory authorities should use a functional assessment protocol that recognizes the watershed perspective to establish permittee compensation requirements.

• Dependence on subjective, best professional judgment in assessing wetland function should be replaced by science-based, rapid assessment procedures that incorporate at least the following characteristics: effectively assess goals of wetland mitigation projects; assess all recognized functions; incorporate effects of position in landscape; reliably indicate important wetland processes, or at least scientifically established structural surrogates of those processes; scale assessment results to results from reference sites; are sensitive to changes in performance over a dynamic range; are integrative over space and time; and generate parametric and dimensioned units, rather than nonparametric rank.

• The Corps of Engineers and other responsible regulatory authorities should take actions to improve the effectiveness of compliance monitoring before and after project construction.

• Compensatory mitigation sites should receive long-term stewardship, i.e., a time frame expected for other publicly valued assets, such as parks.

• The Corps of Engineers and other responsible regulatory authorities should establish and enforce clear compliance requirements for permittee-responsible compensation to assure that (1) projects are initiated no later than concurrent with permitted activity, (2) projects are implemented and constructed according to established design criteria and use an adaptive management approached specified in the permit, (3) the performance standards are specified in the permit and attained before permit compliance is achieved, and (4) the permittee provides a stewardship organization with an easement on, or title to, the compensatory wetland site and a cash contribution appropriate for the long-term monitoring, management and maintenance of the site.

Conclusion 4: Support for regulatory decision making is inadequate.

In addition to using a watershed framework, the federal regulatory authorities can work to improve functional wetland assessment, permit compliance monitoring, staff training, research, and collaboration with state agencies. The committee recommends that the Corps of Engineers, Environmental Protection Agency, and other responsible regulatory authorities take several specific actions.

Recommendations

• To assist permit writers and others in making compensatory mitigation decisions, a reference manual should be developed to help design projects that will be most likely to achieve permit requirements. The manual should be organized around the themes developed in this report. The Corps of Engineers should develop such a manual for each region, based in part on the careful enumeration of wetland functions in the 404(b)(1) guidelines and in part on local and national expertise regarding the difficulty of restoring different wetland types, hydrological conditions, and functions in alternative restoration or creation contexts.

• The Corps of Engineers and other responsible authorities should commit funds to allow staff participation in professional activities and in technical training programs that include the opportunity to share experiences across districts.

• The Corps of Engineers and other responsible regulatory authorities should establish a research program to study mitigation sites to determine what practices achieve long-term performance for creation, enhancement, and restoration of wetlands.

• States, with the participation of appropriate federal agencies, are encouraged to prepare technical plans or initiate interagency consensus processes for setting wetland protection, acquisition, restoration, enhancement, and creation project priorities on an ecoregional (watershed) basis.

Conclusion 5: Third-party compensation approaches (mitigation banks, in-lieu fee programs) offer some advantages over permittee-responsible mitigation.

The committee evaluated several compensatory mitigation mechanisms and developed a taxonomy to evaluate their potential strengths and weaknesses. Mechanisms were characterized by the following five attributes: (1) on-site or off-site compensatory mitigation action; (2) responsible party; (3) timing of the mitigation actions; (4) whether the Mitigation Banking Review Team process is used; and (5) stewardship requirements. The committee does not favor any particular mechanism but has offered recommendations that will, if adopted, assure that permittee-responsible as well as third-party mitigation will secure no net loss of wetlands. In addition, the committee believes that no net loss of wetlands will require a strengthened partnership with the states.

Recommendations

• The taxonomy developed by the committee is recommended as a reference point for discussions about compensatory mitigation. In practice, however, a compensatory mitigation mechanism may not fit neatly into one of the listed categories (e.g., mitigation bank versus in-lieu fee versus cash donation). Accordingly, the committee recommends that when an agency reviews mitigation options, it is most important to focus on their characteristics or attributes (e.g., who is legally responsible, the timing of the mitigation actions, whether the Mitigation Banking Review Team process is used, and whether stewardship requirements are in place).

• Institutional systems should be modified to provide third-party compensatory mitigation with all of the following attributes: timely and assured compensation for all permitted activities; watershed integration; and assurances of long-term sustainability and stewardship for restored, created, enhanced, or preserved wetlands.

• The Corps of Engineers and the Environmental Protection Agency should work with the states to expand their permitting and watershed planning programs to fill gaps in the federal wetland program.

CONCLUSION

The Clean Water Act Section 404 program should be improved to achieve the goal of no net loss of wetlands for both area and functions. The above recommendations will help to achieve this goal. It is of paramount importance that the regulatory agencies consider each permitting decision over broader geographic areas and longer time periods, i.e., by modifying the boundaries of permit decision making in time and space.

1

Introduction

The objective of the Clean Water Act (CWA) is "to restore and maintain the chemical, physical, and biological integrity of the Nation's waters" (Federal Water Pollution Control Act, Public Law 92-500). Accordingly, Congress articulated the more measurable goal of attaining water quality to provide for recreation and to protect fish, shellfish, and wildlife. Toward achievement of this goal, the CWA prohibits the discharge of dredged or fill material into waters of the United States unless a permit issued under Section 404 of the CWA authorizes such a discharge. The CWA vests the U.S. Army Corps of Engineers (Corps), or a state with a U.S. Environmental Protection Agency (EPA)-approved program, with the authority to issue Section 404 permits and to decide whether to attach conditions.

The Corps and EPA define the "waters of the United States" to include most wetlands. Wetlands are included as waters of the United States for purposes of the Clean Water Act because it is recognized that some wetlands may improve water quality through nutrient cycling and sediment trapping and retention. The objective of the Clean Water Act, described above, cannot be achieved if wetlands are not protected. From a legal perspective, defining waters of the United States to include wetlands is also reasonable, at least with respect to wetlands adjacent to traditionally navigable waters (*United States* v. *Riverside Bayview Homes, Inc.*, 474 U.S. 121 (1985)). The U.S. Supreme Court has observed that the broad goal of the Clean Water Act is the improvement of water quality and that adjacent wetlands "play a key role in protecting and enhancing water quality."

As a legal matter, the CWA does not regulate all activities in all wetlands. For example, early in 2001 the U.S. Supreme Court ruled that the assertion of federal jurisdiction over some isolated waters is an unreasonable interpretation of the CWA (*Solid Waste Agency of Northern Cook County v. U.S. Army Corps of Engineers* 2001). This ruling came soon after other court rulings that, taken together, have limited the scope of the federal permitting program. However, numerous states and other nonfederal governments have wetland permitting programs that complement and supplement the federal authority (Want 1994). While the committee's charge is to focus on the CWA Section 404 program, its conclusions and recommendations are applicable to both federal and state regulatory programs.

Over the nation's history, wetlands have been drained and filled for farmland and urban development, mosquito control, and many other activities. However, when wetlands are lost, so are the many functions that they provide within landscapes (Mitsch and Gosselink 2000). Although not all wetlands provide all functions, *wetland functions* can include water-quality improvement; water retention, which helps to ameliorate flood peaks and desynchronizes high flows in streams and rivers; groundwater recharge; shoreline stabilization; and provision of a unique environment, part aquatic and part terrestrial, that supports a diversity of plants and animals, including a majority of the nation's rare and endangered species. In recognition of these functions and their significance to the CWA, the goal of no net loss of wetland area and function was introduced at a national wetland policy forum by the Conservation Foundation in 1988, endorsed by the federal administration in 1990, and supported since. The no-net-loss goal lies behind the federal agencies' efforts to develop Section 404 guidelines that will secure compensation for permitted wetland impacts. The goal was articulated by the agencies in their 1990 Mitigation Memorandum of Agreement (MOA) (see Chapter 4). In its work, the committee accepted the no-net-loss goal as a basis for national wetland policies.

When there is a proposal to discharge dredged or fill material into a wetland, the CWA expects that the Corps, in cooperation with other agencies, will consider the public-interest consequences of issuing a permit. In practical terms, implementation of Section 404 and related programs has followed a general policy that the deliberate discharge of materials must be avoided where possible and minimized when unavoidable. Then if a permit is issued and wetland functions are compromised, some kind of compensatory mitigation may be required to replace the loss of the wetland's functions in the watershed. The Committee on Mitigating Wetland

Losses was established by the National Research Council to evaluate whether compensatory mitigation required of Section 404 permit recipients was contributing to achieving the CWA's overall objective of restoring and maintaining the quality of the nation's waters. In completing its work, the committee assumed that these steps were being followed and focused its efforts on the last step of reviewing compensatory mitigation for unavoidable permitted loss that could not be otherwise minimized.

First, the committee reviewed both the potential and the limits of scientific and technical abilities to replace the function of wetlands in watersheds. Second, the committee examined the likelihood that compensatory mitigation, as provided by the permittee or a third party (e.g., a mitigation bank or an in-lieu fee program), was being executed in a way to secure the goal of the CWA. The committee did not set up an experimental design for comparing in-lieu fee programs, permittee-sponsored and banking mechanisms for securing compensatory mitigation. Such an approach would have required the committee to identify a single mitigation target and then determine which mechanism would most likely meet it. There simply were no data that could be used for such an assessment. As a practical alternative, the committee chose to define the procedural requirements that were most likely to secure mitigation that would meet legal and ecological end points, and the committee compared mechanisms accordingly.

IMPORTANT TERMS

The scientific literature and wetland laws and regulations often attribute different meanings to the same term. Precise definitions are important because confusion about the exact meanings of terms can cloud the arguments being made. To avoid such confusion the committee adopted definitions for the following italicized words. A *wetland* is defined as "an ecosystem that depends on constant or recurrent, shallow inundation or saturation at or near the surface of the substrate" (NRC 1995). *Wetland restoration* refers to the return of a wetland from a disturbed or altered condition by human activity to a previously existing condition (NRC 1992). The wetland may have been degraded or hydrologically altered, and restoration then may involve reestablishing hydrological conditions to reestablish previous vegetation communities. *Wetland creation* refers to the conversion of a persistent upland or shallow water area into a wetland by human activity. Of these, *constructed wetlands*, also referred to as *treatment wetlands*, are created for the primary purpose of contaminant or pollution removal from wastewater or runoff (Hammer 1997). *Wetland enhancement* refers to a human activity that increases one or more functions of an existing wetland. *Wetland preservation* refers to the protection of an existing and well-functioning wetland from prospective future threats. Preservation does not involve alteration of the site.

A *compensatory mitigation project* is the creation, restoration, enhancement, or preservation of a wetland designed to offset permitted losses of wetland functions in response to special conditions of a permit. The mitigation project provides a desired set of hydrological, water quality, and/ or habitat functions in the watershed. (A process for identifying the desired functions for a watershed is described in Chapter 7.) As noted above, *wetland functions* include water quality, water retention, and habitat contributions of wetlands to watersheds.

Functional assessment methods provide useful guidelines for measurement of wetland functions. Such methods that consider how wetland structure (see below), location in the watershed, and the resulting hydrological, geochemical, and biological processes related to that structure and location give rise to certain wetland functions. (Functional assessment is described in Chapter 7.)

The no-net-loss goal is focused on wetland functions; however, the area of a *wetland type* is often used as a proxy for wetland functions. Wetland type describes wetlands according to the U.S. Fish and Wildlife Service's (FWS) classification system (Cowardin et al. 1979; see also Box 1-1). If wetland type is used as a proxy to represent wetland function, compensatory mitigation projects might be expected to result in some number of acres that can be classified as a wetland.

Wetland types in the Cowardin system are differentiated by their

---

BOX 1-1
Wetland Classification System of the
U.S. Fish and Wildlife Service

Cowardin et al. (1979) developed a hierarchical system to classify wetland types for purposes of mapping and inventory. It has at its highest level the "system," of which five are defined (marine, estuarine, riverine, lacustrine, and palustrine). Subsystems further define hydroperiod attributes of the first four systems. Wetland classes are based on substrate type and flooding regime (six classes: rock bottom, unconsolidated bottom, rocky shore, unconsolidated shore, streambed, and reef) or on vegetation types (five classes: aquatic bed, moss-lichen wetland, emergent wetland, scrub-shrub wetland, and forested wetland). Finally, wetlands are classified by their dominance type, based on dominant plants or animals. Various modifying terms are added to describe water regimes, salinity, pH, soil type, or human modifications.

structure. Following the three-part wetland delineation procedure adopted by federal agencies for defining wetlands in the CWA Section 404 program, *wetland structure* can be understood as some combination of hydrology, soil, and vegetation (NRC 1995). When planning a compensatory mitigation project, a wetland's structure and location in the watershed are chosen to secure particular wetland functions. The plan might be based on a functional assessment that relates a wetland's structure and location to its function; alternatively, a compensatory mitigation project plan might seek to secure a particular type of wetland.

A compensatory mitigation site is initiated to satisfy a legal requirement of a permit program. A *mitigation requirement* is a condition of a permit that makes the permit recipient responsible for undertaking and executing a compensatory mitigation site or for paying a third party to take on that responsibility. As a legal matter, the mitigation requirement should establish a measurable outcome, called a *performance standard*, of a mitigation project. Performance standards can be measures of wetland structure or type or a functional assessment score. It may take several years before some measures of functional performance can be achieved at a mitigation project. However, mitigation agreements may avoid performance measures and instead require that, as a measurable outcome, the mitigation project be developed and implemented according to an approved plan. This can be referred to as a project *design standard*. Legal *compliance* with the compensatory mitigation requirement can vary from simply constructing a project according to some approved design (design standard) and/or to being tied to some measure of functional outcome (performance standard).

Wetlands occur in *watersheds*, which are defined in the glossary of this report as "land area that drains into a stream, river, or other body of water." However, a watershed is not area specific, because it can range from a small area near a creek to the entire Mississippi River basin. When positions or management of wetlands within watersheds are discussed, the larger scale is indicated. Discussion of planning for wetland mitigation within watersheds indicates a scale on the order of an ecoregion. The terms landscape and watershed are often used interchangeably.

It follows that there are distinct stages in any compensatory mitigation project; each stage requires an action to be taken that will increase the probability that the compensatory mitigation project will attain its intended results. First, there must be a concept and a general watershed location for the compensatory project. Second, that concept must be translated into a set of site design plans that are expected to secure the target functions over time. Third, the site would be acquired and construction (or other modifications to it) would be undertaken in accord with the design. Inspection of the site would be made to establish whether the construction followed the design plan; the inspection would determine whether design standards had been met and whether some adjustment might need to be made to the project design. Fourth, physical monitoring of the site would continue past design and project construction to determine whether the project was trending toward the desired wetland type or functions; the monitoring would determine whether performance standards were being met. Fifth, there is a regulatory certification that the site has achieved the specified design or performance criteria. At a designated point in the process, there are verifications that the site will be protected and managed in perpetuity. The requirement for establishing mitigation compliance purposes may stop with any of these actions. Also, as noted in Chapters 4, 5, and 8, there can be transfers of legal responsibility for any of these actions from the recipient of the original permit to a third party.

NO NET LOSS AND THE SECTION 404 PROGRAM

According to the FWS, 53% of the conterminous United States presettlement wetland area was lost between the 1780s and 1980s (Dahl 1990). However, there have been dramatic changes in the rate and magnitude of wetland loss in the past 20 years. Wetlands can be lost through direct conversions, such as draining and filling. Wetlands are also lost as the indirect result of other activities that may alter the hydrological regime. Furthermore, not all direct wetland conversions are regulated, and almost none of the indirect losses are regulated. Unfortunately, the available data cannot be used to fully distinguish among these causes. That being said, it is instructive to consider the reported changes in order to put the Section 404 permitting program into perspective and to understand its achievements and possible weaknesses. Doing so means relying on the data published by the FWS and the Corps. Because these two data sets were developed for different purposes and there are important differences in how wetlands are identified, the data cannot be combined (see Box 1-2). However, some useful perspectives can be gained by looking at the two data sets together.

Data for the two most recent FWS reports on the status of and trends in the nation's wetlands were compared (see Table 1-1). The data are for the 10-year period from the mid-1970s to the mid-1980s and an 11-year period from 1986 to 1997. The losses were reportedly due to agricultural and other (nonagricultural) causes. Nonagricultural uses were not reported as separate categories in the earlier report published in 1991; they include silviculture, urban, and rural development uses. Also, the FWS tables include wetland gains through creation and restoration and do not reflect changes that may have occurred in wetland type. In fact, open-water wetlands continue to increase relative to other types in both time periods.

BOX 1-2
U.S. Fish and Wildlife Service Loss Data and
Wetland Delineation for Permitting

FWS relies on air photo interpretation and soil surveys for its national wetland inventory. Therefore, it cannot use the three-part—wetland hydrology, hydric soil, and hydrophytes—delineation system that defines wetlands in the CWA Section 404 program. Instead, FWS must assume that wetland hydrology is present using one of the other two attributes as an indicator. Therefore, it is possible that some areas classified as wetlands by the FWS are not jurisdictional wetlands under Section 404 or the U.S. Department of Agriculture's Swampbuster program. For example, in the southeastern United States, hydrophytes tend to extend across the jurisdictional line "up the hill" relative to either soil or hydrology. If agricultural modifications have occurred on the drier end of the range, areas that may be classified as losses to these causes may not have satisfied jurisdictional criteria. FWS understands this possibility: "The emergent wetlands that continue to be lost are geographically scattered and generally small wetlands: some were already partially drained by surface ditches or completely eliminated through intensified use of existing farmland" (Dahl 2000, p. 45). Under the Food Security Act and CWA Section 404, this represents historical wetland loss, not recent wetland loss. Elsewhere, the authors refer to some of the wetland losses as being "non-jurisdictional" (p. 46).

The best estimate of net acres lost fell from an annual average of about 254,800 acres per year to 58,500 acres per year, a 77% decline (coefficients of variation are included with the original data). Losses to agriculture fell from about 138,000 acres per year to about 15,000 acres per year, almost a

TABLE 1-1   Wetland Losses Due to Agricultural and Nonagricultural Causes

Time Period	Wetland Losses Due to Agriculture	Rate of Wetland Loss Due to Nonagriculture	Total Acreage Lost and Annual Average Loss
Mid-1970s to mid-1980s[a] (10 years)	137,540 acres/year[a] 54% of loss[a]	117,230 acres/year[a] 46% of loss[a]	2,547,700 acres;[b] 254,770 acres/year[a]
1986-1997[c] (11 years)	15,222 acres/year[c] 26% of loss[c]	43,324 acres/year[c] 74% of loss[c]	644,000 acres;[b] 58,545 acres/year[c]

[a]Dahl and Johnson (1991).
[b]Total acreage lost was determined by multiplying the annual average loss by the total number of years evaluated in the study.
[c]Dahl (2000).

90% decline. This decline might be attributed to unfavorable agricultural economic conditions and reforms to federal farm programs that together discouraged wetland conversion and encouraged restoration (Kramer and Shabman 1993; Heimlich 1999). As a result, agricultural losses fell from 54% to 26% of the total loss. While losses to nonagricultural causes increased as a proportion of total loss, the amount of loss to these causes fell from about 117,000 acres per year to 43,000, a 63% decline. This decline is especially remarkable because the 1990s were a time of significant and rapid economic expansion in areas of the nation, such as much of the southeast coast, where wetlands are abundant. In fact, a 1994 report showed that this region accounted for 89% of the total national loss from the mid-1970s to mid-1980s (Hefner et al. 1994). This 63% decline in non-agricultural wetland losses at a time of sharp economic expansion might be attributed to the presence of, and increased understanding about, Section 404 and nonfederal wetland permitting programs. A reasonable conclusion is that these programs may be encouraging those responsible for land development to avoid wetlands. If this is the case, the Section 404 permitting programs may be among the causes for reduced national wetland losses.

FWS reported losses for the 1986 to 1997 period by the more disaggregated categories of urban development (30%), agriculture (26%), silviculture (23%), and rural development (21%). (see Table 1-2). Considering that agricultural and silvicultural activities resulting in wetland losses generally fall outside the scope of Section 404 program and many wetland permit programs (Chapter 4), it might be assumed that the urban and rural development losses were the ones that could have been subject to Section 404 permitting.

However, there are exemptions from permitting that go beyond the silvicultural and agricultural exemptions, and these exemptions continue to expand and contract with administrative and court decisions (Chapter 4). Furthermore, because the FWS criteria for defining wetlands differ from the regulatory criteria for wetlands delineation, it might be expected

TABLE 1-2   Percent Loss by Cause and Acres Lost

Cause of Loss	Wetland Losses 1986-1997 (%)	Annual Average Acreage Lost (Acres)
Urban development	30	17,560
Agriculture	26	15,220
Silviculture	23	13,465
Rural development	21	12,300

SOURCE: Dahl and Johnson (1991); Dahl (2000).

that some of the losses reported may not have been considered wetlands for purposes of permitting. With these caveats in mind, FWS-estimated losses to urban and rural development between 1987 and 1997 were just under 30,000 acres per year.

Data provided by the Corps for its Section 404 permitting program during the 1990s suggest a net gain in wetland area (see Figure 1-1). The Corps' Headquarters, Operations, Construction, and Readiness Division compiles data submitted by the district offices on the area of wetland losses and the compensatory mitigation required in permits. The area of permitted impacts was approximately 24,000 acres per year during the 1990s. Compensatory mitigation required as a condition of these Section 404 permits averaged over 42,000 acres per year. This required mitigation, once implemented, would compensate for the permitted wetland impacts, resulting in a net gain of over 18,000 acres per year. It should be noted that preserved and enhanced wetland acres are counted as equivalent to a newly created or restored wetland in these data for the purpose of esti-mating net wetland loss or gain, but there were no data on how much preservation or enhancement was represented in the total. Likewise, there

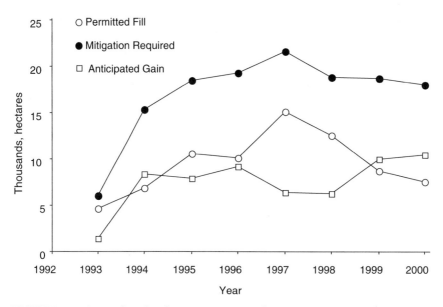

FIGURE 1-1   Area of wetland impacts permitted, mitigation required by the per-mit, and the anticipated gain in wetland area as a result of permits issued by the U.S. Army Corps of Engineers regulatory program from 1993 to 2000. 1 hectare = 2.47 miles. SOURCE: Data from U.S. Army Corps of Engineers Headquarters, Operations, Construction and Readiness Division.

are no data on how much of the mitigation was by created or restored wetlands. Therefore, one cannot draw firm conclusions about such changes in wetland area.

For every 100 acres of permitted fill, 178 acres of wetland were to be restored, created, enhanced, or preserved. However, these data report the mitigation required as a condition of the permit. There are no data on whether the required compensation was initiated. Nor are there data on whether the initiated compensation resulted in a wetland that would be recognized by the jurisdictional criteria used by the Corps or by FWS approach to wetland identification.

Assuming that most mitigation that has been required by the Corps was initiated and resulted in jurisdictional wetlands, the Section 404 program has achieved no net loss of wetland area. This would mean that wetland losses to urban and rural development, as reported by FWS, are occurring outside the scope of the Section 404 program. If this is the case, the program may be discouraging wetland-damaging activities (see above) and is more than replacing the wetlands when such activities are permitted. From this perspective, continued wetland loss to urban and rural development would have to be addressed by expanding the Section 404 program or by some other means. If, by contrast, it is assumed that little of the required mitigation is undertaken in a way that replaces lost wetland area, the acres permitted by the program are about equal to the acres lost to urban and rural development.

The committee is unable to determine the precise extent to which compensatory mitigation is initiated and results in wetlands that would be identified by the FWS inventory process. Indeed, there are no data that would support such an assessment. However, the preceding paragraph does highlight the importance of effectively implementing the required compensatory mitigation when wetland permits are issued. The committee recognizes that the Corps districts now do routinely require mitigation. Hence, this report focuses on increasing the likelihood that this legal requirement will be implemented by those responsible for compensation and that their efforts will result in wetlands that provide important functions in the nation's watersheds.

THE COMMITTEE'S TASK

The Committee on Mitigating Wetland Losses (see Appendix J) was established by the National Research Council, Division on Earth and Life Studies, under the aegis of two boards: the Board on Environmental Studies and Toxicology and the Water Science and Technology Board. The committee's Statement of Task, in the context of Section 404 of the Clean Water Act, is the following:

A multidisciplinary committee will be established to review the scientific and technical and institutional literature on wetland structure and functioning, and options for mitigating wetland loss through restoration, enhancement, creation, and where applicable, in-lieu fee programs. The committee will evaluate the current ability of practitioners to restore various aspects of wetland functioning in a variety of environments and will evaluate options for mitigating wetland loss. The study will address such questions as how wetlands' size and place in the landscape, the ecoregion in which they occur, the kinds of animals and plants that comprise them, their hydrological regime, and other factors affect their structure and functioning in ways that are likely to affect the success of wetland restoration and mitigation of loss. The main criterion for the evaluation will be the degree to which the structure and functioning of the restored wetland match those of naturally occurring wetlands in the same region. The committee will also evaluate other options for mitigating wetland loss, such as in-lieu fee programs. A similar criterion will be used, i.e., to what degree do those options protect or replace the ecological role of naturally occurring wetlands.

The committee will analyze an illustrative set of wetland mitigation projects, including individual projects, mitigation banks, and in-lieu fee programs to the extent that they have ecological goals. As part of its efforts, the committee will consider questions in these three areas:

- Goals for mitigation and criteria for selecting mitigation project type.
- Compliance.
- Mitigation success or failure.

The committee will also consider the following:

- The degree to which experience with success and failure can be extrapolated to other areas and wetland types.
- Whether available information leads to recommendations about circumstances in which compensatory mitigation of various types is more and less likely to succeed.
- What research is likely to improve our success with compensatory mitigation in the near and medium terms.

This report is organized into three sections. Chapters 2 and 3 report on the scientific and technical capacities to create and restore wetland acres and functions in watersheds. Chapters 4, 5, and 6 review the regulatory program that has developed under the authority provided by Section 404. These chapters also comment on the mitigation experience as reported in the literature and as described to the committee during its deliberations. Chapters 7 and 8 point toward the future. These chapters provide technical and institutional suggestions for improving the practice of compensatory mitigation.

2

Outcomes of Wetland
Restoration and Creation

INTRODUCTION

Underlying wetland mitigation is the assumption that it is scientifically possible for humans to recreate the structure and functions of a wetland, either by restoring a site that had previously been a wetland or by creating an entirely new wetland. The purpose of this chapter is to discuss the ecological principles of wetland creation and restoration science and evaluate the current scientific ability of practitioners to restore or create various aspects of wetland functioning in a variety of environments. The chapter is structured to answer several questions posed in the committee's Statement of Task about the ecological basis of wetland mitigation.

Is it Possible to Restore or Create Wetland Structure?

Wetland Types That Have Been Restored and Created

The committee evaluated restored and created wetlands from around the United States, including coastal and inland projects, by reviewing numerous scientific studies (see Appendix A) and by visiting several wetland mitigation sites. The following findings are based on the committee's analysis.

Many types of herbaceous wetlands have been restored or created to a condition that appears to replicate natural wetland structure, such as freshwater emergent marshes (Lindau and Hossner 1981; Niswander and

Mitsch 1995; Wilson and Mitsch 1996; Brown and Bedford 1997) and wet meadows (Brown and Veneman 1998). Mixed results have been reported in the restoration of wet prairies and sedge meadows (Galatowitsch and van der Valk 1996; Ashworth 1997). Certain floristic assemblages, such as sedge meadows visited by the committee in the Chicago area, require extensive planting and intensive management in order to maintain the desired species composition. Thus, the technical ability to attain a prescribed species assemblage may not be affordable or practical in the long term.

Shrub swamps and forested wetlands are more difficult to create or restore because of the time needed to establish mature woody plants (Niswander and Mitsch 1995; Brown and Veneman 1998; King 2000). The committee observed examples of created wetlands where tree saplings had been planted and appeared to be viable, but forest structural characteristics (e.g., stand density, stand height, basal area per tree) were quite different from those of the mature stands they were intended to replace. Planted trees are usually small in diameter, so that basal area per tree is small in comparison to natural forested wetlands. The density (trees per unit area) of planted stands is typically higher than that of natural stands because of either permit specifications or the desire to compensate for mortality. Given sufficient time, planted trees would be expected to attain basal areas comparable to those of trees in natural stands, but densely planted stands would continue to differ from natural stands unless thinned.

Seagrasses and salt marshes are sometimes described as wetlands that are relatively easy to restore or create, based on a long history of mitigation involving eelgrass (*Zostera marina*) and smooth cordgrass (*Spartina alterniflora*; Simenstad and Thom 1992; 1996). Each of these natural vegetation types has the following features that are amenable to restoration or creation:

- One vascular plant species dominates the vegetation; that is, it forms the matrix of the ecosystem.
- The species in question have been well studied for growth requirements and environmental tolerances.
- The matrix species is readily collected and propagated and planted because of its particular clonal growth form (ramets are produced from rhizomes that are easily subdivided).
- These vegetation types grow in relatively wet conditions where environmental variability is buffered by ocean water of relatively constant temperature, salinity, and pH.
- These species are natural colonizers of bare substrates ("early succession" species).

- Transplantation can be done with considerable assurance that environmental conditions will not change substantially after planting (relatively low risk of extreme events, high probability of establishment).
- Microbial and animal components of these ecosystems are readily dispersed by the widely circulating ocean waters that connect distant seagrass beds and/or salt marshes.

If these are indeed the characteristics that make ecosystems easier to restore or create, one can expect greater difficulty in restoring ecosystems that have the following:

- Several codominants.
- Poorly studied species.
- Species that are dispersal limited or that have low reproductive capacity (few seeds, heavy seeds, no vegetative reproduction).
- Dominants that are "late succession" species (not ready colonizers of bare substrates).
- Habitats that have high environmental variability, which allows exotics an opportunity to invade and which pose risks for transplantation efforts (e.g., a marsh plain that is exposed for much of the day can become dry and hypersaline the day after planting, stressing and killing thousands of seedlings (J. Zedler, University of Wisconsin, personal observation, 2000)).
- No aquatic connection to other wetlands and/or no corridors for dispersal.

The restorability of an ecosystem is not necessarily predictable from experiences with smooth and eelgrass cordgrass, even for closely related habitat types. For example, restoration of tall forms of Pacific cordgrass has proved to be difficult because *Spartina foliosa* has a high nitrogen demand and grows poorly in substrates that are coarse in texture and/or low in organic matter (Langis et al. 1991; Zedler 1998). Likewise, seagrass beds in Florida that are dominated by late-succession *Thalassia testudinum* are difficult to restore because this species grows slowly and not densely enough to prevent erosion (Fonseca et al. 1998). Moving upslope to mid- and high-intertidal wetlands complicates restoration and creation efforts even more, as the environment becomes highly variable in soil moisture and soil salinity. In addition, newly planted vegetation becomes susceptible to a broader range of herbivores in that both aquatic and terrestrial animals can attack delicate ramets (Zedler 2001).

Wetland Types That Are Difficult to Restore or Create

Wetland ecosystems that require a specific combination of plant types, soil characteristics, and water supply are difficult to impossible to create from scratch. Examples include vernal pools, fens, and bogs.

Vernal pools

The term "vernal pool" is widely applied to small shallow depressions that hold water for brief periods of time. The vegetation and invertebrates are specific to the regions where such temporary pools form, and the timing of pool formation is also variable. In the forests of the northeastern states, such pools form in spring and fall. In the arid southwestern states, they form during the winter rainy season.

Attempts to restore and create vernal pools in Southern California indicate the importance of locating such efforts where the substrate can support temporary ponding. Restoring or creating vernal pools begins by re-creating the topography where there is appropriate substrate (a clay layer) that seals upon wetting. There, the hydroperiod is critical to restoration or creation of native vegetation and fauna; if the hydroperiod is too long, cattails will invade and outcompete the more ephemeral vascular plants (P. Zedler, San Diego State University, personal commun., 2000). Predators, such as fish, might invade and eliminate the macroinvertebrates (such as fairy shrimp). In a long-term study of California vernal pools that were created by excavating depressions near natural pools, the hydroperiods did not converge with those of the reference systems until year 10 (Zedler al. 1993). Seeds, spores, and resting stages of the plants and animals can be vacuumed from natural pools and used to inoculate restored and natural pools, producing new vegetation highly similar to that of the natural site if the hydroperiod is correct. Even the endangered mesa mint (*Pogogyne abramsii*) has been reestablished at created pools of Southern California. However, few data are available on the similarity of algal and animal components of vernal pools.

Perhaps the greatest difficulty is that vernal pool landscapes cannot be replaced in areas like California, where vernal pools occur on relatively flat topography, which is prime real estate. When the remaining vernal pools are destroyed, there is no place to re-create them except by increasing the density of pools in other remnant landscapes. It is unclear how overall system function is altered by increasing the density of pools while decreasing the landscape area that vernal pools occupy. Another problem is that the uplands surrounding vernal pools also are in short supply and support rare species. Thus, as the available land decreases, vernal pool creation and restoration increasingly conflict with efforts to preserve habitat for other endangered plants and animals.

Fens

Fens are herbaceous wetlands that develop on calcium-rich organic soils and for which groundwater seepage is an important water source. Fens are among the most species-rich wetlands in North America. Their plant diversity is high per unit area, yielding long species lists; approximately 100 vascular plant species might occur in a wetland, with 15 to 30 or more species per square meter (Bedford et al. 1999). Typically, a variety of species share dominance, rather than a preponderance of any one species. Oligotrophic conditions occur because fens are fed by groundwater that is low in nutrients (nitrogen and phosphorus) and high in calcium. Calcium is important, not only for its effect on pH but also because it precipitates out phosphorus, lowers nutrient availability, and thereby reduces the chances that any one species will outcompete its associates.

Drainage of fens changes both the hydrology and the soil chemistry. Exposure to surface-water runoff also changes soil chemistry. In both cases, nutrients are released to the fen, allowing opportunistic species the competitive advantage. Draining also exposes any sulfides to oxygen, forming sulfuric acid, which lowers pH and prevents the restoration of calciphilic vegetation. van Duren et al. (1998) reported slightly decreased pH for drained fens in the Netherlands, where additions of lime (calcium carbonate) were ineffective in reestablishing calciphiles. Vegetation was judged nonrestorable, even when surface-water runoff was entrained in a "treatment wetland" just upstream of fen. Likely constraints on restorability were either inflowing nutrients that escaped treatment or persistent acidic soil.

Bogs

Bogs occur on acidic organic soil ("peat") that develops over millennia from the accumulation of plant decomposition remains. In eight studies summarized by Johnston (1991), natural peat accretion rates ranged from 0.1 to 3.8 millimeters (mm) per year, which indicates an extremely slow rate of development. Bog drainage exposes the organic soil to aeration, accelerating decomposition and fundamentally altering organic carbon compounds in the soil. Agricultural uses of bogs further alter soil chemistry and structure through tillage, fertilizer inputs, and subsidence as soils compact and oxidize. Peat fires can oxidize in days the organic carbon that has taken centuries to accumulate. Although vegetative cover has been reestablished on bogs that have been subjected to such extreme losses and alterations of substrate, restoration of original plant communities is extremely difficult (Mitsch and Gosselink 2000).

The harvesting of live sphagnum moss from bog surfaces is a small but viable industry (Johnston 1988). Research has demonstrated that natural recovery of the moss surface following harvesting takes about 20 years (Elling and Knighton 1984). In contrast, reclamation of wetlands mined for peat has been very difficult because (1) surface mining causes major changes in local hydrology, (2) peat accumulates at a very slow rate, and

(3) the chemistry of old peat is quite different from that of surficial peat layers that support bog vegetation (Updegraff et al. 1995).

The committee concludes that some types of wetlands can be restored and/or created (e.g., freshwater emergent marshes) but that others cannot (e.g., fens and bogs). Not all emergent marshes will be easy to replace. Some types (e.g., species-rich sedge meadows), some hydrological contexts (e.g., late-summer drawdowns), and some functions (e.g., biodiversity support, especially species with narrow ranges of ecological tolerance) will likely be more challenging to reproduce than others (e.g., cattail marshes, continuous flooding, and high plant productivity, respectively). Some types of wetlands will be more difficult to restore or create in certain settings (i.e., where landscape positions, specific substrates, or adjacent land uses are inappropriate).

Are Wetland Functions Replaceable?

Wetlands provide a number of ecological functions. The three most commonly cited wetland functions are related to water quality, hydrology, and habitat, but other functions also exist (e.g., alteration of microclimate, carbon sequestration). Some ecological functions provide human benefits, such as improvement of downstream water quality, whereas others may benefit only nonhuman organisms (i.e., wetland flora and fauna). Knowledge of the existence of wetland functions increases with increasing scientific understanding, but the perceived importance of different wetland functions changes as human values change. For example, the carbon-sequestration function of wetlands has recently assumed increased importance with our increased understanding of the role of atmospheric trace gases in global climate change (Bridgham et al. 1995).

The establishment of wetland structure does not necessarily restore all the *functions* of a wetland ecosystem. For example, denitrification (an ecological process that benefits water quality) requires the presence of nitrate supply, a labile carbon source, anaerobic conditions, and microbial activity. Thus, a site that has wetland structure in terms of its vegetation assemblage might not provide the function of denitrification if these four requirements are not met.

FIVE WETLAND FUNCTIONS

The following sections discuss five major wetland functions that warrant attention in evaluating wetland restoration and creation: hydrological functions, water-quality functions, support of vegetation, support of habitat for fauna, and soil functions. These do not represent all wetland functions but do include a number of important ones that are frequently overlooked in wetland functional replacement. Functional assessment of wetlands proposed for evaluating the development of mitigation sites is discussed in Chapter 7.

Hydrological Function

Hydrology is most often cited as the primary driving force influencing wetland development, structure, function, and persistence. Consequently, establishment of the appropriate hydrology is fundamental to wetland mitigation through either restoration or creation. Hydrological processes influence water quality through nutrient inflow and outflow from the system (Gosselink and Turner 1978; LaBaugh 1986; Carter 1986; Day et al. 1988; Novitzki 1989) and by creating an environment (soil saturation) that allows anaerobic conditions to develop and reducing chemical reactions to operate. Reducing conditions in the soil causes denitrification to occur, organic matter to accumulate, and chemical transformations of phosphorus and iron that influence their solubility (Woodwell 1956; Gosselink and Turner 1978; Richardson et al. 1978; Riekerk et al. 1979; Sharitz and Gibbons 1982; Carter 1986; LaBaugh 1986; Wilcox 1988). Hydrology also influences seed distribution (Gosselink et al. 1990; Sharitz et al. 1990), seed germination, and species establishment and composition (Christensen et al. 1981; Carter 1986; Day et al. 1988; Gunderson et al. 1988; Conner et al. 1990; Gosselink et al. 1990; Sharitz et al. 1990). For many species, seedling recruitment and establishment are a complex process that is only partially understood. Of the seed available at a microsite, germination and establishment of most species are flood/inundation sensitive (Christensen et al. 1981; Day et al. 1988; Duever 1988; Gunderson et al. 1988). Very few species germinate in standing water. Inundation cessation is climatically controlled except where structural devices are used to manage hydrology. Once established, most species tolerate a fairly wide range of hydrological and other environmental conditions (e.g., light, nutrient availability, and soil-water chemistry).

In palustrine nonriverine systems, the hydrological variability that results in species diversity is caused by minor variability in the microtopography (Harper et al. 1965; Christensen et al. 1981; Daniel 1981; Hardin and Wistendahl 1983; USFWS 1983; McDonald et al. 1983; Duever 1988; Gunderson et al. 1988; Titus 1990). Vegetation, in turn, influences hydrology by retarding flow (Gosselink and Turner 1978; Gosselink et al. 1990; Sharitz et al. 1990) and influencing evapotranspiration and affecting soil and water chemistry by leaf litter, transport of oxygen, and other biological processes (Day 1982, 1983; Day et al. 1988).

The difficulty of restoring wetland hydrology increases as the degree of wetland degradation increases (Long et al. 1992). Many wetland restoration efforts have not established the appropriate hydrology (Kusler and Kentula 1990; Pfeifer and Kaiser 1995; Galatowitsch and van der Valk 1996).

One measure of effective restoration or creation is establishment of jurisdictional hydrology. Taking a conservative stance, the U.S. Army Corps of Engineers (Corps) 1987 Wetland Delineation Manual established the 5% criterion (see Box 2-1) as the jurisdictional threshold, a quantitative value that was reaffirmed by the NRC (1995). Since wetland hydrology is fundamental to wetland structure and function, those who implement restoration and mitigation projects have also tended to take a conservative stance, that is, to err toward the wet end of the transition zone (12.5% inundation or saturation during the growing season; see Box 2-1). The structure and character of a wetland that is inundated or saturated 5% of the growing season differ greatly from those of one that is inundated or saturated 12.5% of the growing season. The consequence of this mitigation approach has been the establishment of wetlands that are much wetter than normal for the given landscape position (Cole and Brooks 2000a) or a shift from intermittently inundated or saturated to having open water (Kentula et al. 1992a).

Water-Quality Improvement

Water-quality functions can be mitigated but rarely duplicated. For duplication to happen, the mitigation wetland would have to be of exactly the same wetland type with the same hydrological inputs and the

---

BOX 2-1
Duration and Timing of Inundation or Saturation

Clark and Benforado (1981) divided the hydrological continuum into six functional categories: permanently inundated (inundation >2 meters (m) 100% of the time), semipermanent (inundation <2 m 75-100%), regularly inundated or saturated (25-75%), seasonally inundated or saturated (12.5-25%), irregularly inundated or saturated (5-12.5%) and intermittently or never inundated or saturated (<5%). They suggested that areas saturated less than 5% of the growing season clearly exhibited upland hydrological characteristics and that areas saturated more than 12.5% clearly exhibited wetland hydrological characteristics. Inundation between 5% and 12.5% of the growing season represented the transition zone, with some landscapes in this category exhibiting upland characteristics and others being more characteristic of wetlands.

same chemical (including sediment) composition. It is entirely possible for the restoration or creation site to have water-quality functions superior to those of the impact site. If the impacted wetland is a mineral or organic soil flat (Brinson's 1995 classification), it would make only a passive contribution to water quality (Evans et al. 1993), because its only water input is rain, and the water-quality function is simply to provide an area of runoff where both the surface and the subsurface drainage waters are relatively uncontaminated with pollutants. If a mitigation site is a restored riparian wetland located between a stream and a nonpoint pollution source (either urban or agricultural), the mitigation wetland will have a water-quality function superior to the impact site. However, if the impact site is a riparian wetland while the mitigation site is on a flat, the vast majority of the water-quality function of the impact site is lost. To determine the water-quality function of either the filled area or the mitigation site, it is necessary to make some assessment of both the quality and the quantity of groundwater and surface water entering the wetland (Hill 1996; Hill and Devito 1997; Bedford 1999). Different regions of the United States would need to evaluate both the quantity and the quality of water entering a wetland in order to assess the potential for water-quality improvement.

Support of Vegetation

Wetlands fail to support plant biodiversity when the environment is extremely hostile (e.g., extremely contaminated or hypersaline) or when one or a few species dominate the site. Monotypic vegetation can be formed by native species or exotic species. Cattails (*Typha* species and hybrids) are notorious for overtaking nutrient-rich wetlands (Wilcox et al. 1984), as are giant reed grass (*Phragmites australis/communis*), purple loosestrife (*Lythrum salicaria*), and reed canary grass (*Phalaris arundinacea*).

Invasiveness is a function of both the invader and the habitat it colonizes. Plants that invade wetlands are typically species with high seed production, high germination rates, and the ability to spread vegetatively. Seedling establishment is often the limiting factor, but once established, the clone can expand, such that a clone from a single seedling could come to dominate an entire site. An additional attribute of invasive species is their ability to take up and utilize nutrients from high concentrations in the water or soil supply. In the Everglades, for example, native *Cladium jamaicence* is adapted to oligotrophic waters; it absorbs and stores nutrients in leaf bases. In contrast, the invasive *Typha domingensis* takes up nutrients and distributes them throughout the plant, growing to greater heights and biomass, thereby outcompeting *Cladium* where surface waters are eutrophic (Miao and Sklar 1998).

The habitat being colonized will be more invasible if there are micro-sites available for seedling colonization. Small gaps in the canopy or minor soil disturbance might be all that is needed to allow seeds to establish (Hobbs and Huenneke 1992; Lindig-Cisneros and Zedler 2001). Thus, human- or animal-caused disturbances can lead to establishment events. The combination of canopy gaps and nutrient inflows can virtually guarantee the establishment and spread of invasive species. Wetland restoration and creation sites are very susceptible to species invasions when (1) they are devoid of vegetation, (2) plant canopies have multiple gaps, and (3) their water supplies are eutrophic. All three attributes characterize many mitigation sites.

The ability of a site to support biodiversity is not independent of its ability to improve water quality. It is critical that the relationship between these two functions be understood if mitigation goals are to be set that are ecologically conflicting, such as maintaining high plant biodiversity *and* improving eutrophic water quality. Mitigation sites that receive nutrient-rich surface-water runoff are well situated to perform water-quality-improvement functions, but biodiversity-support functions may suffer in the process. Wetlands that are designed to maximize the water-treatment function typically become monotypes of invasive species within a few years, even if they are initially planted to multiple species (Kadlec and Knight 1996).

Habitat Support for Fauna

None of the compensatory mitigation projects visited by the committee included design and evaluation criteria for animals. Animals are almost never manipulated or introduced into wetlands, in contrast to transplants of higher plants. Even when wetland assessments involve animals, the primary consideration is waterfowl and other birds or identifiable endangered/threatened species. Evaluations do not consider the constraint that most wetland animals are incapable of overland migration if terrestrial corridors are blocked by development, highway systems, or other situations not conducive to overland movement. Many wetland animal species are also dependent on the terrestrial habitat surrounding a prescribed wetland. The importance of considering migratory pathways and upland buffers in the design of a compensatory mitigation plan is discussed in greater detail in Chapter 3.

Soil Functions

Soil performs a number of important functions in a wetland that are usually overlooked in wetland restoration:

- *Rooting medium.* Soil serves as a rooting medium for plants, providing the physical support for above-ground plant structures.
- *Germination medium.* Seed germination requires more specialized conditions than those required to sustain mature rooted plants. Germination of annuals, for example, is often promoted by a moist, temporarily exposed soil that is free of detritus.
- *Seed bank.* Seeds and rhizomes retained in the soil remain viable for months to years.
- *Source of water and nutrients for plants.* Soil is the site of water and nutrient uptake for rooted plants, even rooted plants that are submerged. The release of plant-available forms of nitrogen from unavailable organic forms stored in the soil (i.e., nitrogen mineralization) provides a constant source of nutrition to wetland plants.
- *Habitat for mycorrhizae and symbiotic bacteria.* Roots have complex relationships with soil fungi (mycorrhizae) and bacteria that enable and enhance nutrient uptake. Examples include nitrogen-fixing bacteria living symbiotically in root nodules of legumes and *Alnus* and vascular arbuscular mycorrhizae that associate with *Salix*. Some plants require the presence of specific mycorrhizal species to survive.
- *Water-quality functions.* The soil is the locus of most of the physical, chemical, and biological processes that give wetlands the ability to improve water quality. Sediment retention takes place at the soil surface. The chemical composition of the soil, such as the presence of iron and aluminum hydroxides, affects its ability to sorb phosphorus. Denitrifying bacteria dwell in the soil and depend on soil carbon as an energy source to support denitrification.
- *Habitat for soil macrofauna.* Soil-dwelling fauna sustain wading birds that probe the sediments of mud and sandflats with their long beaks. The role of soil-dwelling fauna in other types of wetlands is less well known.
- *Conduit for groundwater.* Soil permeability affects its ability to convey water. Dense, low-permeability soils may serve as aquacludes, causing water in wetlands to be perched above the regional water table. More permeable soils have higher hydraulic conductivities, allowing wetlands to have greater interaction with groundwater.
- *Source of contaminants.* Contaminants can be released from soils, particularly where the soil is landfill or has a prior history of industrial use. Soils that are high in heavy metals may release toxic forms, such as methylmercury and selenium, when creation of a wetland induces anaerobic conditions.

In wetland restoration and creation projects, soil is generally viewed as merely a rooting medium for the plants that are desired (the first function listed above). The soil that has developed in situ at a wetland creation site is often scraped off to attain a surface elevation that will allow the site to become flooded or intersect the water table. This soil-like material at depth usually has much different texture, structure, chemistry, and biota than the overlying material that was removed. It is nearly devoid of organic matter content, depauperate in nitrogen, may have been compacted by construction activities, and lacks mychorrizal and microbial populations important to plant establishment and water-quality functions. This material cannot provide the same functions as an intact wetland soil, and the plants that can successfully reproduce in such material may not be the desired ones.

The soil organic matter content of created wetlands has consistently been found to be less than that of reference wetlands (Craft et al. 1988, 1999; Gwin et al. 1990; Moy and Levin 1991; Zedler and Langis 1991, Kentula et al. 1992a; Bishel-Machung et al. 1996; Shaffer and Ernst 1999). In a multiple wetland study (n = 95), the average soil organic matter content of newly created wetlands did not change between 1987 and 1993, suggesting that there were no widespread improvements in wetland construction procedures to increase initial soil organic matter content (Shaffer and Ernst 1999). The vertical distribution of organic matter was also different in the profiles of created versus reference wetlands in Pennsylvania, with greater horizonation in the reference wetland soils (soil organic matter content significantly greater at depths of 5 centimeters (cm) than at 20 cm) than in the reference wetland soils (no significant difference in organic matter content at depths of 5 cm versus 20 cm). Although some researchers have reported increases in soil organic matter content in created wetlands over time (Lindau and Hossner 1981; Craft et al. 1988), others have found no significant relationship between soil organic matter content and project age (Bishel-Machung et al. 1996; Simenstad and Thom 1996; Shaffer and Ernst 1999). Low soil organic matter concentrations are associated with reduced levels of function, including poor establishment and growth of vegetation, poor habitat and food chain support for invertebrates and fish, and altered nutrient cycling (Shaffer and Ernst 1999).

Soil bulk density, which is the mass of soil per unit volume, has been found to be greater in created wetlands than in reference wetlands in at least two studies (Craft et al. 1991; Bishel-Machung et al. 1996). Soil texture also differed between created (n = 44) and reference wetlands (n = 20) in Pennsylvania: soils in created wetlands contained more sand and less clay at 20 cm, and reference wetlands were more silty than created wetlands were throughout the soil profile. Created wetlands may lack the redoximorphic features of low chroma and frequent redox depletions that are diagnostic of hydric soils (Confer and Niering 1992; Bishel-Machung et al. 1996).

Soil nutrient concentrations in created wetlands are generally lower than those in natural wetlands. The total nitrogen content of soil is usually positively related to organic matter content; hence, created wetland soils may contain less total nitrogen than do reference wetlands (Langis et al. 1991; Bishel-Machung et al. 1996). Wilson and Mitsch (1996) reported lower concentrations of phosphorus, potassium, calcium, and magnesium in constructed versus natural wetlands. Craft and colleagues (1991) reported lower concentrations of plant-available nutrients in a created salt marsh than in a reference salt marsh, which may have implications for plant primary productivity (Zedler and Langis 1991). Differences in the activity of soil microbes that control nutrient availability may also occur in created versus restored wetlands (Groffman et al. 1996).

Seeds and rhizomes retained in soil can remain viable for many years, germinating when conditions are right. Germination rates are a function of both environmental conditions (e.g., inundation, exposed mineral soils, and temperature) and seed conditions. For example, certain seeds have higher germination rates after a period of cold storage (stratification) or after abrasion of the seed covering (scarification; Thullen and Eberts 1995).

The presence of a seed bank can be either beneficial or detrimental. It is beneficial when desirable plant species naturally revegetate restored wetlands, reducing the cost and effort of planting programs (Mitsch and Wilson 1996; Mitsch et al. 1998). This approach is used most often in the restoration of preexisting wetlands (Galatowitsch and van der Valk 1996). Similarly, soil material with an intact seed bank can be transplanted from a wetland slated for destruction to a wetland creation site to promote revegetation (Brown and Bedford 1997). The presence of a seed bank can be detrimental, however; if it is dominated by weedy or invasive species, it would be viewed as undesirable. On several of the committee's site visits, the removal of soil was recommended for this reason. Planting of wetland vegetation in restored wetlands may not significantly influence the seed bank. Of 136 seed bank taxa found in soils from the shoreline of a reservoir constructed 6 years prior, only 10% were from species that had been planted during the restoration process (Collins and Wein 1995).

The experience at Hole-in-the-Donut in Everglades National Park (see Appendix B) illustrates the importance of restoring the substrate to achieve restoration of native plant communities. The soil that had been artificially created by rock plows during agricultural usage of the site provided substantially different conditions for plant growth than the oolitic limestone that supported the native sawgrass plant communities. Only by scraping off the artificially created soil was it possible to deter invasion by the invasive Brazilian pepper. When that was done, sawgrass regained its competitive advantage and was able to recolonize naturally without human intervention in the form of planting or weed control.

FACTORS THAT CONTRIBUTE TO THE
PERFORMANCE OF MITIGATION SITES

Wetland type and actions taken are two variables that contribute to the performance of mitigation sites. However, regardless of wetland type, the ability of a site to be restored depends in part on the degree to which it was degraded, and the degree to which it was degraded dictates, in large part, the actions that will be required to restore it (Zedler 1999). Degradation is a function of damages to both the watershed and the immediate site. Thus, restoration of a cattail marsh or open pond will be easier than a vernal pool, fen, or bog. And the creation or restoration of a cattail marsh or pond will be easiest at a site that is not too degraded in a watershed that is not too degraded. That is, the site should have a soil substrate (not bare rock as in a gravel pit); it should not be so contaminated that vegetation will not grow (as can be true where oil or pesticides have been spilled); and it should occur in a watershed that still retains the region's natural hydrology (e.g., groundwater should not be substantially depleted), as well as space and a topographic setting appropriate for a wetland to persist in perpetuity. Finally, the actions taken to restore or create a wetland should be appropriate to the type and degree of on-site degradation. If hydrology, soils, vegetation, and fauna have all been degraded, some attention to each of these classes of factors will likely be needed. Chapter 7 outlines appropriate guidelines for restoring or creating wetlands that are ecologically self-sustaining. The committee concludes that in a degraded wetland situation, many wetland functions are difficult to restore to their pre-disturbance condition. The ability to replace wetland functions depends on the particular function, the restoration-creation approach used, and the degree of degradation at the compensation site.

How Is the Likelihood of Achieving
Wetland Restoration-Creation Goals Affected:

By the Hydrological Regime?

Wetlands are transition areas between water (wet 100% of the time) and uplands (seldom wet). In a qualitative sense, wetland hydrology has been defined by the National Research Council (NRC 1995) as recurrent, sustained inundation or saturation at or near the surface at a duration and frequency to support the development of diagnostic wetland features of hydric soils and hydrophytic vegetation. The NRC recognized that this definition is too broad to be applied directly in regulatory practice and so adopted quantitative criteria that focus on frequency, timing, and duration of inundation or soil saturation. In many cases, mitigation and restoration have required establishment of wetlands that are inundated or saturated more than 12% of the time ("seasonally inundated or saturated"; see Box 2-1). In reality, many impacted wetlands satisfy the saturation criteria closer to 5% of the time (Cole and Brooks 2000a). Thus, mitigated sites tend to be wetter than impacted sites (see Chapter 6 for more discussion on this consequence).

Our ability to restore or establish a particular hydrological regime is highly dependent on the desired wetland type. Wetlands that are surface-water-driven are, as a general rule, easier to establish than groundwater-dependent wetlands, but outcomes in both cases are highly dependent on landscape position. These wetland types are typically rainfall- or surface-runoff-driven, although some depressional areas, such as prairie potholes and the southeastern coastal plain, may rely on significant groundwater inflow. Hydrological regime is much more difficult to restore to an original state in highly modified hydrological systems (e.g., dams and levees). Hydrological restoration of prior converted cropland has been a generally successful process in part because the original conversion from wetland to agriculture resulted in relatively minor alteration of the natural hydrology. Wetland hydrology can often be restored to prior converted cropland by disrupting the artificial drainage system. Cooper et al. (1998) reported the restoration of hydrology to a Rocky Mountain fen where damage had been minor (i.e., installation of a drainage ditch). Monitoring data demonstrated that hydrology returned soon after the ditch was blocked. The return of natural water levels, however, might not restore soil chemistry or the ability to support calciphilic species, as discussed earlier. Important groundwater parameters are often inadequately characterized at reference and mitigation sites (Hunt 1996; Hunt et al. 1999). For riparian and riverine wetlands, the ability to restore hydroperiods depends on the degree to which streamflows have been modified (see Coyote Creek case study, Appendix B). Streams may have been channelized to convey storm water, and the need to protect upstream lands might preclude the restoration of natural flood pulses.

In conclusion, surface-water-dominated wetlands are easier to restore or create than groundwater-dominated wetlands, but achieving a natural hydrological regime is always a challenge. Hydrological regime is much more difficult to restore to the original state in highly modified hydrological systems (e.g., with dams and levees).

By Wetland Size?

Large natural wetlands are rarer, and their size imparts additional value to some functions (e.g., habitat for animals with large home ranges). Conversely, small isolated wetlands play a crucial role in the biodiversity of other wetland-dependent fauna, such as amphibians (Semlitsch 2000). There is no question that wetland size has an impact on water quality (Chescheir et al. 1991; Daniels and Gilliam 1996; Gilliam et al. 1997). Larger wetlands have a greater capacity to assimilate constituents in the inflow; however, the larger the wetland, the smaller the pollutant removal per unit area of wetland. Pollutant removal frequently follows a first-order decay curve with regard to residence time in the wetland. For a given input of water, the larger the wetland, the more pollutant will be removed. However, since each additional increment of land removes less pollutant than the previous increment, the removal per unit area of wetland is lower.

Shape may also influence the effectiveness of the wetland for pollutant removal. For functions such as water quality and nutrient retention, edge interface with stream or upland is probably more important than area. For example, pollutant removal by a 50-foot (ft)-wide wetland buffer beside a 100-ft-wide stream would remove far more contaminants than a 100-ft-wide buffer beside a 50-ft-wide stream. That would be true for water and pollutants entering the wetland from land upslope from the stream. However, that might not be true for water entering the wetland from stream bank overflow. However, for water-quality purposes, many small wetlands would be more effective than one large wetland covering the same area. The committee concludes that wetland size affects wetland functions. Thus, replacement area should be proportional to the area required to replace the functions lost.

By Wetland Place in the Landscape?

Riverine and slope wetlands are more difficult to restore or create than are depressional wetlands (Gwin et al. 1999; Shaffer et al. 1999). Position in the landscape affects a number of wetland functions, such as water quality (Johnston et al. 1990; Evans et al. 1993) and biodiversity (Poiani et al. 2000). Wetland place in the landscape is generally not considered a mitigation performance standard. The location of mitigation wetlands may be limited by the availability and cost of land.

Recently, much attention has been focused on the need to define hydrological equivalence in the landscape (Bedford 1996). Wetlands occur in a variety of physical settings, including coastal lowlands, topographic depressions, broad flats on interstream divides, the base of slopes, and topographic highs with little slope (Winter and Woo 1990). Location in the landscape influences geological characteristics such as slope; thickness and permeability of soils; and the composition, stratigraphy, and hydraulic properties of the underlying strata, all of which influence surface and subsurface flows of water. Quite simply, hydrology is the driving force influencing wetland development, structure, function, and persistence. In conclusion, riverine and slope wetlands are more difficult to mitigate than are depressional wetlands.

By the Ecoregion in Which a Wetland Occurs?

Ecoregions are quite diverse, and it was difficult for the committee to generalize about the effect of ecoregion on the likelihood of achieving wetland restoration-creation goals. Created wetlands in areas of variable precipitation, one factor that distinguishes ecoregions, may not meet the jurisdictional definition of a wetland every year. Such areas are more likely to occur in arid ecoregions, although periods of drought and temporal variations in precipitation can occur in any ecoregion. Innate temporal variability of wetlands due to climatic variation is not the same as mitigation failure.

The committee concludes that created wetlands in regions of variable precipitation may not meet the jurisdictional definition of a wetland every year, increasing the risk of noncompliance with performance standards. Thus, the mitigation design should recognize and accommodate hydrological variability and extremes caused by climate. Wetland restoration and enhancement should be preferred over creation in such areas.

By the Kinds of Plants Present?

An important general consideration of wetland design is whether plant material is going to be allowed to develop naturally from some initial seeding and planting or whether continuous horticultural selection for desired plants will be imposed. To develop a wetland that will ultimately require low maintenance, natural successional processes need to be allowed to proceed. For forested wetlands, an initial period of invasion by undesirable species might be temporary if proper hydrological conditions are imposed and if trees shade out early invaders. One strategy is to introduce, by seeding and planting, many of the available species to allow natural processes to sort out the species and communities over time. Selective weeding may be necessary in the beginning or throughout the life of the wetland if aggressive exotic vegetation persists. Preferably, the system can sustain itself through its own successional patterns. Otherwise, labor-intensive management, which is never desirable in a compensatory mitigation wetland, will be needed. In some cases, survival of specific plantings is used to evaluate compliance. Wetlands that are readily invaded by exotics or prove to support undesirable monotypes over long periods require greater attention to planting and an establishment-phase exotic control program (see Boxes 2-2 and 2-3).

BOX 2-2
Seeding Versus Natural Recruitment

Reinartz and Warne (1993) found that early introduction of a diversity of wetland plants may enhance the long-term diversity of vegetation in created wetlands. The study examined the natural colonization of plants in 11 created wetlands in southeastern Wisconsin. The wetlands under study were small isolated, depressional wetlands. A 2-year sampling program was conducted for the created wetlands, aged 1-3 years. Colonized wetlands were compared with five seeded wetlands where 22 species were introduced. The diversity and richness of plants in the colonized wetlands increased with age, size, and proximity to the nearest wetland source. In the colonized sites, *Typha* spp. comprised 15% of the vegetation for 1-year wetlands and 55% for 3-year wetlands, with the possibility of monocultures of *Typha* spp. developing over time in colonized wetlands. The seeded wetlands had a high species diversity and richness after 2 years. *Typha* cover in these seeded sites was lower than that in the colonized sites after 2 years.

In conclusion, wetland mitigation designs should include plantings (e.g., sedges over cattails). Unless actively controlled at the outset, exotic and weedy plant species often dominate restoration sites. Species richness is often low in created wetlands.

By the Kinds of Animals Present?

Natural freshwater wetlands support among the highest levels of regional species' diversity and population densities of fauna in North

---

BOX 2-3
Does Wetland Planting Help Self-Design?

In a multiyear study of the effect of plant introduction on ecosystem function, researchers at the Olentangy River Wetland Research Park in Ohio found that a planted wetland and an unplanted wetland converged in most functions (eight biological measures; eight biophysiochemical measures) in 3 years (Mitsch et al. 1998). Continued studies showed a persistence of the planted vegetation in that basin but dominance by *Typha* in the naturally colonizing basin, with differences in function between the two basins. The planted wetland had more plant communities but had 50% lower net primary productivity, higher summer water temperature, and lower macroinvertebrate diversity than did the naturally occurring wetland 6 years after planting (Mitsch et al. 1999).

America. Several biological and landscape considerations are critical to the sustainability of such species, and a variety of factors must be considered in attempting compensatory mitigation for loss of animal-support functions. Created wetlands are generally not designed to meet the needs of animals found in the impacted wetland; hence, animal species' richness is often low in mitigation sites.

Amphibians are perhaps the best-studied group of organisms from the standpoint of wetland dependency both in terms of association with a particular wetland and the landscape pattern of wetland connectivity. The importance of amphibians as a major component of wetland biodiversity, the significance and necessity of peripheral terrestrial habitat to their existence, and the requirement for wetland interconnectivity in a landscape have been documented for a variety of species and regions in North America (Berven and Gill 1983; Berven and Grudzien 1990; Berven 1990; Pechmann et al. 1991; Dodd 1992, 1993, 1995; Gibbs 1993; Semlitsch et al. 1996; Snodgrass et al. 1999; Madison 1997; Semlitsch 1998, 2000; Semlitsch and Bodie 1998; Lamoureux and Madison 1999). Biological dynamics of animal populations are discussed in Chapter 3.

The committee concludes that for compensatory mitigation of a wetland to be effective for all affected fauna, the biological dynamics must be evaluated in terms of the populations present and the ecological requirements of the species, which include metapopulation aspects that are affected by the relationship of the wetland to other wetlands in the local system.

By Time?

After more than 2 decades of compensatory mitigation, there are now thousands of hectares of restored and created wetlands in the United States. Yet only a few studies have analyzed how various ecosystem components have changed over time, and even fewer describe ecological performance over more than 5 years. Knowing how rapidly an ecosystem matures, or fails to mature, is important for learning how to improve future projects.

A 3- to 5-year monitoring period is a common permit requirement. Five years may be enough time for herbaceous plant cover to reach a peak (see Figure 2-1), but trees obviously take longer than herbaceous plants to reach peak biomass. Biomass, however, may not be equivalent between the restored or created wetland and the natural reference ecosystems. For example, Brown and Veneman (1998) analyzed 68 paired mitigation sites in Massachusetts and, using criteria for establishment for equivalency, reported that plant communities were not equivalent to reference systems after 13 years.

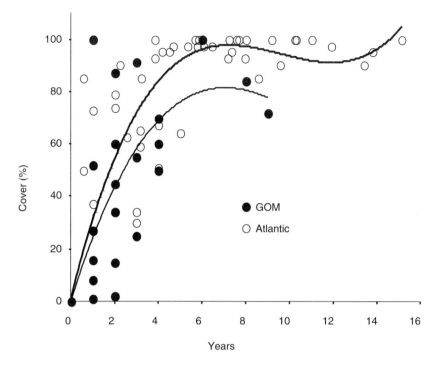

FIGURE 2-1   Percent plant cover on created or restored coastal wetlands on the Atlantic and Gulf of Mexico (GOM) coasts. A second-degree polynomial fit of the data is shown. Note that 100% cover was achieved in most of the Atlantic wetlands within 5 years. SOURCE: Adapted from Matthews and Minello (1994).

The animal components of communities may develop at quite different rates than do plants at mitigation sites. Mobile species may migrate in and colonize with the first floodwaters. Even with fish, however, simple analyses of population structure indicate that the trajectory from colonization to stability is not equal or smooth among population types. For one study, Rulifson (1991) surveyed a created salt marsh constructed as a mitigation site in North Carolina. Although this study was only for a single site, the findings are important. Rulifson found that the number of species captured with one piece of gear (trawl) became more like the number of species in the reference marsh within 20 months and after 42 months for fish collected with shallower-water gear (Wegener Ring). Neither the number of organisms captured with each gear type nor the number of similar species was stable after 4 years. The equilibrium position of various fisheries parameters was not reached within 4 years of the man-

TABLE 2-1   Summary of Results from Study of a Created Salt Marsh
Constructed as a Mitigation Site in North Carolina (1991)

Comparison of Finfish in a Constructed Wetland Compared with Two Reference Sites	Months to Reach Reference Site Conditions	Trend After Equivalency First Reached
Number of organisms		
Trawl	8	Falling
Wegener Ring	28	Rising
Number of species		
Trawl	18	Stable
Wegener Ring	43	?
Species similarity		
Trawl	0-12	Declining
Wegener Ring	0-12	Declining

SOURCE: Adapted from Rulifson (1991).

agement action but then continued to change. In other words, the equilibrium was obtained but not equivalency (see Table 2-1). Zedler (1990) noted that the average number of individual fish species and the average number of individuals caught at a single reference site and a wetland creation project in Southern California were reached within 5 years. However, the similarities of fish and benthic species between the reference site and mitigation sites were only 58% and 54%, respectively.

Results from a variety of trajectory analyses of soil, plant, and animal communities for mitigation sites are given in Appendix C and summarized in Table 2-2. Plant cover and biomass for some marshes may reach

TABLE 2-2   Time Toward Equivalency for Soil, Plant, and Animal
Components in Wetland Restoration Projects Compared with That of
Natural Reference Wetlands

Structural Component	Number of Sites	Range (years)
Soils	17	>3 to >30
Wetland plants		
Cover or biomass	7	3 to >20, or never
Species composition	3	>5 to >10, or never
Below-ground biomass	2	10+, or never
Fish and fisheries	9	>2 to 10
Marsh fisheries	14	1 to >17, or never
Birds	7	<3 to >15, or never

an equilibrium in 3 years, as discussed above, but species composition may take 10 years or longer to stabilize, if at all. Trees, obviously, will not reach an equilibrium until at least the lifetime of the tree. The trajectories for soils, plants, and animals are not the same (e.g., Zedler and Callaway 1999; see Figure 2-2). In contrast to herbaceous vegetation, soil development may be quite slow (Craft et al. 1999), taking from 3 years to 30 years to reach equilibrium, if equilibrium is realizable. The fish and bird components may reach equivalency in as little as 2 years for some species. The trajectory may never reach an equilibrium or the equilibrium conditions may not equal those at the reference sites. However, because there has

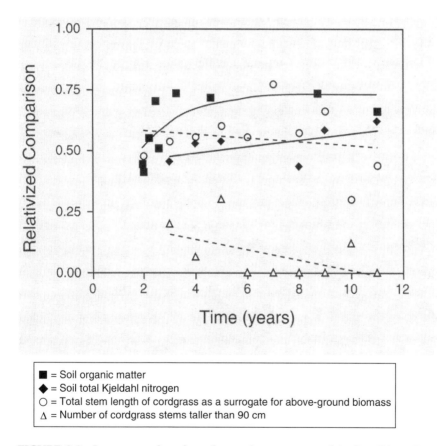

FIGURE 2-2 Long-term data for salt marshes constructed in San Diego Bay. Values are in relation to the adjacent natural salt marsh. The site was required to provide self-sustaining tall cordgrass stems for nesting by an endangered clapper rail. SOURCE: Adapted from Zedler and Callaway (1999).

been virtually no research that compares the relative similarity of undisturbed reference sites, it is hard to make firm conclusions on this point. In their analysis of reference site selection, White and Walker (1997) found that there is so much variation on a regional scale that there may never be a perfect match for a site to be restored. They propose that a conceptual restoration model should be one of interpolation among multiple sites and sources of information, including temporally. Finally, Neiring (in Kusler and Kentula 1990) posits that the end point of the successional process cannot be predicted because it is not an orderly process; that is, we cannot predict what exact species will be at the "end" of the process. In summary, it appears that there is no general trajectory for the development of wetland ecosystems or individual components (structural or functional attributes) of a single watershed.

The significance of these results is not that equivalency among reference and newly managed environments is not reached or that mitigation efforts should not be done. These results demonstrate that (1) ecological equivalency may *not* be reached within a few months or for several years or even decades, depending on the attribute that is of interest; (2) the ecosystem does not move smoothly to an equilibrium or at the same rate for all components; and (3) some components, including ones identified as important in permits currently being issued, may *never* reach equivalency with the natural reference wetland. An obvious conclusion from these results, besides the general paucity of scientific analyses, is that the generally observed 5-year limit on monitoring is insufficient when evaluating whether a site has achieved parity with a reference system. Further, the amount of mitigation required should be based on the amount needed to fully offset the permitted wetland losses. To accomplish that, mitigation ratios (area of mitigation to area lost) will often need to be increased to achieve functional equivalency, rather than simply matching wetland area.

The above review of wetland restoration and creation outcomes has not differentiated projects on the basis of their starting conditions. Some of the projects discussed occurred at sites where damages were relatively minor (e.g., Cooper et al. 1998); others were at highly altered sites with artificial (dredge spoil) substrate (e.g., Zedler and Callaway 1999). As discussed earlier, it is very likely that the ability to achieve desired outcomes is a function of the degree of site degradation. Degradation, in itself, is a complex concept, involving not only the local site but also its watershed. Zedler (1999) suggests that generalizations are not easily drawn among restoration efforts involving sites with different degrees or types of degradation, suggesting instead that ecologists seek predictability of outcomes in a matrix of situations, where degree of degradation is one axis and the type of restoration effort is the other. The general pattern is likely to be that outcomes are more predictable where damages are minimal and restoration efforts most intensive. Badly damaged wetlands and wetland creation sites are less likely to match reference wetlands, especially if restoration efforts are minimal.

In conclusion, the literature and long-term trajectories reported therein suggest that wetland restoration and creation sites do not often achieve functional equivalency with reference sites within 5 years; indeed, up to 20 years may be needed for functional attributes to be determined or assessed correctly.

RECOMMENDATIONS

On the basis of its evaluation of wetland structure and function, the committee makes the following recommendations for compensatory mitigation:

1. Avoidance is strongly recommended for wetlands that are difficult or impossible to restore, such as fens or bogs.

2. The science and technology of wetland restoration and creation need to be based on a broader range of studies, involving sites that differ in degree of degradation, restoration efforts made, and regional variations. Predictability of outcomes should then improve.

3. All mitigation wetlands should become self-sustaining. Proper placement in the landscape to establish hydrogeological equivalence is inherent to wetland sustainability.

4. Hydrological variability should be incorporated into wetland mitigation design and evaluation. Except for open-water wetlands, static water levels are not normal. Because of climatic variability, it should be recognized that many wetland types do not satisfy jurisdictional criteria every year. Hydrological functionality should be based on comparisons to reference sites during the same time period.

5. Because a particular floristic assemblage might not provide the functions lost, both restoration of community structure (e.g., plant cover and composition) and restoration of wetland functions should be considered in setting goals and assessing outcomes. Relationships between structure and function should be better known.

6. The biological dynamics should be evaluated in terms of the populations present in reference models for the region and the ecological requirements of those species.

7. Mitigation projects should be planned with and measured by a broader set of wetland functions than are currently employed.

3

Watershed Setting

As noted in Chapter 2, landscape setting is a critical consideration when planning and constructing a wetland mitigation project. This chapter further discusses the relationship between wetlands, landscape position, and watersheds. After a brief introduction to watershed organization, the relationship of wetland functions to landscape position in the watershed is described, followed by a discussion of how multiple forces within watersheds impinge on wetland functions. It is concluded that wetlands now in the landscape of most watersheds are of a type and location that are an outcome of historical development and regulatory process. The committee concludes that the wetland remnants of the development process may not constitute the best configuration of wetland type for a watershed. This conclusion has implications for the kind of wetland planning that might be required in some of the nation's watersheds and the compensatory mitigation practices in those watersheds.

WATERSHED ORGANIZATION AND LANDSCAPE FUNCTION

A watershed is the land area that drains into a stream or other water body. Within a watershed, both nontidal and tidal stream networks have an orderly arrangement of channels that feed into one another. Stream and river channels are described by stream ordering systems (see Box 3-1). In general, watershed organizational structure governs the flow of water and associated nutrients through these systems, the relationship between hydrological processes and the position of the wetland in the watershed,

> ## BOX 3-1
> ## Stream Order
>
> Stream order is inversely proportional to stream number. There are far more first-order streams than other streams. In many watersheds the first-order streams encompass 60% to 80% of the total watershed area. This relationship between stream order and stream number has been used to estimate the acreage of potential riparian wetlands adjacent to water courses. Brinson (1993) made this estimation and concluded that even though high-order streams had larger wetlands, the number of small-order streams compensated for the small size of the adjacent riparian wetlands. Thus, the total area of wetlands along small streams is similar to that for wetlands along higher-order streams. The difference, however, is that the small wetlands along first-order streams are more subject to developmental stress, a topic discussed later. In a study of the distribution of wetlands in a Maryland coastal plain watershed, Haas (1999) found that most of the wetlands were divided among three main topographic positions: headwater (above the channel head), at tributary junctions, and along the main channel. The wetlands at these sites had different average areas; the headwater wetlands were the smallest, and size tended to increase with stream order.

the distribution of riparian wetlands in the watershed, and the relationship between wetland functions and watershed position. Tidal channels are similarly organized, although with different stream order/area relationships than inland streams (Myrick and Leopold 1963; Rinaldo et al. 1999). In both drainage networks there is a minimum area required to sustain a channel (Montgomery and Dietrich 1988; Rinaldo et al. 1999).

WETLAND FUNCTION AND POSITION IN THE WATERSHED

The hydrological organization of the landscape into watersheds provides a context within which to evaluate the position and possible functions of compensatory mitigation wetlands. Wetlands occur in a variety of physical settings, including coastal lowlands, topographic depressions, broad flats on interstream divides, the base of slopes, and topographic highs with little slope (Winter and Woo 1990). Location in the landscape influences geological characteristics, such as slope; thickness and permeability of soils; and the composition, stratigraphy, and hydraulic properties of the underlying strata, all of which influence surface and subsurface flows of water.

Degradation of wetlands contributes to an overall decrease in watershed ecological function. Watershed scale can include river basins, subbasins or smaller hydrological units or drainage areas, the size of which is dependent on the wetland function(s) of interest. The purpose of this chapter is to demonstrate that these units are hydrologically connected, and thus wetland functions are integrated on a watershed basis. Consequently, wetland mitigation should be considered on a watershed basis. Examples of watershed assessment approaches are presented in Chapter 8.

Several approaches used to determine wetland function focus on the position of wetlands in the landscape. The hydrogeological approach (Winter 1986), hydrogeomorphic approach (Brinson 1993), and hydrological equivalency (Bedford 1996, 1999) all have elements of watershed-scale assessment. Hydrological equivalency and landscape position have been viewed as important components of wetland restoration in recent years (Bedford 1996; Bell et al. 1997).

Restored and created wetlands should be self-sustaining (Mitsch and Wilson 1996); to be self-sustaining, they must be properly sited in the watershed. One way to target mitigation sites to appropriate landscape positions is through the development of basinwide wetland restoration and mitigation plans. The evaluation of watershed position on the functions of existing wetlands and on restored, created, and enhanced wetlands has been aided by the development of new technology. Global positioning system (GPS) technology provides a simple means for locating wetland sites in the landscape. GPS information is easily used with geographical information system (GIS) databases and analysis tools to facilitate watershed analysis. The U.S. Geological Survey supports watershed analysis efforts by providing topographic databases and watershed boundaries that have been coded according to the relative size of the watershed.

How Does Position in the Watershed Affect Hydrology?

One of the most frequently cited functions of wetlands is their ability to reduce the effects of flooding by temporarily storing storm water and gradually releasing it to streams as modulated surface flow (Dennison and Barry 1993) and/or groundwater discharge that constitutes stream base flow. Novitzki (1985) showed that watersheds in the northeastern United States with 4% or greater wetland areas had 50% lower peak flows compared with watersheds without wetland areas. To provide this function, the receiving wetland must occur at a relatively lower topographic elevation within the watershed than the contributing uplands. Typical inland wetlands that provide floodwater storage include riparian or floodplain wetlands.

Riparian wetlands (wetlands immediately adjacent to streams) receive significant groundwater and/or surface-water runoff from a convergent sector of the landscape. This effect of topography on runoff and the development of saturated conditions have been discussed by Dunne and Black (1970) and Dunne and Leopold (1978) and developed into topography-based flow models by Beven (1982). Much of the water carried by stream channels in all regions is first delivered to first- or second-order streams.

In some regions, watershed organizational structure may not be integrated by surface topography, in which case subsurface-water flows dominate the hydrological system. An example of that is the prairie pothole region of North Dakota, South Dakota, and Iowa, where surface-watershed boundaries encompass a small area around each pothole or lake, while in many of these regions the groundwater system is integrated, with small ephemeral potholes that leak groundwater down-gradient to larger and more perennial systems (Winter 1986). Upslope ephemeral potholes recharge groundwater, while the down-gradient potholes constitute groundwater discharge areas. Potholes that remain inundated for prolonged periods of time and still do not attenuate a significant amount of surface runoff are likely dominated by groundwater processes.

How Does Position in the Watershed Affect Water Quality?

As described in Chapter 2, the position of a wetland in a watershed plays an important role in water-quality function. Wetlands improve the water quality of receiving waters by removing nutrients and sediment. To provide this water-quality function, the receiving wetland must also occur at a relatively lower topographic elevation in the watershed than the contributing uplands. Typical inland wetlands that occupy relatively lower landscape positions and provide water-quality functions include riparian or floodplain wetlands, isolated depressional wetlands (such as playas, prairie potholes, and vernal pools), and wetlands at the base of slopes.

Riparian wetlands or buffer zones at the head of stream channels have the greatest opportunity to mediate water quality (due to sediment trapping, denitrification, nutrient uptake, trapping of phosphate sorbed onto soil particles) because their action occurs before the water enters the mainstream channel. There is a large volume of literature describing nutrient cycling and assimilation in many types of wetlands (see review by Vymazal 1995).

The value of riparian wetlands for water quality by preventing nutrients and sediment from entering streams has been shown by many research efforts (Lowrance et al. 1983; Jacobs and Gilliam 1985; Correll and Weller 1989; Groffman et al. 1991; Daniels and Gilliam 1996). Their value for other stream health functions, such as moderating fluctuations in stream temperature, controlling light quantity and quality, enhancing habitat diversity, modifying stream morphology, and enhancing food webs and species richness, may equal their value for pollutant reduction (EPA 1995; USDA 1997). Nitrate reduction by riparian wetlands in sub-surface waters moving toward the stream has been studied more than most other water-quality functions. This function is widely recognized, but the amount of nitrate reduction depends on stream morphology, sediment chemistry, hydraulic conductivity of sediments and soils, carbon content, relative wetness or depth to shallow groundwater, and so forth (Hill 1978; Peterjohn and Correll 1984; Lowrance et al. 1984; Johnston 1993; Bohlke and Denver 1995; Hill and Devito 1997; O'Connell 1999; Prestegaard 2000). The ability of riparian wetlands to remove sediment and phosphorus from surface runoff water may also be diminished by channelized flow (Daniels and Gilliam 1996). Although there are variations in the effectiveness of riparian wetlands for various water-quality functions, in general, they are extremely effective. Because of this, Gilliam et al. (1996) have considered headwater riparian wetlands as the most important factor controlling nonpoint source pollution in humid areas, and it is a national policy of the U.S. Department of Agriculture's Natural Resources Conservation Service to promote use of the general model of riparian buffers presented by Welsch (1991). Wetlands in other watershed locations are important for water quality, but they cannot substitute for the effect of riparian wetlands present on low-order streams.

The value of wetlands for water quality is highly recognized for the Mississippi River drainage basin where a scientific panel from the National Oceanic and Atmospheric Administration (NOAA) recommended significant increases in riparian zones and wetlands to help with hypoxia problems in the Gulf of Mexico (Mitsch et al. 1999). The NOAA committee recommended restoring and/or creating 24 million acres of wetlands and predicted that this increase would reduce the nitrogen input into the Gulf by 40%. This NOAA-appointed committee also recognized that placement of the wetlands in the watershed is of vital importance.

Tremendous water-quality improvement has been documented for constructed storm-water wetlands (Schueler 1992; Brix 1993; Bingham 1994; Brown and Schueler 1997; Malcom 1989) and waste-water treatment wetlands (Reddy and Smith 1987; Hammer 1989; Cooper and Findlater 1990; Moshiri 1993; Corbitt and Bowen 1994; DuBowy and Reaves 1994; Hammer 1997). Constructed wetlands have much potential for assimilating nutrients and improving water quality in a watershed, but treatment wetlands and, to a lesser extent storm-water wetlands, often evolve to dense monoculture stands of *Typha*, *Scirpus*, or *Phragmites*, which will "effectively remove target contaminants from influent waters while providing habitat for a few muskrats, blackbirds and some songbirds but little else" (Hammer 1997). Thus, the use of compensatory wetlands for storm-water treatment may achieve watershed water-quality goals but at the expense of other ecological functions lost or degraded at the impact site. It is important that all lost or impacted functions be mitigated. Storm-water wetlands might best serve watershed water-quality goals, in which case other functions might best be compensated for at another location within the watershed. But such decisions can be effectively addressed only by applying appropriate functional assessment tools on a watershed scale.

In the southeastern United States, there is a large acreage of inland nonriverine wetlands, such as pocosins and pine savannahs, that occur on broad flats and often exist at higher relative elevations in the watershed. The areas are relatively flat, so water moves slowly across the soil surface. They are often located miles from a naturally occurring stream, and excess rainfall can take several weeks to dissipate. The hydrology and degree of wetness are driven by rainfall and evapotranspiration. Although the quality of water discharged from these wetlands is high (Richardson et al. 1978; Richardson et al. 1981; among others), these wetlands do not provide a cleansing water-quality function in the watershed, because they rarely receive natural inputs of poor-quality water (Evans et al. 1993). By releasing storm water slowly, they moderate peak storm flow and provide extended baseflow through the watershed.

Wetlands as Animal Dispersal Corridors in Watersheds

Dispersal of plants and animals is influenced by the proximity and number of wetlands in a geographic area. Connectivity between (Harris 1988) and functional interdependence of wetlands with other landscape units (Bedford and Preston 1988) can also affect animal use because many species (e.g., some amphibians) require an upland-wetland matrix.

Most wetland species of reptiles, amphibians, small mammals, and possibly nonflying invertebrates do not have capabilities for overland migration if terrestrial corridors are obstructed. Birds and flying insects are exceptional in that a disrupted terrestrial landscape can be negotiated without complication, permitting movement to another wetland when necessary.

The functioning of many wetland animal populations on a long-term basis is inherent to the source-sink dynamics of metapopulations that require connectivity in the terrestrial landscape (Gibbs 1993; Burke et al. 1995; Semlitsch 2000). Although populations of many or most wetland animals can fluctuate dramatically in numbers seasonally and annually (Pechmann et al. 1991), most wetland species will remain associated with a particular wetland as long as environmentally suitable conditions persist. However, hydroperiod variability can result in major fluctuations in the numbers of species from year to year (Snodgrass et al. 2000), with the consequence that alternative wetlands must be reached for breeding and feeding opportunities in some years. Many species take advantage of, and actually require, alternative wetlands during periods of drought. To avoid extirpation from natural causes, a variety of isolated wetlands must be accessible by overland routes (see Box 3-2). Species need alternative wetlands in the landscape when a particular wetland experiences a period of environmental duress.

The aquatic and semiaquatic fauna that use wetlands are key components of wetland structure, productivity, and overall functioning. How-

---

BOX 3-2
Ecological Functions of Small, Isolated Wetlands

Of 371 isolated depression wetlands known as "Carolina bays" in South Carolina, most (87%) are smaller than 4 hectares (ha), and 46% are 1.2 ha or smaller. Because they are small, the Carolina bays are more variable than larger wetlands and more likely to dry temporarily during most years. The smaller their size, the lower the probability that species predatory on amphibians, such as fish and dragonfly larvae, will be present during winter and spring when many amphibians are developing. Most fish are restricted to permanent water systems, whereas dragonfly eggs are laid in the warmer months with larvae that persist until the following spring. Thus, if the wetland dries in autumn, neither fish nor dragonfly larvae are present when autumn- and winter-breeding amphibians enter the wetland or while larvae are developing.

Field research in Carolina bays shows that these small isolated wetlands are critical for amphibians. When the bays are of a suitable water depth, they are used for breeding by a wide array of salamanders and frogs. When the smaller wetlands are too dry, the larger bays act as refugia, so that collectively the "metapopulations" of amphibians persist. Additional studies indicate that the maximum dispersal distance for many amphibian species may be less than 1 kilometer (km). The ability of a population to persist is thus limited by the proximity and juxtaposition of small isolated wetlands. As the distance between wetlands increases, the potential for migration and recolonization by amphibians decreases.

Using a GIS with maps of the locations of wetlands of different sizes, Semlitsch and Bodie (1998) showed that if all wetlands smaller than 4 ha were removed, the nearest-wetland average distance would increase from 471 meters (m) to 1,633 m—beyond the critical dispersal distance for most amphibians. The coupling of data on amphibian life histories, dispersal distances, and wetland size and distribution provides convincing evidence that a network of small isolated wetlands is essential for ecosystem function in many regions.

ever, many of the species of animals for which the aquatic portion of a wetland is critical are equally dependent on the surrounding terrestrial habitat. The importance of terrestrial habitat beyond the margin of standard wetland delineation has been unequivocally demonstrated for salamanders and freshwater turtles (Burke and Gibbons 1995; Semlitsch 1998) and is implicit on the basis of the ecology and behavior of other terrestrially dispersing species, including frogs, snakes, and mole crickets (Dole 1965; Semlitsch 1986; Seigel et al. 1995). The issue of including terrestrial habitat in the characterization of wetlands and in evaluating the appropriateness of restored and created wetlands extends to the aspect of terrestrial connectivity between small wetlands in a regional landscape and is an essential feature for assuring the persistence of some wetland species (Semlitsch and Bodie 1998). The biological portion of a functional wetland habitat forms a trophic structure that includes consumers as well as producers; hence, consideration must be given to environmental features of wetlands that are requisite for completion of the life cycle of wetland faunal inhabitants.

On the basis of these facts and principles, the incorporation of animal populations requiring terrestrial movement into the design of compensatory wetlands requires that interwetland distances be taken into account (Semlitsch and Bodie 1998). Local populations can be extirpated and regional species forced to extinction if there are no opportunities for recolonization of wetlands during periods of environmental stress (e.g., extended drought). Also, an undisturbed upland buffer that goes beyond the jurisdictional wetland boundary under the Clean Water Act is essential for some species (Semlitsch and McMillan 1980; Burke and Gibbons 1995; Semlitsch 1998). Therefore, both terrestrial connectivity between wetlands in the landscape and the terrestrial habitat surrounding the prescribed wetland must be considered in designing mitigation wetlands. The ecological requirements for key faunal components of many wetland systems should become a consideration in compensatory mitigation if wetland integrity is to be maintained.

Watershed Position and Self-Sustaining Compensation Projects

A guiding principle in wetland mitigation is that where impacts are permanent, mitigation should be too. However, wetland compensation sites are new features in the landscape, so there must be confidence that the mitigation will protect and preserve desired wetland functions in perpetuity. Permanence means locating, designing, and managing the site for its long-term sustainability in a changing landscape. Permanence also means establishing the institutional means for assuring protection and management of the site over time.

Permanence is promoted when the enhanced, restored, and created wetlands are self-sustaining. A self-sustaining wetland does not require machines or human intervention in order to exist. Water inputs to the wetland come from natural sources (surface water, groundwater, precipitation) without the use of pumps and other water-control structures. Once established, vegetation should be maintained by natural regeneration and competitive selection, as opposed to using herbicides, replacement plantings, and weeding to promote certain plant species over others. In practice, created wetlands are rarely self-sustaining. The committee saw many examples of created wetlands in which costly management practices were implemented during the 5-year monitoring period typically required by design specifications, practices that maintained the wetland in a state that would not be ecologically sustainable should those practices cease (see Appendix B).

To be self-sustaining, wetlands must be properly sited in the landscape. An approach to increase the likelihood of establishing sustainable hydrology is to identify reference wetlands on a landscape or basinwide scale for a wide range of wetland types. Brinson and Rheinhardt (1996) state that the advantages of using a reference wetland approach are (1) making explicit the goals of compensatory mitigation through identification of reference standards from data that typify sustainable conditions in a region, (2) providing templates to which restored and created wetlands can be designed, and (3) establishing a framework whereby a decline in functions resulting from adverse impacts or a recovery of functions following restoration can be estimated both for a single project and over a larger area accumulated over time. Key hydrological parameters that need to be quantified include location, frequency, duration, and timing of saturation or inundation.

Although there are differences in quantifying wetland hydrology, there are also promising new approaches. Bedford (1996) suggests that the numbers needed to quantify hydroperiod and other hydrological variables on a long-term basis are only available for a small number of wetland types. Hunt et al. (1999) suggest that linkages between hydrology and wetland structure have been difficult to quantify, especially when the hydrology is driven by groundwater flow processes. Tweedy (1998) demonstrated that simulation models could be used to predict many of the hydrological parameters for nonriverine wetlands on broad flats (see Figure 3-1), especially such parameters as water-table depth, with reasonable reliability. Suhayda (1997) used simulation modeling to evaluate the impacts of barrier islands on wetland hydrology in Louisiana. Hunt (1996) suggested the use of reference wetland simulations to establish reference wetland hydrological parameters for jurisdictional purposes. A similar approach would seem practical for relating reference wetland hydrology

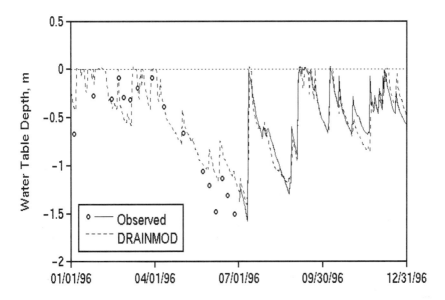

FIGURE 3-1   Comparison between observed and DRAINMOD (hydrological model)-simulated water-table depths for a wetland restoration site in Craven County, N.C., 1996. SOURCE: Tweedy (1998). Reprinted with permission of the author.

to mitigation sites. Short-term monitoring of the reference site could provide the data necessary for model calibration. The calibrated model could then be used to establish hydroperiod relationships based on simulation analyses of long-term climate records.

Constructed, enhanced, or restored wetlands may be particularly vulnerable to external influences because they are still immature and may not have developed resilience to chronic change, catastrophic disturbance, and surrounding population growth and development that bring increased nutrient and contaminant loading and more frequent hydrological changes. For this reason, a site should be able to "evolve" with the landscape over time.

Numerous sites observed by the committee were not positioned in landscape locations that would ensure sustainability. This observation was judged to be due in part to preference of on-site, in-kind mitigation. Some sites were properly located but were threatened by future developments in the watershed, demonstrating that landscape position alone is not sufficient. The problems associated with watershed development include altered hydrology, trash accumulation, and invasive plants and animals. Once a watershed is developed, it may be impossible to provide conditions that are favorable to the mitigation site.

Other external factors that may impinge on the long-term ecological sustainability of a mitigation wetland include deleterious influences of natural pest species (e.g., intense grazing by herbivores such as migratory or resident geese) and large-scale disturbances such as hurricanes, fire, sea-level rise, and climate change (see Box 3-3). The committee recognizes that it is impractical to expect individual permittees to design for and be

---

BOX 3-3
Sea-Level Rise and Wetlands Placement

Sea-level rise, caused by both natural processes and increased concentrations of greenhouse gases, threatens the sustainability of coastal wetlands because (1) sea level will continue to increase for the foreseeable future, (2) a large rise in sea level will cause a net loss of wetlands, and (3) coastal development will block the natural inland migration of wetlands (Titus 1999). Global sea-level rise estimates vary dramatically, but 1.8 to 1.9 mm/yr may be a reasonable median rate, including correction for postglacial rebound (Douglas 1995, 1997). Based on current projections of greenhouse gas emission rates, with no future remedial reductions, sea level may rise from 31 to 110 cm by 2100 (IPCC 1990). A 50-cm rise in sea level would involve inundation of 24,000 square kilometers (km^2) in the United States (Neumann et al. 2000). The areas most vulnerable to sea-level rise are in the mid- and south-Atlantic states and along the Gulf Coast, where land subsidence is also a concern, although parts of New England, San Francisco Bay, and Puget Sound also are vulnerable (Neumann et al. 2000). For example, it is estimated that 21% (22,000 acres) of Delaware's coastal emergent wetlands would be inundated (MARA Team 2000). Inundation would not be the only threat; storm frequency, intensity, and surge levels also would increase.

The contingency of climate change and sea-level rise argues for landscape-scale planning and implementation of wetland restoration, creation and enhancement, and preservation. The consequences of increased temperatures and reduced precipitation may need to be designed into mitigation projects particularly vulnerable to changes in flooding duration and frequency in wetlands such as prairie pothole and peatland wetlands. Drier wetlands, such as depressional, slope, flats, and river and lake fringe wetlands (Brinson 1995), may need additional design features to ensure protection of the proper hydrological regime. Inland migration of coastal emergent marshes, mangroves, forested wetlands, and seagrass and other submergent vegetation systems may need to be accommodated to some extent through, for example, "managed retreat" and reduction of armored shorelines. Strategic restoration of coastal marshes by breaching of dikes and levees may need to be advanced to accommodate an increased tidal prism and coastal erosion in estuaries. Marsh sediment accretion rates must be maintained by the preservation and enhancement of sediment sources and transport patterns and rates. Dams further impair sustainability of downstream wetlands by eliminating sediment transport that could counteract rising sea levels.

held accountable for such long-term, uncontrollable factors. However, watershed assessment and prioritization provide a framework for identifying and avoiding future high-risk areas for mitigation sites. The contingency of climate change and sea-level rise argues for landscape-scale planning and implementation of wetland creation, restoration, and preservation.

WATERSHED-SCALE PATTERNS OF WETLAND LOSSES

As discussed in Chapter 1, wetland losses have occurred with changes in runoff and erosion due to urbanization and agricultural land uses. Other factors that result in both direct fill/destruction and indirect impacts and wetland losses include channelization, groundwater withdrawal, and flood-control practices.

Losses Due to Urbanization

Urbanization of watersheds is often extensive in headwater regions. In older, built-out urban areas, headwater wetlands and wetlands along first-order streams may have been put into storm sewer networks. This loss of streams and springs is well documented in some regions (e.g., Williams 1977). The loss of wetlands in this context can be evaluated by a comparison of wetlands and their distribution in urban and adjacent nonurban watersheds.

Losses Due to Agricultural Uses

The position of the stream channel head in the landscape is controlled by runoff processes and surface topography (Dietrich et al. 1986). An increase in overland flow tends to move a stream channel upslope because less area is required to initiate the channel head. This upslope migration of stream channels has been documented in agricultural areas and has often resulted in the loss of headwater wetlands and some first-order stream wetlands. Thus, the pattern of wetlands in a watershed often reflects previous land-use practices. For example, channel incision in Wisconsin, Maryland, and Pennsylvania has resulted in significant loss of wetlands in headwater positions and along first-order streams (Prestegaard 1986; Prestegaard and Matherne 1992).

Many wetlands have also been lost due to land drainage for agricultural or other land uses. For examples, sedge meadows, wet prairies, and other wetlands were easily drained for agriculture in central Wisconsin (Curtis 1959), Iowa, and elsewhere in the Midwest (Prince 1997). In this process, unchannelized portions of the landscape are channelized into existing watersheds, thus extending the stream network. Channelization of downstream portions of these river networks often deepens the existing river channel and destroys adjacent riparian zones and wetlands (Prestegaard et al. 1994). Inadvertent channel network changes have also occurred as a result of agricultural and urban land uses. Increased runoff from agricultural lands has generally caused a headward migration of stream channels in many areas. This leads to incised stream channels in many headwater regions (Costa 1975) and loss of headwater wetlands (McHugh 1989; Prestegaard and Matherne 1992). Thus, channelization practices have led to the loss of both prairie pothole wetlands that were not originally part of watershed systems and riparian wetlands along the original river courses.

Losses Due to Groundwater Withdrawals

Groundwater withdrawals have particularly affected wetlands and riparian zones along higher-order streams in arid and semiarid regions. An example is provided by Stromberg et al. (1996), who demonstrated the effect of groundwater withdrawals on riparian zones and riparian wetlands in arid regions.

Wetland Losses Due to Flood-Control Practices

Wetland losses have also occurred as a result of flood-control practices. For example, levees restrict connections between the river and the adjacent flood plain, affecting riparian wetlands. Levees, reservoirs, and other flood-control structures also serve to modify the timing of flood events, either by minimizing the size or modifying the frequency and duration of flood flows. Infrequent flooding can modify floristic communities in flood-plain areas, often allowing the development of forests in formerly herbaceous wetlands (Bren 1992). The importance of flooding, particularly in large (downstream) river systems, has been emphasized as the flood-pulse concept (Bayley 1995; Bornette and Amoros 1996; Middleton 1999).

A WATERSHED TEMPLATE FOR WETLAND
RESTORATION AND CONSERVATION

Several authors have argued for a hydrogeological or hydrogeomorphic template for wetland mitigation and development (Moore and Bellamy 1974; Bedford 1996; 1999). This would suggest compensation projects that would be selected based on set functional priorities of the watershed. In practice, some in-lieu fee programs (see Chapter 4) have already stated such a watershed orientation for selecting compensation projects (Scodari and Shabman 2000). In addition, a watershed perspective may suggest preservation (Kentula 1999; Winston 1996) as an integral part of maintaining wetland heterogeneity in watersheds.

CONCLUSIONS

1. Watershed organizational structure governs the flow of water and associated nutrients through a watershed, the relationship between hydrological processes and the position of a wetland in the watershed, and the relationship between wetland functions and watershed position.

2. Wetland location and position in the landscape influence surface and subsurface flows of water.

3. Equivalency of hydrological conditions and landscape position with reference systems and impact sites are viewed as important components of wetland restoration and creation.

4. Restored and created wetlands should be self-sustaining; to be self-sustaining, they must be properly sited in the watershed.

5. The position of a wetland in a watershed plays an important role in water-quality function.

6. Dispersal of plants and animals in a watershed is influenced by the proximity and number of wetlands in a geographic area and the functional interdependence of wetlands with other landscape units.

7. Numerous mitigation sites observed by the committee were not positioned in landscape locations that would ensure sustainability.

RECOMMENDATIONS

1. Site selection for wetland conservation and mitigation should be conducted on a watershed scale in order to maintain wetland diversity, connectivity, and appropriate proportions of upland and wetland systems needed to enhance the long-term stability of the wetland and riparian systems. Regional watershed evaluation should greatly enhance the protection of wetlands and/or the creation of wetland corridors that mimic natural distributions of wetlands in the landscape.

2. Riparian wetlands should receive special attention and protection because their value for stream water quality and overall stream health cannot be duplicated in any other landscape position.

4

Wetland Permitting: History and Overview

This chapter describes the evolution of compensatory mitigation requirements in the Clean Water Act (CWA) Section 404 program, including agency guidance on the use of mitigation banks and in-lieu fees. Also noted is the somewhat limited role that CWA compensatory mitigation plays in the attempt to achieve no net loss of the nation's remaining wetland base. The chapter concludes with a brief overview of the CWA Section 404 permitting process.

EVOLUTION OF COMPENSATORY MITIGATION REQUIREMENTS IN THE CWA SECTION 404 PROGRAM

The U.S. Army Corps of Engineers (Corps) makes its decisions regarding mitigation requirements within a framework of multiple statutes, regulations, guidance, and policy documents (see Box 4-1). These provisions include general mitigation requirements (i.e., ones that derive from sources other than the CWA and that may apply to all federal agencies in general); general Corps policies for evaluating permit applications (i.e., requirements that apply to all permit programs the Corps administers); and CWA-specific mitigation requirements (i.e., obligations that apply solely in the Section 404 context and that flow from the CWA, the Section 404(b)(1) guidelines, and policy documents). After examining these requirements, agency guidance regarding the use of mitigation banks and in-lieu fees is reviewed.

BOX 4-1
**Timeline of Significant Federal Actions Regarding Wetland
Permit and Mitigation Requirements**

1890s Rivers and Harbors Act enacted (the earliest regulation of activities in waters of the United States)
1934 Fish and Wildlife Coordination Act enacted
1968 Corps Public Interest Review promulgated
1969 National Environmental Policy Act enacted
1972 Federal Water Pollution Control Act (FWPCA) enacted
1973 Endangered Species Act enacted
1975 U.S. Environmental Protection Agency (EPA) Section 404(b)(1) Guidelines Promulgated
1977 FWPCA amended (as Clean Water Act)
1980 EPA Section 404(b)(1) Guidelines revised
1990 U.S. Congress instructs the Corps to pursue the goal of "no overall net loss" (Section 307, Water Resources Development Act)
1990 Corps and EPA mitigation Memorandum of Agreement calls for sequencing and establishes preferences for on-site, in-kind mitigation
1993 Corps and EPA Joint Memorandum (Interim Guidance) on Mitigation Banking issued
1995 Interagency Mitigation Banking Guidance issued
1998 Transportation Equity Act for the 21st Century enacted, expressing congressional preference that mitigation for highway projects be supplied by mitigation banks
2000 In-Lieu Fee Guidance issued

GENERAL MITIGATION REQUIREMENTS

Fish and Wildlife Coordination Act

Mitigation to offset the impacts of dredging and filling projects is not a new concept for the federal government. Indeed, in 1934 the Fish and Wildlife Coordination Act required federal agencies that construct or permit dams to consult with the then-existing Bureau of Fisheries to make provisions for fish migration (Public Law 73-121). Subsequent amendments to the act now require federal agencies that engage in or permit projects that modify bodies of water to consult about habitat loss with the Fish and Wildlife Service (FWS). Thus, prior to making Section 404 permit decisions, the Corps must discuss with the FWS a proposal's impact on wildlife resources. The act calls on the FWS to advise federal agencies about proposed projects' impacts on fish and wildlife habitats and to recommend compensatory mitigation measures. Agencies are not, however, required to follow the FWS's recommendations (*Sierra Club* v. *Alexander*, 484 F. Supp. 455 (D.C.N.Y. 1980)).

National Environmental Policy Act of 1969 (NEPA)

The NEPA also requires federal agencies to consider mitigation measures before taking action, including the granting of federal permits, that may have adverse environmental consequences (Public Law 91-190). The Council on Environmental Quality (CEQ), which is responsible for overseeing federal compliance with the NEPA, has promulgated regulations that are binding on federal agencies (40 CFR §§ 1500-1517 (2000)). CEQ defines mitigation to include (a) avoiding the impact altogether by not taking a certain action or parts of an action; (b) minimizing impacts by limiting the degree or magnitude of the action and its implementation; (c) rectifying the impact by repairing, rehabilitating, or restoring the affected environment; (d) reducing or eliminating the impact over time by preservation and maintenance operations during the life of the action; and (e) compensating for the impact by replacing or providing substitute resources or environments.

The Corps must discuss mitigation options in its NEPA documentation when examining alternatives to the proposed action. Similar to the Fish and Wildlife Coordination Act, NEPA largely imposes procedural requirements. Accordingly, federal agencies must consider the need for mitigation to compensate for federal actions (including the granting of a permit), but NEPA does not mandate that the agencies perform or require mitigation (*Robertson* v. *Methow Valley Citizens Council*, 490 U.S. 332 (1989)).

Endangered Species Act (ESA)

In contrast to the Fish and Wildlife Coordination Act and the NEPA, the ESA has more substantive mitigation requirements (Public Law 93-205, as amended). For example, when a federal agency proposes to take an action (including the granting of a permit), the agency may need to consult with the FWS or the National Marine Fisheries Service (NMFS) to ensure that the proposed action will not violate the ESA. After consultation, the FWS or NMFS may issue a biological opinion that contains "reasonable and prudent alternatives" that the agency must follow to comply with the ESA. Additionally, any ESA permit that authorizes the taking of a protected species must specify how the applicant will "minimize and mitigate the impacts of such taking."

Food Security Act (FSA)

Although the mitigation requirements of the FSA are limited to agricultural activities and are not directly applicable to the CWA Section 404 program, they deserve some mention. To discourage farmers from converting wetlands into agricultural areas, the FSA, through its swamp-buster program, penalizes landowners who plant agricultural commodities in converted wetlands (Strand 1997). Such landowners may become ineligible for certain federal agricultural loans and payments. A landowner may retain eligibility for federal benefits, however, by performing compensatory mitigation: restoring, enhancing, or creating wetlands. The FSA presumes that the mitigation will be provided on "a 1-for-1 acreage basis," although more mitigation may be required if needed to offset lost wetland functions and values (Public Law 104-127). The FSA requires that such mitigation be "in the same general area of the local watershed as the converted wetland" and that a conservation easement be placed on the mitigation site.

GENERAL CORPS MITIGATION REQUIREMENTS

Long before assuming its responsibilities under the CWA, the Corps administered a regulatory program under the Rivers and Harbors Acts (RHAs) of 1890 and 1899 for work conducted in traditionally navigable waters (Strand 1997). For example, Section 10 of the RHA of 1899 declared excavating or filling such waters to be illegal without a Corps permit. For many years the Corps based its permit decisions "primarily upon the effect of the proposed work on navigation." Environmental impacts were generally not considered.

In 1967 the Secretary of the Interior and the Secretary of the Army entered into a Memorandum of Understanding (MOU) that articulated how the Corps would implement its obligation to consult with the FWS under the Fish and Wildlife Coordination Act when it made its RHA decisions regarding dredging, filling, excavating, and other related work in traditionally navigable waters (Fed. Regist. 33(Dec. 18):18672-18673). The MOU recognized that permits to conduct work in these waters may, as a result of the consultation, include conditions the Corps determined "to be in the public interest."

The following year the Corps codified the MOU in its regulations and announced a significant shift in its RHA permit decision criteria (Fed. Regist. 33(Dec. 18):18671). Rather than focusing on navigational impacts, the Corps would now evaluate "all relevant factors, including the effect of the proposed work on navigation, fish and wildlife, conservation, pollution, aesthetics, ecology, and the general public interest." Under this public-interest review, the Corps could deny RHA permit applications based on environmental impacts or impose permit conditions to alleviate those impacts. Courts subsequently affirmed the Corps's authority to consider the environmental impacts of its permitting decisions (*Zabel* v. *Tabb*, 430 F.2d 199 (1971)).

In 1977 revisions to its regulations (by which time the Corps had assumed CWA Section 404 responsibilities), the Corps again noted its obligation to consult with the FWS under the Fish and Wildlife Coordination Act (Fed. Regist. 42(July 19):37137). The regulations also expressly provided that an "applicant will be urged to modify his proposal to eliminate or mitigate any damage to such [fish and wildlife] resources, and in appropriate cases the permit may be conditioned to accomplish this purpose." In 1982 the Corps's regulations expanded the authority to add conditions when necessary to satisfy a legal requirement (e.g., the ESA) or to meet a public-interest objective, which now encompassed all environmental impacts (Fed. Regist. 47(July 22):31794). The last significant revision to the Corps's general mitigation policies occurred in 1986 when the Corps emphasized that if a permit applicant declined to provide compensatory mitigation needed to ensure that the project was not contrary to the public interest, the district engineer must deny the permit (Fed. Regist. 51 (Nov. 13):41206). Furthermore, the Corps pointed out that this general statement of mitigation policy was separate from and did not supercede any compensatory mitigation required by the CWA Section 404(b)(1) guidelines.

The factors that the Corps now considers in its public-interest review are found at 33 CFR 320.4. In its review, the Corps must specifically evaluate the proposed activity's likely effect on wetlands. See Appendix I for a list of the factors that the Corps must take into account in its permit decision-making process.

CWA SECTION 404 MITIGATION REQUIREMENTS

The terms "mitigate" and "mitigation" do not appear in CWA Section 404. Nor does Section 404 expressly authorize the Corps to require mitigation of permit applicants. Nevertheless, by virtue of the interplay between Sections 404(b)(1) and 403(c), the statute does provide implicit authority for the Corps to require permit applicants to avoid and minimize wetland impacts. Section 404(b)(1) requires EPA, in conjunction with the Corps, to develop the criteria that the Corps uses in its Section 404 permit decisions. These criteria, known as the 404(b)(1) guidelines, must be based on criteria identified in CWA Section 403(c). These criteria are applicable to ocean-based discharges (Federal Water Pollution Control Act (Public Law 92-500)). Section 403(c) requires consideration of alternative disposal sites, including land-based sites, and minimization of adverse environmental impacts. Accordingly, EPA's Section 404(b)(1) guidelines must also consider alternatives (i.e., avoidance) and minimization of unavoidable impacts.

The EPA issued its Section 404(b)(1) guidelines in 1975 (Fed. Regist. 40(Sept. 5):41292-41298)). In their first iteration the guidelines stressed the need to avoid wetland impacts. If no less environmentally damaging practicable alternative existed and if a project would not cause unacceptable adverse impacts on aquatic resources, the Corps could issue a permit. The guidelines also called for the impacts of a permitted project to be minimized. No mention was made of restoration, enhancement, or creation of wetlands, although the 1975 guidelines stated that "[c]onsideration shall be given to preservation of submersed and emergent vegetation."

The current Section 404(b)(1) guidelines, promulgated in 1980, reaffirmed the avoidance and minimization requirements in greater detail (Fed. Regist. 45(Dec. 24):85336-85357). Subpart H (230.70-77) describes a number of actions that the Corps should consider as permit conditions to minimize adverse effects—for example, actions concerning the location of discharge, composition of discharge material, control of material after discharge, method of dispersal, and use of appropriate equipment and technology. Included in the minimization discussion is a reference to compensatory mitigation: "Habitat development and restoration techniques can be used to minimize adverse impacts and to compensate for destroyed habitat." Thus, in the Section 404(b)(1) guidelines, compensatory mitigation, such as the restoration of wetlands, is a subset of minimization.

Additional support for compensatory mitigation can also be found elsewhere in the Section 404(b)(1) guidelines. The guidelines require that a permitted activity not cause or contribute to significant degradation of the waters of the United States, either individually or cumulatively. When determining whether a proposed activity will result in significant degradation, the Corps will consider to what extent compensatory mitigation will offset the activity's adverse effects.

The Section 404(b)(1) guidelines were developed in accordance with the Administrative Procedure Act's public notice and comment procedures and are binding regulations that appear in the Code of Federal Regulations. Frequently, agencies turn to less formal documents, such as MOAs or regulatory guidance letters (which are typically not subjected to public notice and comment) to interpret the requirements of the CWA and the Section 404(b)(1) guidelines (Gardner 1991). These documents are issued to provide guidance to agency personnel and the public to explain how the agencies intend to apply the statute and regulations in the field. Much of the detail of the mitigation policies for the Section 404 program is found in these guidance documents.

A 1990 MOA between EPA and the Department of the Army explains how mitigation determinations should be made (Fed. Regist. 55(Mar. 12):9210). The MOA notes that the mitigation requirements of CEQ's regulation and the CWA Section 404(b)(1) guidelines are compatible and, as a practical matter, may be condensed to three general types of mitigation: avoidance, minimization, and compensatory mitigation (see Figure 4-1). The MOA emphasizes that this mitigation must be applied in a sequential fashion: an applicant must first avoid wetlands to the extent practicable; then minimize unavoidable impacts; and, finally, compensate for any remaining impacts through restoration, enhancement, creation, or, in exceptional cases, preservation. With respect to compensatory mitigation, the MOA expresses a preference for on-site, in-kind mitigation, with restoration as the first option considered.

The Corps and EPA mitigation MOA's sequencing requirement is limited in several important respects. First, the MOA applies only to individual permits, not general permits such as nationwide permits. Significantly, the Corps has reported that up to 85% of authorized projects in waters of the United States proceed under a general permit (Davis 1997).

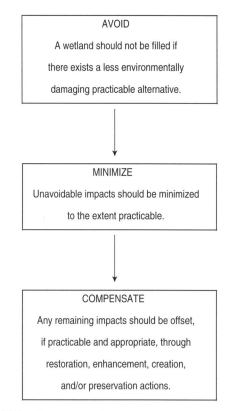

FIGURE 4-1   Mitigation sequencing.

Second, subsequent guidance documents have relaxed the rigorousness of the avoidance requirement for individual permits when the wetland impact site is of low environmental value or when the permit applicant is a small landowner (Fed. Regist. 65(Mar. 9):12518-12519).

MITIGATION BANKING

The Corps and EPA define a mitigation bank as "a site where wetlands and/or other aquatic resources are restored, created, enhanced, or in exceptional circumstances, preserved expressly for the purpose of providing compensatory mitigation in advance of authorized impacts to similar resources" (Fed. Regist. 60 (Nov. 28):58605). Mitigation banks provide off-site mitigation. Mitigation banks may be established by permittees who anticipate having a number of future permit applications or by third parties who develop wetland credits for sale to permittees needing to provide compensatory mitigation. Although the Corps and EPA MOA expressed the agencies' preference for on-site compensatory mitigation, it does acknowledge that "[m]itigation banking may be an acceptable form of compensatory mitigation under specific criteria designed to ensure an environmentally successful bank." The Corps and EPA MOA promised additional guidance on mitigation banking; that was forthcoming in the form of a Corps and EPA joint memorandum to the field issued in 1993.

The 1993 memorandum announced interim national guidance for mitigation banking in the Section 404 program (Fed. Regist. 60(Mar. 14):13711, also Corps Regulatory Guidance Letter 93-02). The interim guidance identified several benefits of relying on mitigation banks rather than individual mitigation projects. First, because mitigation is done in advance of project impacts, it reduces "temporal losses of wetland functions and uncertainty over whether the mitigation will be successful in offsetting wetland losses." Second, the agencies suggested that it may be ecologically advantageous to have consolidated mitigation sites instead of smaller isolated projects. Third, the agencies believed that mitigation banks were more likely to marshal together the financial resources and scientific expertise necessary for effective mitigation projects. Fourth, the guidance also stated that the economies of scale resulting from mitigation banks should lead to "cost-effective compensatory mitigation opportunities" for permit applicants. The guidance emphasized that mitigation banks could provide compensatory mitigation only and thus could be available to offset wetland impacts only where the permit applicant had complied with the sequencing requirement (i.e., avoid first, then minimize impacts before compensating).

The 1993 interim national guidance explained that the Corps, EPA, and other relevant agencies should enter into an agreement with the entity establishing the bank. Such agreements should address the bank's location, its goals and objectives, the bank's sponsor and participants, a development and maintenance plan, an evaluation and assessment methodology, procedures for crediting and debiting mitigation gains, the geographic service area, monitoring provisions, remedial actions, and long-term protection of the site. The interim guidance authorized the establishment and use of third-party banks (i.e., banks operated by entities other than the permittee, such as an entrepreneurial bank) but noted that permittees must remain legally responsible for ensuring mitigation compliance.

In 1995 the Corps, EPA, FWS, Natural Resources Conservation Service (NRCS), and National Oceanic and Atmospheric Administration (NOAA) published interagency federal guidance on the establishment, use, and operation of mitigation banks (Fed. Regist. 60(Nov. 28):58605). The 1995 guidance, which supplanted the Corps and EPA's 1993 interim guidance, was subjected to public notice and comment prior to its issuance. Nevertheless, the preamble to the 1995 guidance notes that it is an interpretive rule and not a regulation that has the force of law.

The 1995 guidance identified two additional benefits that mitigation banking offers. First, consolidation of mitigation at a bank makes agency compliance monitoring more effective. Rather than visiting many individual mitigation sites, a regulator need visit only a single bank site. Second, the agencies suggested that the availability of mitigation banks could make compensatory mitigation appropriate and practicable in cases where the Corps had not previously required any compensatory mitigation, especially for authorized activities under general permits. In this way, the agencies reasoned, mitigation banking could contribute to the attainment of the goal of no net loss of wetlands.

The 1995 guidance elaborated on the mitigation banking approval process in greater detail. To initiate the process, the bank sponsor, the entity responsible for establishing the bank, should submit a prospectus to the Corps. The prospectus is a document that describes the bank sponsor's proposal on how the bank will be established and operated. An interagency group, the Mitigation Banking Review Team (MBRT), which may consist of representatives from federal, state, tribal, and local agencies, reviews the proposal. For mitigation banks in the Section 404 program, the Corps representative serves as MBRT chair.

The MBRT review process may lead to the development of a formal agreement, the banking instrument. The 1995 guidance expands on the 1993 interim guidance with respect to what information should be included in a banking instrument: ownership of the bank site; bank size and types of wetland classes that will be included in the bank, with a site plan and specifications; a description of the baseline (existing) conditions at the bank site; wetland impacts suitable for compensation; financial assurances; and compensation ratios. The 1995 guidance also provides greater detail about the timing of credit withdrawal and financial assurances. For example, the guidance allows limited early withdrawals of mitigation credit prior to meeting mitigation performance standards. The guidance cautions, however, that early withdrawals should be done only when there is a high likelihood that the bank will achieve its requirements. Furthermore, the MBRT must have approved the banking instrument and plans; the bank site must have been secured; the bank sponsor must have provided financial assurances (e.g., performance bonds and escrow accounts); initial physical and biological work must be finished no later than the first full growing season after the early withdrawals; and the banking instrument may impose higher compensation ratios.

With respect to financial assurances, the 1995 guidance states that a bank sponsor should secure sufficient funds for remedial actions in the event that the mitigation project fails or founders. Moreover, the bank sponsor must provide financial assurances for monitoring and maintaining the bank through its operational life (while the credits are generated and debited) and for long-term monitoring and maintenance. Most significantly, the 1995 guidance also shifts legal responsibility for compliance of the mitigation site from the permittee to the bank sponsor.

In 1998, Congress expressed its preference that mitigation banks be used to offset wetland impacts from federally funded transportation projects (Public Law 105-178). The Transportation Equity Act for the 21st Century states that the Corps shall give mitigation banks preference "to the maximum extent practicable," if banks are approved in accordance with the 1995 guidance and sufficient credits are available.

IN-LIEU FEES

In addition to authorizing individual mitigation projects and mitigation banks to satisfy a permittee's compensatory mitigation obligation, the Corps has sanctioned the use of in-lieu fee mitigation. The Corps and EPA define an in-lieu fee as a payment "to a natural resource management entity for implementation of either specific or general wetland or other aquatic resource development projects, [which] . . . do not typically provide compensatory mitigation in advance of project impacts" (Fed. Regist. 60(Nov. 28):58605). In the 1995 guidance on mitigation banking, the agencies suggested that the sponsor of an in-lieu fee account should enter into an agreement, "similar to a banking instrument," to define the conditions when in-lieu fee mitigation is appropriate.

For several years after the 1995 mitigation banking guidance, policy statements about the use of in-lieu fees appeared in the *Federal Register* in the Corps's preamble discussion regarding its nationwide permit program. Traditionally, the Corps did not require any compensatory mitigation to offset the impacts of activities authorized by nationwide permits (NWPs). In 1996, however, the Corps concluded that compensatory mitigation may be appropriate for some NWPs; permittees could satisfy the mitigation requirement through credits from mitigation banks or in-lieu fee payments "to organizations such as The Nature Conservancy, state or county natural resource management agencies, where such fees contribute to the restoration, creation, replacement, enhancement, or preservation of wetlands" (Fed. Regist. 61(Dec. 13):65874-65922). In a 2000 notice in the *Federal Register*, the Corps expressed its preference that compensatory mitigation for NWPs come from consolidated mitigation approaches, which include mitigation banks and in-lieu fees (Fed. Regist. 65(Mar. 9):12818-12899). At that time, however, the Corps declined to express a preference between mitigation banks and in-lieu fee arrangements.

In November 2000 the Corps, EPA, FWS, and NOAA issued interagency guidance on the use of in-lieu fees to offset wetland fill impacts (Fed. Regist. 65(Nov. 7):66914). That guidance reiterated the Corps and EPA mitigation MOA preference for on-site, in-kind mitigation but recognized that such mitigation may not always be available, practicable, or environmentally preferable. With respect to compensating for impacts from individual permits, the guidance provides that in-lieu fee arrangements may be used if there is a formal agreement that is developed, reviewed, and approved through the interagency MBRT process.

For impacts from general permits, the 2000 guidance offers more detail. As a general rule, the agencies prefer on-site mitigation to off-site mitigation. When, however, off-site mitigation is permitted, the agencies state that the "use of a mitigation bank is preferable to in-lieu fee mitigation where permitted impacts are within the service area of a mitigation bank approved to sell mitigation credits, and those credits are available." The preference for mitigation banks does not apply when (1) the mitigation bank does not provide in-kind mitigation and the in-lieu fee arrangement offers in-kind restoration, or (2) the mitigation bank provides only preservation credits and the in-lieu fee arrangement offers in-kind restoration. The 2000 guidance requires that in-lieu fee sponsors who wish to offset impacts from activities authorized by general permits enter into a formal agreement with the Corps. The in-lieu fee agreement should contain provisions very similar to those in mitigation banking agreements.

THE CLEAN WATER ACT AND THE GOAL OF NO NET LOSS

The CWA vests the Corps (or a state with an EPA-approved program) with the authority to issue a Section 404 permit and to decide whether to attach conditions. Section 404 permit conditions may include a requirement to provide compensatory mitigation. However, it is important to note how the Corps and EPA—the federal agencies charged with administration of the CWA Section 404 program—interpret the program's goals.

The agencies state that a goal of the program is to seek no overall net loss of wetland functions and values (Fed. Regist. 55(Mar. 12):9210). The no-net-loss goal is a statement of policy or an interpretive rule that the agencies articulated in their 1990 mitigation MOA. Congress subsequently established, through the Water Resources Development Act of 1990, a no-net-loss goal for the Corps's water resources development program. However, the no-net-loss goal does not appear in Corps or EPA regulations, and the statutory goal does not specifically apply to the Corps's regulatory program.

In the 1990 mitigation MOA, the Corps and EPA recognize that no net loss may not be satisfied in every Section 404 permit action but emphasize that a goal of the CWA Section 404 program "is to contribute to the national goal of no overall net loss of the nation's remaining wetland base." Thus, the agencies acknowledge two important limitations of the CWA Section 404 program with respect to wetland mitigation. First, the program is not designed to remedy historical losses of wetlands; rather, it focuses on existing or remaining wetland functions and values. Second, the program is not expected to achieve the goal of no net loss of existing wetland functions and values by itself. Accordingly, before examining the permit process and the technical aspects of wetland mitigation, it may be instructive to consider the somewhat limited role that wetland mitigation in the CWA Section 404 program plays in efforts to achieve the goal of no net loss.

The CWA does not vest the federal government with the authority to assert jurisdiction over all wetlands and all wetland-damaging activities. The geographic scope of the CWA is limited by two main sources: the language of the CWA itself and the U.S. Constitution. The CWA provides the Corps with jurisdiction over "waters of the United States." The Corps had interpreted this phrase to include isolated waters, including isolated wetlands, that provided habitat to migratory birds. In *Solid Waste Agency of Northern Cook County* v. *U.S. Army Corps of Engineers*, however, the U.S. Supreme Court ruled that the Corps's "migratory bird rule" was an unreasonable reading of the plain language of the CWA. The Supreme Court's decision leaves open the possibility that Congress may amend the CWA to make it clear that the act should regulate activities in isolated waters, including isolated wetlands. If Congress chooses to do so, however, the decision suggests that such an exercise of federal power may be constitutionally suspect unless there is a sufficient nexus to Congress's power over interstate commerce.

Regardless of where the line is drawn for geographic jurisdiction, another limitation of the CWA Section 404 program concerns activities that trigger a permit (and perhaps a mitigation) requirement. Jurisdiction of the CWA Section 404 program is activity specific; the program only regulates the discharge of dredged or fill material (Want 1994). Agencies have no CWA jurisdiction over other activities that destroy wetlands. Some activities that result in wetland impacts are not regulated by the Section 404 program and thus are not subject to its mitigation requirements. For example, the mere draining of a wetland does not trigger Section 404. Nor may the Corps require a permit for draining, dredging, or excavation activities that result in only incidental fallback of dredged material (*National Mining Association* v. *U.S. Army Corps of Engineers*, 145 F.3d 1399 (D.C. Cir. 1998)). While it may be difficult and expensive for developers to excavate and drain wetlands in a manner that avoids triggering Section 404, it is technically feasible for a developer to legally drain hundreds of acres of wetlands outside the scope of the CWA. Indeed, the case that prompted the Corps to issue the so-called Tulloch rule involved the draining of approximately 700 acres of pocosin wetlands in North Carolina (Gardner 1998). Moreover, even some activities that constitute the discharge of dredged or fill material are not subject to CWA regulation because of the statutory exemption for normal agricultural, silvicultural, and ranching activities. The Corps is not required to report losses on activities they do not regulate; therefore, loss of these wetlands from such unregulated activities does not figure into the Corps's calculations when it declares that it requires more acres in mitigation than it permits to be filled.

Even if an activity is regulated by CWA Section 404, that fact alone does not lead to the conclusion that mitigation will be required to offset the loss of functions and values. As will be explained in more detail below, the Corps issues two types of Section 404 permits: individual permits and general permits. Individual permits, which include standard permits, are issued on a case-by-case basis. Applications for activities that require a standard permit are subjected to more rigorous review, and the sequencing requirement may apply. The Corps must issue a public notice prior to making its permit decision and must decide what level of compensatory mitigation, if any, is appropriate in each particular case. The Corps and EPA report that approximately 15% of activities authorized under the Section 404 program proceed under an individual permit; most activities are authorized by general permit (Davis 1997). Section 404 authorizes the Corps to issue general permits (which may be nationwide permits, regional permits, or programmatic permits) for any category of activity, if the activity will cause only minimal individual and cumulative adverse environmental impacts (Strand 1997). In contrast to individual permits, general permits authorize activities to occur in wetlands with little agency oversight. Indeed, some activities authorized by general permit allow the permittee to proceed without notifying the Corps. Initially, the Corps required little or no compensatory mitigation to offset wetland impacts for activities authorized by general permit (33 CFR Parts 320-330, Nov. 13, 1986). Significant changes, however, have been implemented in the permitting program. First, the types of activities and size of fill eligible for a general permit have been limited. As a result, more fills require individual permits and permit recipients are expected to provide compensatory mitigation. Second, the Corps now more frequently imposes compensatory mitigation requirements for general permits.

In sum, some isolated wetlands are beyond the scope of the Section 404 program. Moreover, some wetland-damaging activities are not subject to CWA Section 404 regulation. Most activities that do trigger Section 404 are authorized by general permit and thus may not require compensatory mitigation. For those activities that require an individual Section 404 permit, however, the Corps frequently imposes a compensatory mitigation condition. It should be noted that the Section 404 program may lessen impacts to the aquatic environment better than would occur with no such requirements, because developers are encouraged to avoid or reduce impacts through the sequencing process of individual permit requirements or by further reducing project impacts to allow use of the simpler general permit program (33 CFR Parts 320-330, Nov. 13, 1986). The data presented and the discussion of those data in Chapter 1 illustrate this possibility.

SECTION 404 PERMIT PROCESS

Section 404 of the CWA authorizes the Department of the Army to issue permits for discharge of dredged or fill material into waters of the United States, including wetlands. The Corps categorizes Section 404 permits as either standard or general permits. Individual permits include standard permits and letters of permission, while general permits include regional permits, nationwide permits, and programmatic permits for projects that should result in only minimal impacts to the aquatic environment.

Standard Permits

The most common form of individual permit is the standard permit. Standard permits are issued after a "case by case evaluation of a specific project involving the proposed discharges in accordance with the procedures of this part and 33 CFR part 325 and determination that the proposed discharge is in the public interest pursuant to 33 CFR part 320" (Fed. Regist. 51(Nov. 13):41206). A review of an individual permit application may include a preapplication meeting in which the Corps and other resource agencies meet with the applicant and discuss the project. Preapplication meetings can help streamline the permitting process by alerting the applicant to potentially time-consuming concerns that are likely to arise during the project evaluation. Figure 4-2 illustrates of the complexities of the Section 404 process for California.

Upon receipt of an application, Corps staff have 15 days to determine if it is complete (33 CFR 325.1(d)), and staff must contact the applicant within the 15-day time period if additional information is necessary. Once an application is deemed complete, the next step is to determine what form of permit review is appropriate for the proposed project and to issue a public notice within 15 days. The Corps then considers public comments on the notice. The district engineer may determine that a public hearing is necessary when it would provide additional information not otherwise available that would enable a thorough evaluation of pertinent issues.

An integral part of the individual permit application process is the analysis required for evaluating compliance with Section 404(b)(1) guidelines and their sequencing requirement, as discussed earlier. The project must also be evaluated to ensure that it is not contrary to the "public interest" (33 CFR Part 320.4). At the same time, the Corps prepares its documentation pursuant to NEPA.

To assist with its internal evaluations, the Corps consults with the FWS pursuant to the Fish and Wildlife Coordination Act. The Corps may also be required to initiate consultation with the FWS and/or NMFS for project impacts to species or habitat protected under the ESA. For the purpose of evaluating permit applications, the Corps considers the scope of analysis under ESA to be limited to the boundaries of the permit area plus any additional area outside Corps jurisdiction where there is sufficient federal control and responsibility (USACE 1999a).

The Corps makes its decision to authorize or deny the permit application based on its evaluation of the data collected. If a proposed project does not satisfy Section 404(b)(1) guidelines, the Corps must deny a permit. Similarly, if the proposed project is not in the public interest or would violate the ESA, the Corps must deny a permit. Although many proposed projects are modified through the permit process, most permit applicants receive Corps approval (Davis 1995). That authorization, however, may contain conditions pertaining to compensatory mitigation. Under the Corps's administrative appeals process, once a standard permit is issued, its terms and conditions may be appealed to the division engineer.

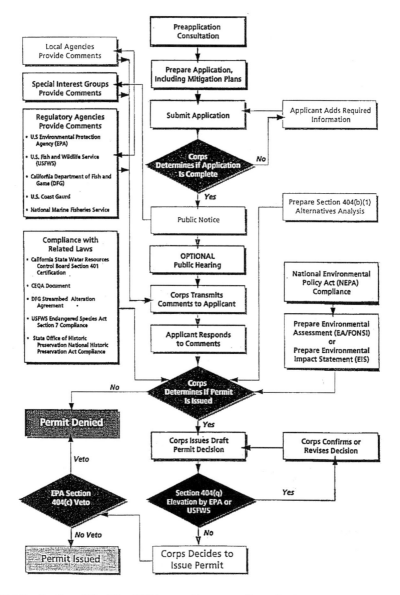

FIGURE 4-2   Section 404 of the CWA permit process flow chart.

General Permits

The most commonly used form of general permit is the nationwide permit (NWP). Like all general permits, NWPs are issued for classes of activities that should result in only minimal individual and cumulative adverse effects to the aquatic environment (Long et al. 1992). The Corps publishes proposed NWPs for public comment in the *Federal Register* and considers input from the public before deciding to issue an NWP. Initially, the Corps published a list of current NWPs in an appendix to 33 CFR Part 330 but no longer does so. Instead, current NWPs are available at the Corps district offices and may be found on the Corps website at www.usace.army.mil. Figure 4-3 illustrates how the NWP process operates.

The term "minimal," as it is used in the CWA and regulations, is not quantified, leaving the determination of what constitutes a minimal impact to the interpretation of the Corps regulatory staff. The thresholds in the NWP program provide some guidance as to what level of impact the Corps considers acceptable, and this threshold has become increasingly lower since the program was first authorized. For example, NWPs authorized on November 22, 1991, included NWP 26, which authorized the filling of up to 10 acres of nontidal wetlands. In 1996 the threshold for use of NWP 26 was reduced to 3 acres (Fed. Regist. 61(Dec. 13):65874-65922). More recently, the Corps eliminated NWP 26 and replaced it with NWPs for which the impact threshold does not generally exceed 0.5 acres of discharge into nontidal waters.

As noted earlier, NWPs may authorize activities to occur in wetlands with little agency oversight. Indeed, some activities authorized by general permit allow the permittee to proceed without notifying the Corps. These are commonly referred to as "nonreporting" NWPs and include activities for which the notification impact thresholds are not exceeded. If a project's impacts fall below the notification impact threshold, the project is automatically authorized and the Corps does not require that the applicant provide written documentation. No mitigation is required for impacts authorized by nonreporting NWPs. Nonreporting NWPs make it difficult for the Corps to determine overall program impacts.

Many NWPs now require a prospective permittee to provide the Corps with a preconstruction notification (PCN). For NWPs that trigger a PCN, the Corps may now require compensatory mitigation (Fed. Regist. 65(Mar. 9):47). Table 4-1 lists current NWPs and notes whether they are nonreporting NWPs or ones that require a PCN.

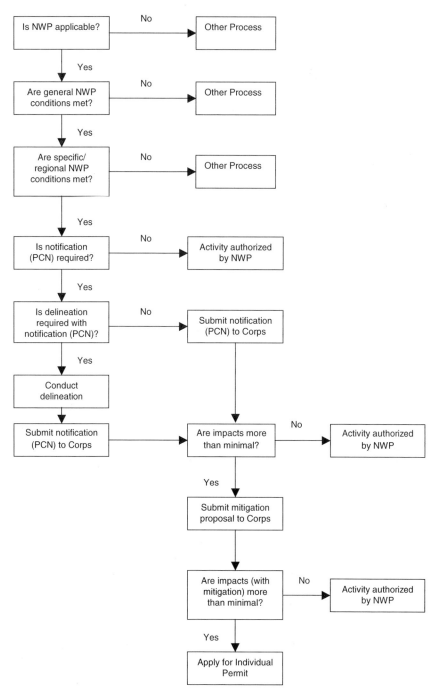

FIGURE 4-3   Approach to the nationwide permit process.

TABLE 4-1 Listing of Current Nationwide Permits

Nationwide Permit	SEC	MAX	PCN	DEL	PLAN
1. Aids to navigation	10				
2. Structures in artificial canals	10				
3. Maintenance	10/404		C		
i. Currently serviceable structure	10/404				
ii. Sediment and debris removal [all waters]	10/404	200	A		
iii. Upland restoration [all waters]	10/404		A		
4. F&W harvesting, enhancement, and attraction devices and activities	10/404				
5. Scientific measurement devices	10/404		C		
6. Survey activities	10/404				
7. Outfall structures and maintenance	10/404		A		
i. Outfall structures	10/404		A	D	
ii. Maintenance excavation	10/404		A	D	B
8. Oil and gas structures	10				
9. Structures in fleeting and Anchorage areas	10				
10. Mooring buoys	10				
11. Temporary recreational structures	10				
12. Utility line activities	10/404	1/2A (Total)	C	D	
i. Utility lines [all waters]	10/404		C, 500	D	
ii. Substations [nontidal waters only]	10/404	1/2A	C, 1/10	D	
iii. Foundations [all waters]	10/404		C	D	
iv. Access roads [nontidal waters only]	10/404	1/2A	C, 500	D	
13. Bank stabilization	10/404		C, 500	D	
14. Linear transportation crossings	10/404		C	D	A, C
a. (1) Public—[nontidal waters, excluding wetlands adjacent to tidal]	10/404	1/2A	C, 1/10	D	A, C
a. (2) Public—[tidal waters, including adjacent nontidal wetlands]	10/404	1/3A, 200	C, 1/10	D	A, C
a. (3) Private—[all waters]	10/404	1/3A, 200	C, 1/10	D	A, C
15. U.S. Coast Guard-approved bridges	404				
16. Return water from upland contained disposal areas	404				
17. Hydropower projects	404		A		

Activity	SEC	MAX	PCN	DEL	PLAN
18. Minor discharges	10/404		C	D	
19. Minor dredging	10/404				
20. Oil spill cleanup	10/404				
21. Surface coal mining activities	10/404		A	D	C
22. Removal of vessels	10/404				
23. Approved categorical exclusions	10/404				
24. State-administered Section 404 programs	10				
25. Structural discharges	404				
26. Reserved					
27. Stream and wetland restoration activities	10/404		C		B
28. Modifications of existing marinas	10				
29. Single-family housing [nontidal waters of the U.S.]	10/404	1/4A	A	D	B
30. Moist soil management for wildlife	404				
31. Maintenance of existing flood-control facilities	10/404		A	D	B, M
32. Completed enforcement actions	10/404				
33. Temporary construction, access, and dewatering	10/404		A		
34. Cranberry production activities	10A/404		A	D	R
35. Maintenance dredging of existing basins	10			D	
36. Boat ramps	10/404		A		
37. Emergency watershed protection and rehabilitation	10/404		A	D	
38. Cleanup of hazardous and toxic waste	10/404		A		
39. Residential, commercial and institutional development [nontidal waters only]	10/404	1/2A, 300*	C, 1/10	D	A, C
40. Agricultural activities [nontidal waters only]	404	1/2A, 300	C, 1/10	D	C
41. Reshaping existing drainage ditches [nontidal waters only]	404		C, 500	D	
42. Recreational activities [nontidal waters only]	404	1/2A, 300*	C, 1/10	D	C
43. Storm-water management facilities [nontidal waters only]	404	1/2A, 300*	C, 1/10	D	A, C, M
44. Mining activities	10/404	1/2A	A		A, R

NOTES: SEC—(10) Applicable to Section 10 work, activities, structures; (404) applicable to Section 404 fill discharges; (10/404) Applicable to both Section 10 and Section 404; MAX—A maximum acreage or linear restriction applies (* restriction applies to intermittent and perennial streams only); PCN—(A) Corps notification always required, (C) required under certain conditions; DEL—(D) Special aquatic sites must be delineated; PLAN—(A) Avoidance statement required, (B) baseline drawings required, (C) compensatory mitigation required, (M) maintenance plan required, (R) restoration plan required.

INSPECTION AND ENFORCEMENT

Once the Corps issues an individual or a general permit and the permittee has commenced construction of the permitted project, the Corps may visit the construction site to determine whether the avoidance and minimization requirements are being followed. If the permit requires that the permittee provide compensatory mitigation, the Corps may require the permittee to provide periodic monitoring of the physical features of the site. If the compensation is made through a permittee-sponsored or commercial mitigation bank or a fee payment program (see more discussion of the difference in Chapter 5), the inspection process focuses on the off-site mitigation area. In these cases, the Corps also may expect that monitoring reports be filed.

Corps headquarters expects that its staff will inspect a relatively high percentage of compensatory mitigation sites to ensure compliance with permit conditions, the banking instrument, or the conditions in the fee agreement. However, to minimize the number of field visits and the associated expenditure of limited staff resources, Corps field offices may ask the responsible mitigation providers to certify that the mitigation is being done in accordance with agreed-to conditions when they submit their monitoring reports (USACE 1999a).

Once the mitigation site matches predetermined criteria that may be included in a permit's performance standards, a banking instrument, or some other form, the Corps will sign off on the mitigation, deeming that the requirements have been satisfied. Many permits allow for this sign-off or regulatory certification after 5 years. Although the Corps may find that a site satisfies the legal requirements of a permit and will therefore provide its regulatory certification, the mitigation site may not achieve the desired functional effectiveness (Josselyn et al. 1990). Once the sign-off has occurred, there typically is no legal requirement on the permittee to maintain the mitigation site.

The Corps has primary enforcement jurisdiction over violations of permit conditions, including conditions relating to compensatory mitigation (USACE/EPA 1990). If the Corps discovers a violation of a permit condition, it may issue a compliance order, initiate civil judicial action, and/or suspend or revoke the permit (see Figure 4-4). In some cases, especially with mitigation banks, the party responsible for the compensatory mitigation may be asked to post financial assurances against the possibility that the mitigation will not achieve the required results. The Corps can determine whether these financial assurances will be returned to the responsible party or be used to repair the site. However, as with site inspection, enforcement actions are not a high priority for the use of limited Corps staff time and budget. This problem is discussed later in the report.

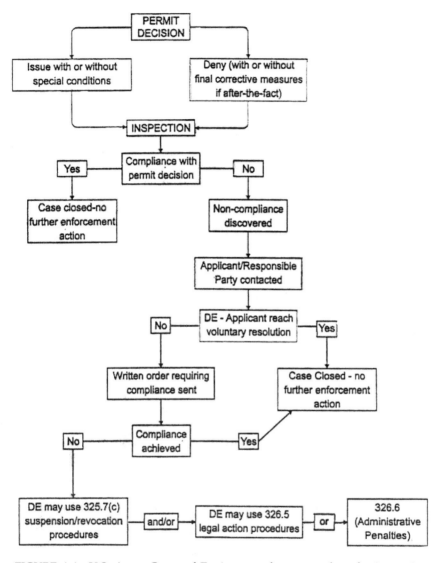

FIGURE 4-4 U.S. Army Corps of Engineers enforcement chart for inspection and noncompliance.

5

Compensatory Mitigation Mechanisms Under Section 404

The previous chapter provides an overview of the statutory and regulatory framework for compensatory mitigation. Agency guidance suggests that there are three general, discrete mechanisms for providing compensatory mitigation: permittee-responsible mitigation, mitigation banks, and in-lieu fees. In practice, however, there are many variations within and among these mechanisms. This chapter expands on the discussion in the previous chapter to further compare and categorize compensatory mitigation mechanisms. In developing these categories, the committee considered the following areas of possible differences as reflected in regulations, guidelines, and field practice:

- Where the compensatory mitigation is located: on-site or off-site.
- Who is legally responsible for meeting regulator-defined criteria for compensatory mitigation compliance: the permittee or a third party.
- What compensatory mitigation actions are required before the permittee is allowed to proceed with the activity authorized by the permit.
- Whether the criteria for compensatory mitigation actions have been approved through the interagency Mitigation Banking Review Team (MBRT) process.
- Whether requirements are imposed for the stewardship of the compensatory mitigation site.

Each area is discussed and then used to develop a taxonomy of six mechanisms for securing compensatory mitigation. Permit-specific and single-user mitigation banks are two forms of permittee-responsible mitigation; commercial mitigation banks, in-lieu fee programs, cash donation programs, and revolving fund programs are third-party mitigation mechanisms (see Table 5-1).

Central to developing a taxonomy is the definition of mitigation "action." A compensatory mitigation project consists of distinct actions, including a general design concept, identification of a general watershed location for the project, development of site design plans, development of ecological performance standards (target wetland functions), site acquisition, construction in accordance with design standards, monitoring to determine whether the design is trending toward the target wetland functions, achievement of performance standards, and regulatory certification that a site meets required mitigation requirements. Another distinct stage is an action to assure that the site is protected and managed in perpetuity. With each step the actions taken increase the assurance that the compensatory wetlands will contribute to the ecological values of the watershed.

Much of this chapter is devoted to describing compensatory mitigation mechanisms where a third party (some entity other than the permit recipient) is responsible for each of these mitigation actions. It is important to note, however, that the typical permittee will perform the mitigation itself or hire an agent to perform it, and the permittee remains responsible for the mitigation. For example, the U.S. Army Corps of Engineers (Corps) Headquarters, Operations, Construction, and Readiness Division reports that 75% of the compensatory mitigation required in 1998, under NWP 26, which is no longer in effect, was expected to be implemented by the permit recipient. Nine percent of this mitigation was done at a mitigation bank, and the balance was to be provided through other mechanisms (e.g., in-lieu fees and in-kind exchanges). It is not clear from the data description whether the banks were single-user or commercial; however, in either case, permittee-responsible mitigation is the dominant means for offering mitigation.

LOCATION OF THE COMPENSATORY MITIGATION ACTION

On-site mitigation is restoring, enhancing, creating, or preserving wetlands adjacent to an impact site. The Corps and the U.S. Environmental Protection Agency (EPA) define on-site mitigation as mitigation in "areas adjacent or contiguous to the discharge site" (USACE/EPA 1990). When on-site mitigation is not warranted, the agencies state that it should be "in close proximity [to] and, to the extent possible, [in] the same watershed" of the project that is adversely affecting wetlands. The agencies' preference for on-site mitigation is not absolute, however, and does not preclude the use of off-site mitigation when it provides greater environ-

TABLE 5-1   Taxonomy of Compensatory Mitigation Mechanisms

	Permittee-Responsible Mitigation		Third-Party-Responsible Mitigation
	Permit Specific	Single-user Mitigation Bank	Commercial Mitigation Bank
Location of compensatory mitigation action	On- or off-site	On- or off-site	Off-site
Responsible party	Permittee	Permittee	Sponsor, typically a private firm that provide capital for project initiation. Recovers cost and earns a market rate return by selling mitigat credits to permittees.
Relationship of mitigation actions to permitted activities (timing)	Mitigation action required concurrent with or soon after the project begins. Mitigation action required before permitted activity can begin varies by permittee.	Permitted activity cannot commence unless there are available credits; number of credits available is generally commensurate with the level of aquatic functions at a bank's site. Limited early withdrawals from the bank may be possible.	Permitted activity canno commence unless there a credits available for sale the permit recipient. Number of credits available is generally commensurate with the level of aquatic function at a bank site. Limited early sales of credits ma be possible.
MBRT review	No	Yes	Yes
Stewardship requirements	None required, although requirements may be imposed on an ad hoc basis.	Yes. Banking instrument should contain provisions for long-term management endowment and transfer of site to government agency or appropriate stewardship organization.	Yes. Banking instrumen should contain provisio for long-term managem endowment, and transfe site to government agen or appropriate stewards organization.

-Lieu Fee	Cash Donations	Wetland Revolving Fund
f-site	Off-site	Off-site
•e administrator, typically •nservation organizations government agencies that ive entered into a formal •OA with the Corps and llect cash payments for •itiating mitigation actions.	Typically conservation organizations or government agencies that use the cash donation for an ongoing action, but there is no in-lieu fee MOA.	State agency, which initially provides capital for mitigation projects; recovers costs through collection of fee payments from permittees.
•rmitted activity cannot •mmence until a fee has •en paid to the fund •ministrator. Compensation tions are taken after equate funds are collected.	Permitted activity cannot commence until a fee has been paid. Funds are typically applied to an ongoing mitigation project.	Permitted activity cannot commence until a fee has been paid to the fund administrator. Mitigation actions, up to and including certification of credits, may be implemented before the permitted activity occurs.
•s, to compensate for •pacts associated with •dividual permits. No, to •mpensate for impacts •sociated with general •rmits. Requirements of •e MBRT review may •t be similar to banks.	No	No
•s. In-lieu fee agreement •ould contain provisions for •g-term management •dowment and transfer off- •e to government agency or •nservation organization.	None required, although requirements may be imposed on an ad hoc basis.	The fund assumes responsibility for long-term stewardship. Management oversight may be transferred to a conservation entity.

mental benefits than on-site mitigation (Fed. Regist. 60(Nov. 28):58605-58614). As a general rule, the compensatory mitigation undertaken under all but the permit-specific mitigation mechanism is located off-site. Off-site mitigation mechanisms tend to "consolidate" at a single site the compensatory mitigation required to offset the impacts of numerous permitted activities that may be scattered across the landscape.

LEGAL RESPONSIBILITY FOR THE MITIGATION

Initially, a permittee is legally responsible for satisfying permit conditions relating to compensatory mitigation. A permittee may choose to implement the mitigation project or hire an environmental or engineering firm to do so. Alternatively, the compensatory mitigation condition in the permit may allow use of the permittee's own single-user mitigation bank. A permittee may develop a "single-user mitigation bank" by making an up-front investment in creating mitigation credits (see below for a definition) and then draw credits from the bank to satisfy mitigation conditions for future permits. A permittee who expects to have a significant number of future permits and has access to a source of funds to capitalize the bank might develop a single-user bank. In either scenario the permittee remains the party responsible for fulfilling the compensatory mitigation actions. If a permittee fails to comply with the compensatory mitigation stated in the permit conditions, the Corps may issue a compliance order, initiate a civil judicial action, and/or revoke or suspend the permit. In the case of a single-user mitigation bank, the permittee may have posted a financial assurance, and the Corps may require that the escrowed funds be used to undertake the action required in the permit.

In the case of a "commercial mitigation bank," legal responsibility is shifted from the permittee to the bank sponsor (Shabman et al. 1998). The permittee is required to secure a certain number and type of wetland credits in a certain general location as a condition of the permit. By making a payment to the commercial bank that has made an investment to create credits for sale, the permittee satisfies its compensatory mitigation requirement when it purchases the requisite credits. At the time of credit purchase, responsibility for the mitigation site shifts to the commercial bank.

The mitigation banking guidance defines a credit as "a unit of measure representing the accrual or attainment of aquatic functions at a mitigation bank; the measure of function is typically indexed to the number of wetland acres restored, created, enhanced or preserved" (Fed. Regist. 60(Nov. 28):58605-58614). In concept, credits are realized after inspection and monitoring have established that functional performance standards have been met. To ensure that a bank sponsor is legally responsible for the mitigation credits, the Corps and EPA state that "the bank sponsor should sign such [Clean Water Act Section 404 and/or Rivers and Harbors Act Section 10] permits for the limited purpose . . . of confirming that those [mitigation] responsibilities are enforceable against the bank sponsor if necessary." In addition, the banking guidance calls for the posting of financial assurances so that the Corps may require that the escrowed funds be used to rectify conditions at the bank site that may prevent it from meeting the performance standard.

Legal responsibility also will be shifted under in-lieu fee arrangements (a fee is paid in lieu of permittee-responsible mitigation). In this case, once the permittee provides the required funds to the entity administering the in-lieu fee financial account, the permittee has satisfied its compensatory mitigation obligations. The in-lieu fee financial account holds the funds until they are adequate for mitigation actions to begin. In practice, the actions taken by the in-lieu fee programs had no binding requirement to meet formal design or performance standards established at the time the permit was issued (Scodari and Shabman 2000). Instead, there was an expectation that the in-lieu fee administrator would undertake actions that would lead to functional wetland sites. Indeed, prior to the October 2000 interagency guidance on in-lieu fees, the legal responsibilities of in-lieu fee administrators for meeting either design or performance criteria as a condition of accepting the fees was unclear. As a practical matter, it appears that in-lieu fee administrators did not accept such legal responsibility. For example, the Nature Conservancy's in-lieu fee agreement with the Corps Sacramento district states that the Conservancy"does not guarantee any specific results, actions or effects on any lands acquired, managed or restored under this agreement but will use good faith efforts to meet the objectives. The Corps recognizes that The Nature Conservancy cannot guarantee specific results for mitigation efforts." The October 2000 guidance provides that in-lieu fee agreements should now "clearly state" that legal responsibility for undertaking specified mitigation actions for the ecological performance of mitigation site conditions rests with the organization accepting the in-lieu fee.

Some Corps districts authorize permittees to make cash donations on an ad hoc basis to satisfy their compensatory mitigation obligations (Gardner 2000). Such donations are not technically in-lieu fee arrangements because there is no formal general agreement between the Corps and the entity accepting the fee. These cash donations are often used to defray the expenses of an ongoing mitigation project; in contrast, most formal in-lieu fee arrangements (at least prior to the October 2000 guidance) involved the collection of funds for future mitigation projects. It is not clear what design or performance criteria the recipient of the cash donation must meet. As in the case of commercial mitigation banks and in-lieu fees, however, these ad hoc cash donations shift the legal responsibility for site conditions from the permittee to the recipient of the funds.

RELATIONSHIP OF MITIGATION ACTIONS TO PERMITTED ACTIVITIES (TIMING)

Permittee-responsible *permit-specific* mitigation actions will only be implemented concurrent with or after initiation of a permitted development project. At best, permit-specific compensatory mitigation actions (the dominant mechanism for compensatory mitigation) required as a permit condition include a regulator-approved plan for site location, design, and construction. However, it is also possible that the compensation plan may be just a conceptual idea that is a "promise" of future detailed design and follow-on construction.

If a permittee makes a payment to a mitigation bank, the permitted activity may move forward. Mitigation banks, whether single-user or commercial, are often described as "advance" mitigation. Consider that the Corps and the EPA define a mitigation bank as "a site where wetlands and/or other aquatic resources are restored, created, enhanced, or in exceptional circumstances, preserved expressly for the purpose of providing compensatory mitigation in advance of authorized impacts to similar resources" (Fed. Regist. 60(Nov. 28):58605-58614). As noted above, a credit from a bank presumes that the credits are compensation wetlands that have met performance standards.

More realistically, the MBRT process may ask that only certain actions be taken before some of the bank credits can be accepted as compensatory mitigation for wetland impacts. There are several distinct stages in a compensatory mitigation project, beginning with conceptual design, moving through site identification and acquisition, to detailed project design and initiation, and ending with a certification of success based on specified success criteria. At each stage of the process, the assurance of ecological performance increases. Limited debiting from the bank may be made before there is any construction activity, especially if there are financial assurances in place. However, the 1995 federal mitigation banking guidance contemplates that most bank credits will not be released until the mitigation project is actually constructed or completed. For example, as practiced in Florida under state regulations and federal guidance, mitigation banks are typically allowed to debit about 15% of their credits upon perpetual preservation of a mitigation site and implementation of the short- and long-term financial obligations. In these cases, 85% of the credits can be used only after project implementation and certification that performance standards have been met. As a result, mitigation banks have been developed when there was a private-sector (permittee or third-party) source of funds to capitalize the investment in developing credits for future use or sale.

Permission to use some credit sales in advance of certification of performance allows for some early return on the invested funds. If the prices for the credits are high enough, the sale of a small number of credits may yield enough up-front revenue to encourage private investment in credit development. However, the conditions necessary for encouraging such investment are not always present (Shabman et al. 1994). (More discussion of the financial conditions surrounding private credit sales is found in Chapter 8.) Nonetheless, if a permittee satisfies its mitigation obligation with a payment to a mitigation bank, actions required of a bank sponsor are likely to offer greater assurance of long-term performance than is the case with permit-specific compensation.

Permittees also may commence with activities authorized by a permit if they have made a payment to an in-lieu fee program. In the 1995 mitigation banking guidance, the agencies define in-lieu fees as an arrangement where a permittee pays money "to a natural resource management entity for implementation of either specific or general wetland or other aquatic resource development projects, [which] . . . do not typically provide compensatory mitigation in advance of project impacts" (Fed. Regist. 60(Nov. 28):58605-58614; Fed. Regist. 65(Nov. 7):66914-66917). Viewed in the context of the preceding paragraphs, this definition suggests that an in-lieu fee is paid to a program where there are no specific locations chosen, no sites acquired, and no detailed plans in place for the compensatory mitigation action; only after adequate funds are acquired will these actions be undertaken. However, the guidance also says that the Corps may approve payments of in-lieu fees for compensatory mitigation if "they meet the requirements that would otherwise apply to an off-site, prospective mitigation effort and [provide] adequate assurances of success and timely implementation." This language is confusing because these requirements would suggest that there must be a plan in place for mitigation actions at a specific site.

The October 2000 guidance on in-lieu fees offers some clarification and timetables for implementation of mitigation actions funded through in-lieu fees. Before the Corps approves the use of in-lieu fee mitigation, the in-lieu fee administrator and the Corps should enter into a formal agreement that describes "potential site locations, baseline conditions at the sites, and general plans that indicate what kind of wetland compensation can be provided" and a "schedule for conducting the activities that will provide compensatory mitigation or a requirement that projects will be started within a specified time after impacts occur." With respect to timing, the guidance suggests that actions to include "[l]and acquisition and initial physical and biological improvements should be completed by the first full growing season following collection of initial funds." In limited circumstances the guidance does allow initial physical and biological improvements to commence in the second full growing season. If these requirements were to be applied, then the in-lieu fee programs would be similar to permit-specific compensation conditions that require only that mitigation plans be in place when the permitted activities are initiated and that plans be implemented expeditiously. However, as noted elsewhere, some requirements, such as timing, may not be stated in all permits.

A significant difference between mitigation banks and in-lieu fee programs, such as third-party compensation, is the different ability of each to financially capitalize mitigation actions. In a commercial mitigation bank, the private-sector bank sponsor invests money prior to Corps approval of the use or sale of mitigation credits. In an in-lieu fee arrangement, the entity administering the fund does not invest its own money. However, up-front capitalization need not be confined to private-sector entrepreneurs: government agencies may also provide the initial funding. The state of North Carolina, which is implementing a wetland revolving fund, offers an important illustration (discussed again in Chapter 8). The North Carolina system operates in one sense as an in-lieu fee: the permittee pays a fee into a wetland restoration fund that has been recognized by a Memorandum of Agreement (MOA) between the Corps and the state, with the state serving as the fund administrator. However, the state also initially capitalized that fund to set up watershed plans and compensation site designs and to implement projects before fees are collected. The goal of the fund is to have wetland restoration projects in place before impacts are permitted (Scodari and Shabman 2000). Therefore, the desire is to have many compensatory mitigation actions completed in advance of project impacts, as envisioned in a mitigation banking system.

Of course, in all cases execution of mitigation actions by the responsible party will depend, in part, on inspection and enforcement. For permittee-specific compensation, inspection may include only the Corps's review of a conceptual mitigation plan prior to permit issuance. There may be little follow-up to see if the plan is refined and executed. Third-party mitigation is generally provided by entities that have a financial (bonds and future business) or other (their reputation with the agency) stake in meeting their mitigation obligations. Each time the third-party mitigation site is used by a permittee (i.e., credits are purchased), that use must be approved by the Corps. Thus, until credits are fully sold, the third party has regular oversight by the Corps. Although the Corps may still need to exercise careful oversight on these mitigation providers, inspection and enforcement challenges may be less severe. Confidence in the final result may be increased when the compensatory mitigation has been reviewed by the MBRT or if the Corps has entered into a formal MOA with an in-lieu fee mitigation provider.

THE MBRT PROCESS

As discussed in the previous chapter, a mitigation bank sponsor must proceed through the MBRT process to develop the banking instrument, a formal agreement that describes how the bank is to provide compensatory mitigation. The MBRT process can be time consuming and can impose significant approval costs on the prospective banker (Shabman et al. 1994; Rolband et al. 2001). The 1995 mitigation banking guidance calls on the Corps to enter into formal agreements with in-lieu fee administrators, "similar to a banking instrument," although the guidance does not specify the level of involvement of other agencies or the detail required by the MBRT review. When a permittee provides permit-specific compensatory mitigation, it does not proceed through the MBRT process; rather, the Corps approves the mitigation proposal that is the condition of the permit, although it may consider comments from other agencies.

The October 2000 guidance on in-lieu fees creates two separate processes for in-lieu fee arrangements. If the in-lieu fee arrangement is to offset impacts from individual permits, the in-lieu fee administrator should go through the MBRT process. It is unclear whether this guidance means that in-lieu fee arrangements that compensate for individual permits should secure the capital necessary to take some mitigation actions in advance of impacts or whether it simply requires interagency involvement in the approval of the MOA that sets up the in-lieu fee program. For in-lieu fee arrangements designed to offset impacts from activities authorized under general permits, the process appears to be less formal than the MBRT process. A formal agreement between the in-lieu fee administrator and the Corps is still necessary, but the Corps need only consult with other federal agencies; apparently, those agencies are not necessarily expected to be parties to the formal agreement.

STEWARDSHIP REQUIREMENTS

Once the Corps determines that the responsible party has met its design or performance obligation for a site, the agency signs off on the mitigation project, and the compensatory mitigation condition is deemed satisfied. Although Corps and EPA guidance stresses that compensatory mitigation should be self-sustaining, it is clear that many sites may require management and corrective actions after sign-off. Moreover, a mitigation site may require an entity or organization that is committed, both by its mission and financially, to the site's long-term stewardship.

Although a particular individual permit may impose stewardship conditions, such as an endowment fund for corrective action after sign-off and the transfer of an easement or title to a conservation entity, nothing in Corps and EPA regulations or guidance documents requires that permittee-responsible mitigation account for the long-term stewardship of compensatory sites. Indeed, the 1990 Corps and EPA mitigation MOA, which provides the most comprehensive guidance with respect to permittee-specific mitigation, is silent on the matter.

In contrast, a mitigation bank's banking instrument should provide for long-term stewardship. For example, the 1995 mitigation banking guidance states that the bank operator is responsible for securing adequate funds for "long-term management" of the bank site. Additionally, to assist in protecting the bank site in perpetuity, the banking instrument should specify that title or a conservation easement will be transferred to a government agency or nonprofit conservation organization.

Prior to the October 2000 in-lieu fee guidance, agencies had not clearly articulated the long-term stewardship responsibilities of in-lieu fee administrators. On the other hand, most in-lieu fee administrators were entities that were philosophically oriented to long-term stewardship of protected lands. The guidance suggests that these in-lieu sponsors should secure adequate funds for site maintenance and arrange for the site to be protected in perpetuity by conveying an easement or title to a government agency or nonprofit conservation organization.

A TAXONOMY

The committee found it instructive to develop a taxonomy of compensatory mitigation mechanisms (Table 5-1) to discuss differences among the alternative mechanisms for achieving compensatory mitigation. There are two main categories: permittee-responsible mitigation and third-party-responsible mitigation. Permittee-responsible mitigation includes permit-specific mitigation and single-user mitigation banks, and third-party-responsible mitigation includes commercial mitigation banks, in-lieu fees, cash donations, and revolving funds. The differences in mechanisms turn primarily on the factors listed on the left side of the table and discussed in detail in this chapter. Other descriptors, such as in-kind or use of restoration, creation, enhancement, or preservation, are not necessarily specific to a type of mitigation mechanism. Instead, differences in mitigation practices are attributed to site-specific conditions and not to the mitigation mechanism used.

RECOMMENDATION

The committee suggests that this taxonomy be used as a reference point for discussions about compensatory mitigation. In practice, however, a compensatory mitigation mechanism may not fit neatly into one of the listed categories (e.g., mitigation bank versus in-lieu fee versus cash donation). Accordingly, the committee recommends that when an agency reviews mitigation options, it is most important to focus on their characteristics or attributes (e.g., who is legally responsible, the timing of the mitigation actions, whether the MBRT process is used, and whether stewardship requirements are in place).

6

Mitigation Compliance

There are six generic stages of the mitigation process, and these sequential actions that must be undertaken to assure that compensation wetlands (whether creation, restoration, or enhancement) will secure the expected watershed functions. First, there needs to be a concept and a general watershed location for the project. Second, that concept is translated into a set of site design plans expected to secure the target functions over time. Third, the site for a project is acquired and construction (or other modifications to the site) undertaken in accordance with the design. Fourth, inspection of the site is made to determine whether construction followed the design plan and whether design standards have been met. Fifth, physical monitoring of the site is executed for a period of time to determine whether the design is trending toward the target wetlands functions. At this point, the monitoring would determine whether performance standards are being met. Sixth, regulatory certification would concur that the site has achieved the specified performance criteria. Included at this stage are actions to ensure that the site is protected and managed in perpetuity.

If permittees or third parties are to be held responsible for the mitigation they provide, the permitting agency needs to take steps to ensure that the required mitigation actions are being taken. This might be termed the compliance challenge. The problem of defining design or performance standards and then enforcing compliance with the standards has long been recognized in the regulatory program. The committee relied on an interpretation of the extant literature to explore the practice of establishing and enforcing compliance. In reviewing and synthesizing the published studies, there were several interpretative challenges. First, the studies may be dated, and their results may not reflect the rapidly changing requirements of the program described in Chapter 4. Second, some studies may not be related to the Clean Water Act's (CWA) Section 404 program but may instead evaluate nonfederal programs. The committee recognized this possibility in drawing its conclusions. Third, these studies may not indicate whether the responsible party was the permittee or a third party; however, it is suspected that most studies were for permittee-responsible mitigation, because third-party mitigation is still the exception and not the rule. Fourth, often it cannot be determined if a mitigation was on-site or off-site or whether the action taken was restoration, creation, or enhancement. The committee is therefore reluctant to draw specific conclusions about mitigation in the current Section 404 program based on these studies. However, the committee also drew on its field visits, on testimony from presenters at its meetings, and on the collective experiences of committee members. By cautiously integrating these various perspectives with the literature, compliance can be characterized.

MITIGATION PLANNING

Mitigation plan development begins with a functional assessment of the impact site and continues through the selection and development of a mitigation site leading to the replacement of the impacted site's functional values. While this is the expected scenario, testimony provided at committee meetings indicates that, in many cases, permit files sometimes lack a mitigation plan, and at times, mitigation may not be required to replace wetland impacts. Performance standards were often unspecified or vague and not directly related to the measurement of the sites' overall performance (Zedler 1998). The committee heard testimony that in some cases mitigation plans do not specify the most basic requirements for a wetland: water source, water quality, water retention, water quantity, soil, and topography, structure (flora and fauna), and location. Absent such basic considerations, adequate performance is unlikely.

Area To Be Lost and Proposed Mitigation

Mitigation plans, when clearly written, specify the area of wetland to be lost and the measures proposed for reducing the impact of that loss. The literature suggests that mitigation plans (particularly for older projects) are not always required for each permit (Table 6-1). On a national basis there is an anticipated gain of 78% in wetland area as a result of mitigation. However, results of independent scientific reviews suggest

TABLE 6-1   Required Mitigation as Restoration, Creation, and Enhancement for Permits Issued under Permitting Programs[a]

Location	No. of Permits	Area Impacted (Ha)	Years	% of Permits Requiring Mitigation	% of Area Impacted	Source
Alabama	18	18	1981-87	100	100	Sifneos et al. 1992a
Arkansas	7	703	1982-86	71	98	Sifneos et al. 1992b
California, Statewide	324	1,176	1971-87	NA	107	Holland and Kentula 1992
Orange County	70	168	1979-93	13	97	Sudol 1996
Southern CA	75	112	1987-89	92	140	Allen and Feddema 1996
Sacramento and San Francisco Bay	30	168	1987-90	NA	144	DeWeese 1994
San Francisco Bay	36	NA	1977-82	64	NA	Race 1985
Florida, Corps	NA	NA	NA	NA	246	GAO 1988, as cited in Torok et al. 1996
Jacksonville District	NA	26,280	1981-87	41	10	Sifneos et al. 1992a
Louisiana Florida, St. John's River Water Management District	680	NA	1984-89	48	98	Lowe et al. 1989[b]
New Jersey Section 404 Program	NA	333	1985-92	NA	100	Torok et al. 1996[b]
State FWPA	NA	58	NA	NA	147	Torok et al. 1996
Mississippi	10	1,095	1981-87	50	100	Sifneos et al. 1992a
Mississippi, Corps Vicksburg District	NA	NA	NA	NA	1	GAO 1988, as cited in Torok et al. 1996
Ohio	32	371	1990-95	68	93	Sibbing 1997[c]
Oregon	58	74	1977-86	NA	57	Kentula et al. 1992a
Tennessee	50	34	1992-96	100	100	Morgan and Roberts 1999

TABLE 6-1 Continued

Location	No. of Permits	Area Impacted (Ha)	Years	% of Permits Requiring Mitigation	% of Area Impacted	Source
Texas	45	2,995	1982-86	NA	69	Sifneos et al. 1992b
Washington	35	61	1980-86	NA	74	Kentula et al. 1992a
Washington	45	40	1992-97	100	598	Johnson et al. 2000
Wisconsin	NA	40	NA	NA	62	Owen and Jacobs 1992, as cited in Torok et al. 1996
National Totals	NA	76,500	1993-00	NA	178	USACE 2000

[a]No field examinations were made, and functional equivalency is not assumed.
[b]Wetland creation only; wetland enhancement may not be included.
[c]Permits based on 25 with data and 8 not restored or created; acreage is for 32 permits and includes enhancement. NOTES: FWPA = Federal Water Pollution Act. NA = not available.

a range of a net loss in 8 of 19 reviews with data to gains as great as 598% in one review (Johnson et al. 2000).

MITIGATION DESIGN STANDARDS

If the functions and values of jurisdictional wetland habitat are negatively affected, a net wetland loss will occur if these functions and values are not replaced. Spelling out the particular requirements for replacement in the Corps permit is the critical first step in the permitting process. A recent review (Streever 1999a; see Table 6-2, Appendix E) indicates that such requirements vary widely among Corps districts. These requirements range from physical to biological criteria, and most often include a standard related to plant dominance or abundance.

Because hydrological processes determine many wetland functions, design standards often seek to grade the topography down to the groundwater source, connect the site to a local stream channel, control the water source (e.g., with tide gates or berms), or other features. Some designs require connecting the site to adjacent rivers and wetlands in the watershed.

Sometimes a desired wetland type is the standard to be achieved (e.g., a sedge meadow or an emergent wetland). Many mitigation proposals state that wetland complexes will provide a desired habitat. Some plans

TABLE 6-2 Review of Corps Permits Issued Nationwide

Example No.	Performance Standard	Location	Type	Year	Size, acres
1	75% survival of planted *Juncus roemerianus*, 4,800 plants per acre after three growing seasons	Alabama	Salt marsh	1985	40
2	Sustain 85% or greater cover by obligate and/or facultative wetland plant species; less than 10% cover by nuisance plant species; "proper hydrology condition."	Florida	Forested and herbaceous freshwater	1991	21.9
3	Hydrology must meet wetland definition of 1987 Corps wetland manual, with saturation to surface of the soil for 12.5% (31 days) of the growing season; at least 50% of woody vegetation must be facultative or wetter, with woody vegetation stem counts of 400 per acre or canopy cover of 30% or greater by woody vegetation; at least 50% of all herbaceous must be facultative or wetter with aerial cover of at least 50% in emergent wetland areas (excluding "scrub/shrub or sapling/forest vegetation").	Virginia	Forested freshwater	1995	27.4
4	Emergent and aquatic bed portions of mitigation site not to be inundated with salt or brackish water; less than 10% cover by invasive species during any monitoring event; staged vegetation requirements as follows: Year 1: 100% survival of planted stock, 50% cover in emergent areas Year 2: 80% survival by planted stock, 20% cover by native shrub species, 70% cover in emergent areas Year 3: 70% survival and 40% cover by native shrub species, 80% cover in emergent areas Year 5: 60% cover by native shrub species, 100% cover in emergent areas	Washington	Emergent, scrub/shrub, and forested freshwater	1998	9.4

5	Less than 5% cover by nuisance and exotic plant species; planted and nonnuisance wetland plant species to have areal cover of 50% in the first year, 70% in second year, and 80% in third year, with provisions for remedial planting to meet percentage requirements.	Florida	Freshwater marsh and wet prairie	1990	10
6	Must meet the regulatory definition of wetlands; specified portions of the mitigation area must meet the definitions of palustrine forested, palustrine scrub/shrub, and palustrine emergent wetland types as documented in *Classifications of Wetlands and Deepwater Habitats of the United States*; cover by hydrophytic vegetation; vegetation not to consist of common reed (*Phragmites australis*) or purple loosestrife (*Lythrum salicaria*); all performance standards must be met for 3 consecutive years.	New York	Forested, scrub/shrub and emergent palustrine wetlands	1998	25
7	Areal cover in 90% of planted area equivalent to natural reference marsh; benthic invertebrates and fish with 75% similarity to natural reference marsh; upper soil horizon with 1% organic matter by dry weight.	Alabama	Salt marsh	1988	25.3
8	Vernal Pool Habitat Suitability Index less than or equal to 0.55 with 60% of pools more than 0.7 (VPFI = $a/(a+b)$, where a = number of species that the pool and the "vernal pool species list" share and b = number of species in the pool not on the "vernal pool species list" (the list includes those species typically found in the region's vernal pools); hydrology assessed as suitable on the basis of presence of wetlands plants.	California	Vernal pools	1996	27
9	Hydrogeomorphic approach.				

SOURCE: Streever 1999a. Reprinted with permission from the *National Wetlands Newsletter*, copyright 1999, Environmental Law Institute, Washington, D.C.

include percentages of facultative or obligate wetland species. Others propose levels of species diversity or abundance (see Box 6-1). The permit typically includes a map of the proposed communities with planting lists.

In some cases, performance is stated as a measure of primary productivity, such as algal and macroinvertebrate community richness, species diversity, taxonomic composition, and trophic relationships, that might reveal the system's functionality (FDER 1992; FDEP 1994, 1996). Water chemistry parameters and indexes of pollution geared to wetland condition might also be useful indicators of desired endpoints (Box 6-1).

Some permits require the addition of topsoil with seed banks. In others, the risk of introducing invasive species leads to a restriction of the use of topsoil. Although a common design requirement is to spread wetland topsoil over the site, primarily to provide plant propagules, the soil characteristics desired are rarely stated. Because characteristics of soil usually do not develop quickly (Craft et al. 1999), many restored or enhanced wetlands do not have the carbon or mychorrhizal contents of natural wetlands, and so soil characteristics may not be used as part of a performance endpoint.

Other common permit performance requirements include the percent survival of planted vegetation; percent cover of native versus weedy or exotic species; similarity of a site to a reference site; similarities to habitat

---

BOX 6-1
Performance Standards Used by the
Chicago District and the Corps

In the Chicago region, wetland restoration evolved along with prairie restoration efforts. In a state with about 90% loss of historical wetlands (Dahl 1990), wet prairies were among the few remnants of unplowed, unforested land. Prairie remnants were studied for their biodiversity, especially distributions of plants (Swink and Wilhelm 1994) and insects (Panzer et al. 1995). Species restricted to the least disturbed sites were considered good indicators of natural conditions. The endpoint requirements were as follows:

- Evaluation of aerial coverage by plants (90% in 3 months)
- Floristic Quality Index
- Native mean wetness (less than 0)
- Relative importance value of native species
- Cover (less than 0.5 square meters)
- Macroinvertebrate Biotic Index
- Habitat evaluation procedures (not usually measured)
- Water quality (usually not measured, except for siltation or sedimentation)

types described in other documents; or presence of certain wildlife species, particularly species with special status designations, such as species listed pursuant to the federal Endangered Species Act. Few permits specify animal populations or communities; however, habitat for selected species (e.g., waterfowl and endangered species) is sometimes specified.

PROJECT IMPLEMENTATION

Mitigation Permit with Special Conditions

Once a Section 404 mitigation plan is completed, it is provided to the Corps for review and approval. When the Section 404 permit is issued, the items prescribed in the mitigation plan are included as a special condition of the permit. These special conditions form the legal requirements of the permit. Because the permit requirements are legally binding, it is important that the permit conditions be clear, complete, and comprehensive so that the desired mitigation outcome is achieved. If the special conditions are not included in the permit or if they do not clearly describe mitigation milestones to be achieved, it is possible for regulatory certification to be obtained by the permittee even though the mitigation does not produce a mitigation site that replaces the impact area's functions and values (see Coyote Creek case study, Appendix B).

Mitigation Specified But Not Carried Out

The committee learned that in some cases specified mitigation was not initiated as required in the Corps permit. Eight studies provided information as to whether required mitigation was initiated for a mixture of programs (Table 6-3). In addition, numerous studies on mitigation required by permits revealed that as much as 34% of the mitigation was never installed (FDER 1991b; Allen and Feddema 1996; Sudol 1996; Robb 2000). In southeast Florida, Erwin's (1991) study of 40 mitigation wetland creation and restoration projects found that only about half of the required 430 hectares of wetlands had been constructed.

The committee found that compliance inspections are rarely conducted by the Corps and that this is policy. In a Memorandum for Commanders, Major Subordinate Commands, and District Commands dated April 8, 1999, Major General Russell Furman provided the Standard Operating Procedures (SOPs; Appendixes F and G) and described how the regulatory program would be executed across the United States. In addition to describing policy and program administration, the SOPs prioritized regulatory activities by the percentage of staff time devoted to them. Activities to be emphasized are described as "above the line," and de-

TABLE 6-3   Mitigation Initiated for Permits Requiring Mitigation

Location	Source	No. of Permits Considered	Initiated, %
Orange County	Sudol (1996)	57	96
Southern California	Allen and Feddema (1996)	75	92
Florida	Erwin (1991)	97	66
Indiana	Robb (2000)	345	62 (completed)[a]
			14 (not attempted)
			20 (incomplete)
Massachusetts	Brown and Veneman (1998)	114	74
New Jersey	Torok et al. (1996)	80	28
Ohio	Fennessy and Roehrs (1997)	14	100
Washington	Johnson et al. (2000)	45	93

[a]For 1988-1993 permits as of 1995; additional ones may be under construction in later years.

emphasized activities are described as "below the line." The SOPs specifically note that "below-the-line" activities should be accomplished only after the "above-the-line" activities are fully executed. The SOP lists 10 activities under Permit Evaluation, plus another eight under Mitigation. Under Permit Evaluation, activity number 1, resource permit evaluation for timely decisions, is above the line. Extensive negotiation with other agencies to reach consensus (number 7) and multiple site visits and meetings of extensive preapplication (number 10) are below the line. Under Mitigation, compliance inspections for all mitigation (number 6) and multiple visits to a mitigation site (number 8) are below the line. Of the five activities under Enforcement (number 2) implement self-reporting and certification for compliance is above the line. The committee found that the cumulative effect of these policy decisions indicates that evaluating and issuing permits takes priority over careful evaluation of mitigation projects.

In addition to the SOPs, testimony provided to the committee by Corps staff from Chicago, Los Angeles, and San Francisco indicated that the workload for Corps regulatory staff is exceedingly high. Regulators from all the Corps Districts providing testimony to the committee indicated that there are consistently more Section 404 permit applications than there is time for Corps staff to perform adequate reviews. That problem, coupled with guidance provided in the SOPs, indicates that priority is given to issuing permits (which often require mitigation), yet mitigation development and follow-up inspections to determine if mitigation commitments are being met is not a priority, and the activity is not encouraged. Indeed, the SOPs suggest that since compliance inspections and site visits fall "below the line," these activities are not adequately performed because there is insufficient time for staff to perform the activities "above the line." The committee believes that compliance inspections should be an "above-the-line" activity to ensure that the programmatic goal of no net loss of wetland functions and values is met.

If the Corps recognized mitigation compliance and increased compliance as a priority activity, mitigation would more likely be carried out as specified in Section 404 permits. The committee recognizes that increasing compliance efforts would result in increased staff workloads requiring additional regulatory staff.

COMPLIANCE WITH PERMIT CONDITIONS

The literature shows that many mitigation sites are not performing as specified in Corps permits (Allen and Feddema 1996; Sudol 1996; FDER 1991b; Race 1985). These same studies also show that where mitigation is performing as specified, many of those sites do not support functions and values equivalent to similar reference sites. In some cases, the standard to be met by an individual compensatory mitigation project may stop with a requirement to secure some wetlands structure or to design the project in a particular way. In this case, the premise is that restoring hydrology will facilitate the development of other wetland functions. However, permits do not always call for hydrological measurement.

If mitigation is not carried out to the level specified in the permit, then the Corps can take enforcement action and require a permittee to perform the agreed-upon mitigation. However, if the permit does not specify mitigation, or if the permit is not clear as to the level of mitigation that must be performed or what parameters must be met for the mitigation to be considered complete, it becomes difficult for the Corps to determine if the project is in compliance. Mason and Slocum (1987), for example, found that compliance rates were twice as high when the permits contained specific conditions compared with those that had no specific conditions. For this reason, it is important that Section 404 permits specifying mitigation contain specific language about the expected mitigation outcome (mitigation goal).

The mitigation goal statement should be followed by specific objectives that consist of specific statements about the intended mitigation outcome (Streever 1999b). Performance standards are then developed from the mitigation goal statement and objectives. When these performance standards are included in the Section 404 permit as special conditions, they become legally binding upon the permittee.

Design Standards and Detailed Performance Standards

With detailed assessment of the impacted sites and/or reference systems selected as targets, the committee could set detailed performance standards. But neither data set is typically available. Thus, projects are designed without adequate knowledge, and performance criteria are general and few in number (Streever 1999b). Ecologists, hydrologists, and other scientists who study mitigation sites find many shortcomings in comparing mitigation sites with reference systems (see Chapter 2). Thus, it seems that regulators need to agree that either (1) design standards constitute reasonable performance criteria, or (2) detailed assessment of functions lost must be matched by detailed assessment of mitigation site performance and penalties developed for failure to achieve performance standards.

A consistent set of procedures to identify wetlands is required in order to permit wetland filling under the guidelines of the CWA. The Corps created preliminary guides to regional wetlands and developed techniques for identifying wetlands (USACE 1978a,b,c,d; Reppert 1979; USACE 1987; NRC 1995). The resulting schemes were based on a triad of wetland characteristics: hydrological conditions, soil characteristics, and plant communities. Lists of wetland plant species and hydric soils were created for all parts of the country (USDA 1982, 1985, 1987, 1991). Hydrological requirements were codified (such as number of days of flooding and depth to groundwater) and, to some extent, adapted to various regions. Hydrological data were not available for many wetland sites; therefore, procedures were developed for estimating hydrological conditions from soils and other features (NRC 1995). More detail on the history of the federal wetland manuals and current and past practices in wetland delineation is presented in NRC (1995).

Basic to all wetland restoration and creation projects is the need to set goals for each site's hydrological conditions. Hydrology is most often cited as the primary driving force influencing wetland development, structure, function, and persistence (Gosselink and Turner 1978; Carter 1986; LaBaugh 1986; Day et al. 1988; Novitzki 1989; Wilcox 1988; Gosselink et al. 1990; Sharitz et al. 1990; FDER 1991a; Reaves and Croteau-Hartman 1994; Bedford 1996, 1999; Morgan and Roberts 1999). Consequently, establishment of the appropriate hydrology is fundamental to wetland mitigation whether through restoration or creation (NRC 1992, 1995; Brinson 1993; Bedford 1996; Mitsch and Wilson 1996; Shaffer et al. 1999; Cole and Brooks 2000b). In a survey of 175 federal, state, private, and environmental professionals working in wetland restoration, hydrology was considered one of the most difficult structural features of a wetland to establish and the most important component of a project (Holman and Childres 1995).

One measure of mitigation compliance is the restoration of jurisdictional hydrology. An explicit hydrological standard is the percentage of the growing season that soils need to be saturated. Clark and Benforado (1981) suggested that areas saturated less than 5% of the growing season clearly exhibited upland hydrological characteristics and that areas saturated more than 12.5% clearly exhibited wetland hydrological characteristics. The 1987 Corps wetland delineation established the 5% criterion as the jurisdictional threshold, a quantitative value that was reaffirmed by the NRC (1995). However, there are major differences in depth to water table between a wetland that satisfies the 5% standard and one that meets the 12.5% standard (see Figure 6-1). These differences in wetness lead to very different ecological communities (Scherrer et al. 2001).

Because the permittees responsible for the mitigation need some time frame that clearly defines the length of their mitigation responsibility, hydrological performance standards may be based on 5 years or less of water-table monitoring. However, the hydrological regime in nonriverine, intermittently saturated freshwater wetlands varies not only seasonally but also year to year (see Figure 6-2). During a short monitoring period,

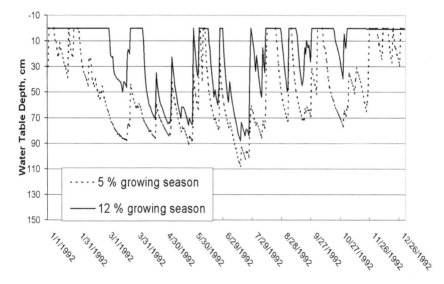

FIGURE 6-1 Water-table position and duration of root zone saturation for wetland site that satisfies the jurisdictional hydrology criteria (5% of growing season) as compared with wetland site that satisfies the criteria (12% of the growing season). Simulation modeling (DRAINMOD) was used to determine values. SOURCE: Skaggs (1978). Reprinted with permission; copyright 1978, Water Resources Research Institute of the University of North Carolina, Raleigh.

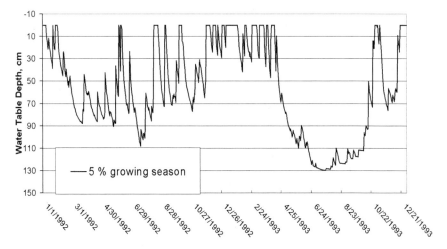

FIGURE 6-2   Year-to-year variations in water-table depth and duration of root zone saturation for a wetland site that satisfies jurisdictional hydrology criteria at least 5% of the growing season. Year-to-year extremes are typical for intermittently saturated wetlands. Values determined from simulation modeling using DRAINMOD. SOURCE: Skaggs (1978). Reprinted with permission; copyright 1978, Water Resources Research Institute of the University of North Carolina, Raleigh.

water levels might not meet hydrological standards for several consecutive years, even though the wetland could satisfy criteria over the long term. Depending on the date when the 5-year monitoring period began and ended, there could be six 5-year periods where the wetland did not satisfy hydrological criteria (see Figure 6-3). If this were a mitigation site and the 5-year monitoring period occurred during one of these six periods, the mitigation project would not comply with performance standards. Recognizing this potential shortcoming, practitioners tend to err toward the wet end of the range, creating wetlands that are much wetter than normal for the given landscape position (Cole and Brooks 2000b).

In many cases this approach has resulted in the creation of open-water areas as compensation for loss of intermittently inundated or saturated wetlands (Kentula et al. 1992a). The stable-water pond has come to typify mitigation efforts in many parts of the country (Cole and Brooks 2000b). Mitigation projects that stress the wet end of the range will not replace the functions provided by much drier impact sites. For example, use of a mitigation site as a stormwater storage, attenuation, or treatment wetland may compromise biodiversity goals.

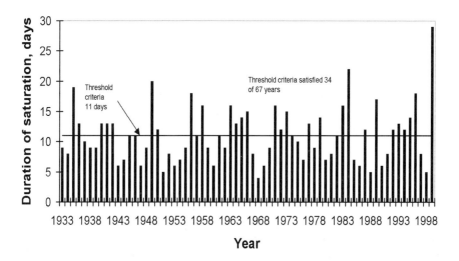

FIGURE 6-3   Year-to-year variation of the longest period that wetland hydrological criteria satisfied. Results obtained from long-term simulation modeling using
DRAINMOD. NOTE: There are several 5-year periods where criteria are not satisfied 3 out of 5 years. SOURCE: Skaggs (1978). Reprinted with permission; copyright 1978, Water Resources Research Institute of the University of North Carolina, Raleigh.

Breaux and Serefiddin (1999) examined 110 compensatory wetland
mitigation projects in California (permitted from 1988 to 1995) and determined that the most commonly measured parameter was vegetation (type
or cover) (Table 6-4). Two of the most commonly assumed wetland val-

TABLE 6-4   Parameters Measured in 110 Compensatory Wetland
Mitigation Projects in California from 1988 to 1995

Parameter	% of Sites Measured
Vegetation	72
Hydrology	22
Wildlife	38
Water quality	7
Soils	3
Invertebrates	3
Flood storage	Not mentioned

SOURCE: Adapted from Breaux and Serefiddin (1999).

ues, flood storage and water-quality improvements, were supposed to be examined in less than 10% of the permits.

The committee concludes that current permitting procedures do not always result in permit conditions that are clear and enforceable and lead to the development of viable mitigation that compensates for the functions and values of the permitted impact. Instead, permits typically contain performance standards that measure only one or several easily measured parameters of a mitigation site, and in many cases, these parameters do not reflect the overall viability of the mitigation site. Recommendations relevant to this conclusion are provided in Chapter 8.

MITIGATION RATIOS

Mitigation ratios are the proportional requirements for replacing wetlands that are permitted for fill. A point that is frequently raised in assessments of mitigation is that the ratios (the number of required mitigation acres to the permitted acres) are too low (Morgan and Roberts 1999; Allen and Feddema 1996). Ratios vary across permits, often because the logic behind the ratios differs. Higher ratios might be required for sites and wetland types that are difficult to restore. Higher ratios might be also used if there is a long time expected between the permitted activity and the achievement of the desired endpoint for the compensation site. Ratios have been used to reflect the functional values of the impact site, that is, the ratio would be higher for a pristine wetland than for a severely degraded wetland. An example of ratio guidelines used by the California Department of Fish and Game incorporates this principle in its guidelines for mitigating impacts to streams and associated habitat (see Appendix D). Mitigation ratios are 1:1 for low-value habitat (e.g., unvegetated streams), whereas ratios can be as high as 5:1 for impacts to endangered species habitat (e.g., mature willow riparian inhabited by least Bell's vireo).

The Corps and the Environmental Protection Agency (EPA) mitigation Memorandum of Agreement (MOA) states that "mitigation should provide, at a minimum, one-for-one functional replacement (i.e., no net loss of values), with an adequate margin of safety to reflect the expected degree of success associated with the mitigation plan . . . [T]his ratio may be greater where the functional values of the area being impacted are demonstrably high and the replacement wetlands are of lower functional value or the likelihood of success of the mitigation project is low. Conversely, the ratio may be less than 1 to 1 for areas where the functional values associated with the area being impacted are demonstrably low and the likelihood of success associated with the mitigation proposal is high."

Whether required ratios were met was examined in nine studies for four nonfederal (not Section 404) programs (see Table 6-5). The mitigation ratio requirements were never fully met, but the ratio for the mitigation implemented was higher than 1:1 in three of the nine studies. These results present another way of examining the compliance rate of mitigation when viewed on a programmatic scale. The average percentage was 69%, implying that for these nine mitigation efforts, a 1.5:1 ratio of mitigation:loss acreage would be needed to equal the area lost (if all other permit conditions are met, including functional equivalency). The committee concludes that some mitigation will not be fully implemented. The reasons that mitigation projects do not meet expectations are partially dependent on performance and design criteria, program oversight, and execution. These are discussed in the next sections.

In all programs surveyed by Allen and Feddema (1996) in Southern California, the area mitigated was about equal to the area lost due to permitting. That was accomplished by varying the amount of mitigation required by habitat type so that different replacement risks were built into the permit requirement (see Table 6-6). The committee heard testimony to the effect that pressure to mitigate for one rare wetland type with a high mitigation ratio resulted in a more common wetland type with a lower ratio. In effect, the regulatory program may reassemble the landscape with a different habitat mix than the wetlands being lost.

TABLE 6-5   Mitigation Ratios Required and the Actual Ratios Met, Based on Post-Construction Evaluation (assumes complete compliance in meeting permit conditions)

Location	Ratios Required	Ratios On Site	% Compliant	Source
California				Fenner (1991)
San Diego County	1.51:1	1:0.93	62	Allen and Feddema (1996)
Southern Sacramento	1.40:1	1:0.96	69	DeWeese (1994)
San Francisco	1.44:1	1:0.29	90	DeWeese (1994)
Orange County	1.03:1	1:0.18	17	Sudol (1996)
Indiana	2.48:1	1:1.1	44	J. T. Robb (personal communication 2000)
Ohio	1.5:1	1:1.26	84	Fennessy and Roehrs (1997)
Ohio	1.72:1	1:0.66	38	Wilson and Mitsch (1996)
Ohio	1.5:1	1:0.93	62	Sibbing (1997) (includes enhancement)
Tennessee	1:1	1:0.87	87	Morgan and Roberts (1999)

NOTE: Data not directly comparable among locations because of different types of surveys.

TABLE 6-6   Mitigation Ratios (Area Basis) and Achievement Rates (%) for Different Wetland Types in Southern California

Wetland Type	Required Ratio	Actual Ratio	Rate
Riparian habitat	1.39:1	0.97:1	70
Freshwater	0.9:1	0.61:1	67
Saltwater	2.1:1	2.1:1	100
Unidentified riparian	1.59:1	0.40:1	67
Riparian woodland	2.40:1	1.6:1	67
Total	1.39:1	1.096:1	69

SOURCE: Adapted from Allen and Feddema (1996).

Mitigation ratios are considered further in Chapter 8 under the topic of Permit Conditions. The adjustment of ratios is one of the principal tools for addressing risk and temporal loss with the ultimate goal of achieving permit compliance.

MONITORING OF MITIGATION PROJECTS

Once a Section 404 permit is issued by the Corps and the agreed-upon mitigation and corresponding performance standards are outlined in the permit special conditions, it is the responsibility of the permittee to conduct ongoing monitoring of the site to ensure that the performance standards are being achieved. The Corps should review the monitoring reports submitted by the permittee and conduct periodic inspections of the site to ensure compliance with the permit.

Table 6-7 lists results of seven reports on the monitoring frequency for six states. The highest monitoring rate was for a California study in which 324 sites had at least one monitoring site visit; two-thirds of these 324 sites had no second site visit required. One study (Louisiana) reported that monitoring occurred on only 10% of the sites. Six of seven studies reported a monitoring rate of about 50% or less. This result has important implications for the success of a mitigation program. Several studies have shown that permit compliance rate may be quite low when monitoring is sparse or does not occur. Mason and Slocum (1987) evaluated 32 wetlands in Virginia and, at the time of their analysis, documented that permits with specific conditions for creating wetlands were 86% compliant, whereas when there were no specific permit conditions, then permit compliance was 44%. When time limits for completion were specified, 100% of the mitigation efforts were compliant, compared with

TABLE 6-7   Frequency of Monitoring for Permits That Required
Mitigation

State	Source	No. of Permits	No Monitoring at:
Alabama	Sifneos et al. (1992a)	18	61% were not monitored
California	Holland and Kentula (1992)	324	40% had incomplete acreage; 2% had no completion date; two-thirds did not require follow-up
Louisiana	Sifneos et al. (1992a)	93	90% were not monitored
Mississippi	Sifneos et al. (1992a)	5	80% were not monitored
Oregon	Storm and Stellini (1994)	58	53% had no site visit recorded
Washington	Storm and Stellini (1994)	17	67%; monitoring was done at three of the nine sites requiring monitoring
Washington	Kentula (1986)	35	49% had no site visit

50% without deadlines or time limits. Morgan and Roberts (1999) concluded that "some applicants apparently believe that they will not be held accountable for their projects." Morgan and Roberts (1999) implied that this self-interest in reporting data to agencies might have an influence on the accuracy of the evaluation and made a plea that "we strongly recommend that consultants responsible for site development not be allowed to submit the monitoring reports for their own projects." Zentner (1988) found that the "lack of monitoring was a common element of unsuccessful projects."

In addition, the permit may sparsely quantify the necessary requirements. Lowe et al. (1989) estimated that 86% of the 29 permits he surveyed "did not contain enough details or were not clear enough to ensure success of the created wetland, nor were they drawn tightly enough to enable the District to enforce the terms to correct problems. Two of the most often noted deficiencies in the permit conditions were the absence of success and maintenance criteria and the lack of provisions for corrective action should the created wetland fail." DeWeese (1994) noted that the omission of monitoring reports meant that the need for remedial action went unnoticed and therefore was not done or was not done as well. These sentiments were shared by Race and Fonseca (1996):

> [O]ur survey of past mitigation projects nationwide indicates that the success rate of permit-linked mitigation projects remains low overall. In addition, there is continuing difficulty in translating mitigation concepts into legal principles, regulatory standards, and permit conditions that

are scientifically defensible and sound. Based on the record of past poor performance, we assert that continued piecemeal revision efforts focused on technical or scientific details are not likely to make compensatory mitigation more effective. There is need to acknowledge the extent to which non-scientific, real-world complications plague current policies and practices. To prevent continued loss of wetlands under compensatory mitigation, decisive action must be taken by placing emphasis on improving compliance, generating desired acreage, and maintaining a true baseline.

Based on design requirements, Morgan and Roberts (1999) found that 72% of the mitigation sites inspected were smaller than required. Morgan and Roberts (1999) also found that many sites were not constructed to topographic specifications, resulting in sites that could not attain appropriate hydrological requirements to create the intended results.

Accordingly, the Corps/EPA Mitigation MOA (USACE/EPA 1990) recognizes that "[m]onitoring is an important aspect of mitigation, especially in areas of scientific uncertainty. Monitoring should be directed toward determining whether permit conditions are complied with and whether the purpose intended to be served by the conditions are actually achieved." Thus, a mitigation site must be physically monitored to determine compliance with design standards as well as compliance with performance standards (whether the site exhibits a trend toward achieving the target wetland functions).

MONITORING DURATION

The literature and testimony provided to the committee indicate that monitoring periods commonly last between 3 and 5 years following mitigation site construction. However, for many created and restored systems, particularly those such as woody riparian systems that require long periods of time for plant establishment, a short-duration monitoring (3-5 years) might not be long enough to determine whether mitigation goals will be acheived (see Coyote Creek case study, Appendix B). The mitigation guidelines from the St. Paul District (Eggers 1992) recommend monitoring beyond 5 years if necessary. For forested wetlands, Morgan and Roberts (1999) recommend revising the monitoring period to a time frame that better reflects the time needed to achieve the desired outcomes.

A 5-year monitoring window is a common permit requirement. For example, Tennessee, like most states, requires that most mitigation projects be monitored annually for 5 years (Morgan and Roberts 1999). This requirement is especially true of tree-dominated mitigation sites that may take 50 years or more to mature (Morgan and Roberts 1999).

Obviously, the specific needs and performance requirements of a particular mitigation site will influence monitoring frequency and duration.

THE COMPLIANCE RECORD

In the mid-1980s, some scientists involved in wetland restoration and creation believed that mitigation was effective (Harvey and Josselyn 1986). Others emphasized a need for more research, enforcement, and monitoring (Kusler and Groman 1986; Race 1986). Since then, follow-up studies of compensatory restoration and creation projects have identified significant shortcomings with respect to the functioning of mitigation sites (see Chapter 2). Shortcomings are evident in every region of the country. The discussion focuses on plant communities, because that is what is most commonly monitored. The committee notes, however, that while vegetation may be easily measured, it is a poor indicator of function (Reinartz and Warne 1993).

One of the most comprehensive investigations involved a review of 61 permits for 128 projects in six counties around Chicago, Illinois (Gallihugh and Rogner 1998). That study found that 17% of the wetland vegetation proposed was established, and an additional 22% had established wetlands but with vegetation other than that proposed. Fifty-two percent of the wetlands had excessive or unplanned open water, and 9% had insufficient hydrology. The wetland area lost was 117 hectares (ha) and the approved wetland mitigation amounted to 144 ha. So, in theory, there was a mitigation ratio of 1.2:1, resulting in a net gain of 27 ha. In actuality, the study found that 29 ha were not established and at least 99 ha were found to have unsatisfactory hydrology (too wet or too dry).

In a smaller sampling, Wilson and Mitsch (1996) evaluated five wetland projects in detail to estimate their ecological and legal outcomes (legal compliance was determined by the authors in consultation with the regulators). Only two of the five mitigation cases showed the mitigation projects were in full legal compliance, but four of the five were on a trajectory toward legal compliance with permit requirements (Table 6-8). Overall, 24.4 ha of wetlands were lost, and about 16 ha were actually created or restored. Because of the failure of one large site, only 38% of the desired wetland area was established at the time of their study. For the four wetlands that were compliant, there was an overall mitigation ratio of 1.4:1.

Evaluations of ecological equivalency between mitigation sites and reference sites are rarely conducted as part of a programmatic review. Kentula et al. (1992b) found that 65% of the permits that they surveyed in Oregon and Washington required a functional assessment, but these assessments lacked detail. A salt marsh creation project in southern Califor-

TABLE 6-8   Permit Requirements and Compliance for Five
Replacement Wetlands Investigated in Ohio

Location, County Replaced	Wetland Type Lost	Required	Implemented	Wetland Area, hectares			
				Lost	Required	Area Implemented	% of Required
Portage	Emergent	Emergent/ woody	Emergent/ woody	0.4	0.6	0.6	100
Delaware	Emergent/ woody	Emergent/ woody	Emergent/ woody	3.7	5.4	~4.0	74
Franklin	Emergent	Emergent/ submergent	Emergent/ submergent	15	28	32	11
Jackson	Emergent	Emergent/ scrub-shrub	Emergent/ scrub-shrub	4.8	7.2	7.5	105
Gallia	Emergent	Emergent/ woody	Emergent	0.5	0.8	0.7	88
Total				24.4	42	~16.0	38

SOURCE: Data from Wilson and Mitsch (1996).

nia was evaluated using several indicators of function, and the results show the importance of measuring more than vegetation to characterize ecological performance (Table 6-9). Four constructed salt marshes were studied for 5 years or longer, and 11 attributes were compared with nearby natural sites. Three attributes of plant health (biomass, height, and nitrogen content) varied from 42% to 84% equivalency, two benthic invertebrate parameters varied from 36% to 78% equivalency, and four soil parameters (organic content, sediment nitrogen, pore water nitrogen, and nitrogen fixation) varied from 17% to 110% of that in the reference marsh. A high score for one parameter does not guarantee a high score for another. A recent compilation of the 10-year data set indicated that this project did not comply with permit conditions, which included tall vegetation suitable for nesting by an endangered bird (Zedler and Callaway 1999).

A systematic approach to measuring ecological equivalency involves the application of Brinson's (1993) Hydrogeomorphic Method (HGM), which is broadly applicable to most wetlands. Sudol (1996) used the HGM method to assess 40 mitigation projects covering 97 hectares of impacts for 104 hectares of proposed mitigation in Orange County, California.

TABLE 6-9   Index of Functional Equivalency for Four Constructed Salt Marshes in Relationship to Natural Sites in Paradise Creek, Southern California

Parameter	% Equivalency to Reference Wetland
Organic matter content	51
Sediment nitrogen as inorganic nitrogen	45
Sediment nitrogen (total Kjeldahl)	52
Pore-water nitrogen as inorganic nitrogen	17
Nitrogen fixation in top 1 cm	51
Nitrogen fixation in rhizosphere	110
Biomass of vascular plants	42
Foliar nitrogen concentration	84
Height of vascular plants	65
Epibenthic invertebrate numbers	36
Epibenthic invertebrate species list	78
Average of all comparisons	57

NOTE: Only the most vigorous stands of vegetation in the mitigation site were sampled for comparison with the reference site.

SOURCE: Zedler and Langis (1991). Reprinted with permission from *Ecological Restoration*; copyright 1991, University of Wisconsin Press, Madison.

Fifteen habitat functions were compared in 7 reference sites. Forty-two percent of the compensatory mitigation wetland area met the terms of the permit requirements. However, not one mitigation effort was completely successful in terms of the HGM analyses. Fourteen mitigation efforts were partially successful. The overarching reason given to explain the lack of success was that there was insufficient restoration and creation of the necessary hydrological conditions.

Brown and Veneman (1998) examined 70 mitigation permits in Massachusetts and made field visits to 68 permittee-responsible sites that underwent restoration and also to a subset of sites that apparently were mitigation banks. Various environmental parameters were measured in both mitigation wetland and an appropriate reference site, although full information on how the comparison was made is not available (discussed in Chapter 2). Some of their results are in Table 6-10. The compensatory wetland had fewer species among all sites, but the mitigation bank sites had more species than the reference sites. Although plant cover and other indices of plant community health were similar in reference and mitigation wetland, the species were not the same. The differences in plant species may explain why the use by amphibians, mammals, and birds (but not reptiles) was higher in the reference sites.

TABLE 6-10 Ecological Parameters in Paired Replacement and Reference Wetlands in Massachusetts

Parameter	All Reference Wetlands	All Replacement Wetland	Reference Wetlands	Mitigation Bank Wetlands
Plant				
Number of species	11.4 ± 2.7	7.8 ± 3.8	15.5 ± 3.4	17.3 ± 2.4
Number of facultative or wetter species	10.2 ± 2.7	7.8 ± 3.8	Same in both	Same in both
Cover all species	More in reference sites	More in reference sites	Same in both	Same in both
Cover wetland species	More in reference sites	More in reference sites	Lower in reference sites	Lower in reference sites
Wetland index value	2.3 ± 0.33	2.3 ± 0.68	2.1 ± 0.7	2.0 ± 0.4
Similarity of species	Dissimilar	Dissimilar	Dissimilar	Dissimilar
Water-quality function value	N.A.	N.A.	0.86 ± 0.8	0.95 ± 0.32
Sediment stabilization	N.A.	N.A.	0.86 ± 0.075	0.86 ± 0.05
Wildlife utilization (Jaccard Similarity Index)				
Amphibians	N.A.	N.A.	More in reference sites	More in reference sites
Mammals	N.A.	N.A.	More in reference sites	More in reference sites
Reptiles	N.A.	N.A.	Similar	Similar
Birds	N.A.	N.A.	More in reference sites	More in reference sites

NOTES: The mean and ± 1 standard deviation are shown, when given in the source document. A. For 68 sites of all sizes; B. for 12 sites that were for "variances," which appear to be mitigation banks. These variance sites have much better site planning and oversight than the nonvariance (mitigation bank) wetlands. N.A., Not available.

SOURCE: Adapted from Brown and Veneman (1998).

Estimates of functional equivalency and compliance rates from various studies are summarized in Table 6-11. The percentage of permits meeting permit compliance ranges from 3% to 100%, and the percentage of permits resulting in various measures of ecological functionality (equivalency) ranges from 0 to 67%. For studies with estimates of both permit compliance and ecological equivalency, the averages are 55% and 23%, respectively. The average for all eight estimates for ecological equivalency is 14%. Some of these sites may improve with time (see discussion of trajectories in Chapter 2).

Eighteen studies used the number of restoration or creation sites that met permit conditions as an indicator of permit compliance (see Table 6-12). Ten of the 18 studies had compliance rates of greater than 50%. Sudol (1996) surveyed 70 (required as part of 80 permits) mitigation sites in Orange County, California, to determine permit compliance for permits issued between 1985 and 1993. There were 128 ha of impact, of which 19 ha met all conditions of mitigation. Of the 80 permits, 30 projects were considered compliant with permit conditions. Of the remaining projects,

TABLE 6-11 Comparison of the Percentage of Permits Meeting Their Requirements and Percentage of Those Permits Meeting Various Tests of Ecological Functionality or Viability

Location	No. of Permits	% Permit Compliance	% Meeting Viability/ Function[a]	Source or Notes
California	57	18	0	Sudol (1996)
Florida	29	79	45	Lowe et al. (1989)
Florida	63	6	27	FDER (1991a)
Freshwater	34	3	12	
Saltwater	29	10	45	
Florida	N.A.	N.A.	4	Erwin (1991)
Florida, St. Johns	N.A. (1992)	43	27	OPPAGA (2000)
Water Management District (WMD)	N.A. (1999)	78	67	OPPAGA (2000)
Ohio	10	100	0	Fennessy and Roehrs (1997)
Oregon	17	47	18	Storm and Stellini (1994)
Unknown	29	21	3	Mockler et al. (1998)

[a]Criteria used: Sudol (1996) classified sites as a complete success and not irrigated; Lowe et al. (1989) classified site viability as good or poor; DeWeese (1994) rated sites as successful if they scored 7 or higher on a scale of 0 to 10; FDER (1991a) rated sites based on hydrology, soils, vegetation, and fauna.
NOTES: N.A., Not available.

TABLE 6-12   Compliance (Based on Permit Number) for Fully Implemented Mitigation Plans

Location	No. of Permits	% in Compliance[a]	Source
California			
Orange County	57	13	Sudol (1996)
Southern Sacramento	75	42	Allen and Feddema (1996)
San Francisco	30	50	DeWeese (1994)
Florida	29	79	Lowe et al. (1989)
Florida	42	10	Erwin (1991)
Florida (Northeastern)	201	86	Miracle et al. (1998)[b]
Florida			
Southwestern WMD	33	33	OPPAGA (2000), for 1988 to 1989
Southwestern WMD	254	82	OPPAGA (2000), for permits since 1995[c]
St. Johns WMD	N.A.	78	OPPAGA (2000), for 1999
Suwannee River WMD	N.A.	100	OPPAGA (2000)
Florida Department of Environmental Protection			
Southeastern District	N.A.	67	OPPAGA (2000), no date
Northeastern District	N.A.	87	OPPAGA (2000), no date
Illinois	N.A.	4	Gallihugh and Rogner (1998)
Massachusetts Department of Environmental Protection	84	49	Brown and Veneman (1998)
Ohio	14	100	Fennessy and Roehrs (1997)
Ohio	5	80	Wilson and Mitsch (1996)
Virginia	32	N.A.	Mason and Slocum (1987)
1.a	N.A.	86	When with specific permit conditions
b	N.A.	44	When without specific permit conditions
2.a	N.A.	100	Permits with time limits
b	N.A.	50	Permits without time limits
Washington	17	53	Storm and Stellini (1994)
Washington	43	35	Johnson et al. (2000)
Unknown	29	21	Mockler et al. (1998)

[a]Compliance based on 100% compliance for Allen and Feddema (1996); scale of 8 out of 10 (75% in compliance) for DeWeese (1994).

[b]After 5 years; some mitigation was still in the monitoring stage (and compliant).

[c]Includes mitigation efforts that achieved success and that were "trending toward" success, as defined by the evaluators.

NOTES: Based on field inspection or monitoring reports and when all permit conditions were met. Unverified mitigation attempts were considered noncompliant. N.A., Not available.

6 were not compliant, 2 were never attempted, and 13 permits were never completed. The remaining projects were only partially successful in meeting the permit requirements. Eight studies of five state permitting programs examined whether attempts at mitigation met permit conditions (see Table 6-13). Compliance with the permit was measured by the area of created or restored wetlands. The results indicate that the compensatory wetland mitigation area was often less than the permitted area.

Mitigation sites have also been evaluated using more subjective measures. DeWeese (1994) evaluated 30 projects in California to determine mitigation compliance using a 10-point scale based on professional judgment. About half of the sites were at least 5 years old. DeWeese ranked only 1 of the 30 sites high, 13 were average or above average, and 6 were below average (Table 6-14). Two sites had low value, and two others were judged to have no value. The average value was 4.7 for 30 mitigation projects. There was no net loss in area as a result of permitting, but there was a net loss in ecological functionality.

Allen and Feddema (1996) examined 75 sites in California using three simple criteria evaluated subjectively: weed invasion, plant cover, and vegetation status. They found a 69% "success rate" and estimated that 77 ha replaced 81 ha lost, for a net loss of 5%. Storm and Stellini (1994) found a mixed result in their review of 17 compensatory mitigation projects in Washington (Table 6-15). Fifty-three percent could not be verified as being in compliance with the permit, 29% were verified as being out of compliance, and two-thirds were not ecologically equivalent to the wetlands lost through the permitting program. Monitoring was done for only

TABLE 6-13   Compliance (Area Basis) for Mitigation That Was Attempted Based on Field Inspection or Monitoring Reports

Location	No. of Permits	Impacted Hectares	% Area Gain (loss)	Source
California				
Orange County	68	128	(92)	Sudol (1996)
San Diego County	N.A.	102	(8)	Fenner (1991)
Southern Sacramento	75	80	(8)	Allen and Feddema (1996)
San Francisco	30	168	44	DeWeese (1994)
Florida	29	269	(32)	Lowe et al. (1989)
Indiana	31	14	10	J. T. Robb (personal commun. 2000)
Ohio	5	24	(33)	Wilson and Mitsch (1996)
Tennessee	50	38	(13)	Morgan and Roberts (1999)

TABLE 6-14   Ranking of Compliance for 30 Sites in San Francisco Bay That Were Issued Section 404 Permits

Permits (No.)	Compliance (%)
3	100
6	85 to 95
6	75 to 85
12	45 to 75
1	1 to 14
2	0

NOTE: DeWeese (1994) ranked sites (scale 1 to 10) to determine their ecological success rate.

one-third of that required. These subjective measures of equivalency suggest that compliance is much lower than expected.

The historical and national perspective of mitigation in the United States that has been presented comes from a wide range of studies, and we do not think that our review is a parochial or geographically restricted one. A summary of key data reviews is in Table 6-16. Between 70% and 76% of the mitigation required in the permits is implemented, and about 50% to 53% of the implemented mitigation projects did not meet the permit requirements. In addition, the estimate of functional equivalency of mitigation wetland was about 20% of that intended. These estimates (based on the mean or median value in the tables) suggest that there is a substantial net loss in wetland area from wetlands permitting program. In terms of the ecological equivalency of these wetlands, there is a low value of the wetlands actually built.

TABLE 6-15   Results from an Analysis of Compliance for 17 Mitigation Projects with Field Investigation in Western Washington

Condition	Number	Percent
Not verified as being in compliance	9	53
Out of compliance	5	29
Not functioning ecologically	11	65
In compliance with regulatory requirements	3	18
Monitoring done/required	9/17	53
Monitoring done when required	3	33

SOURCE: Storm and Stellini (1994).

TABLE 6-16 Summary of Data from Previous Tables on Wetland Permit Implementation, Compliance, Ecological Success, and Monitoring Frequency

Parameter	No. of Studies	No. of States	Range	Mean	Median
1. % Mitigation attempted for required mitigation permits (778+ permits; Table 6-3)	8	7	28 to 100	76	70
2. % Compliance for mitigation required, based on field inspections					
a. % area gain (loss) (Table 6-13)	8	5	(92) to 44	(17)	(32.5)
b. % permits issued (Table 6-12)	19	6	0 to 100	58	53
3. % Ratio required and ratios met (post-construction; Table 6-5)	9	4	17 to 90	61	62
4. % Functional equivalency of completed mitigation (Table 6-11)	9	4	0 to 67	21	18
5. % Sites insufficiently monitored for permitted mitigation (Table 6-7)	7	6	40 to 90	63	61

NOTE: The average of a tie for the median value is given.

Record Keeping

The committee found that in many cases, mitigation was required to offset impacts; however, some project files did not contain a mitigation plan or other explicit agreements on the size and type of mitigation to be provided. Because mitigation projects extend over many years, mitigation plans need to be on record, with mitigation requirements clarified, so that all subsequent parties involved in the evaluation of the mitigation site know exactly what was required.

Many of the problems of tracking wetland loss and gain resulting from mitigation implementation could be addressed by improved record-keeping on the part of the Corps. Through improved compliance inspections described in Chapter 8, information should be included in a national database, such as the Corps Regulatory Analysis and Management System (RAMS) database. If each Corps project manager followed data-entry quality-assurance measures, then wetland losses and gains could be tracked more accurately on a national scale.

CONCLUSIONS

1. It appears that the performance standards sought in compensatory mitigation have not often been well defined.

2. Wetland restoration and creation trajectories do not suggest equivalency with reference sites within the commonly used 5-year monitoring period.

3. The literature and testimony provided to the committee indicate that the national goal of "no net loss" for permitted wetland conversions is not being met.

4. The gap between what is required and what is realized is not precisely known; however, the evidence strongly suggests that the required compensatory mitigation called for by wetland permits to date will not be realized.

5. Permit follow-up is sparse or too infrequent, and a higher post-monitoring rate will increase permit compliance rates. Compliance monitoring is commonly known to be nonexistent after 5 years. Better documentation and monitoring will increase compliance rates.

6. The sparse compliance monitoring is a direct consequence of its designation as a "below-the-line" policy standard. Raising compliance monitoring to "above the line" will greatly enhance mitigation success. Chapter 8 further discusses mitigation compliance issues; specifically, recommendations 4 and 6 in that chapter addresses concerns outlined here.

RECOMMENDATIONS

The committee makes the following recommendations relative to future mitigation compliance to ensure that the nation has an accurate reporting of wetland losses and gains:

1. The wetland area *and* functions lost and regained over time should be tracked in a national database. This database may or may not be the Corps Regulatory Analysis and Management System (RAMS) database.

2. The Corps should expand and improve the quality-assurance measures for data entry in the RAMS database.

3. Mitigation goals must be clear and those goals carefully specified in terms of measurable performance standards in order to improve mitigation effectiveness. Performance standards in permits should reflect mitigation goals and be written in such a way that ecological viability can be measured and the impacted functions replaced.

7

Technical Approaches Toward Achieving No Net Loss

This chapter describes lessons that can be drawn from the preceding chapters to increase the likelihood that mitigation requirements will in fact move forward and serve water quality and other functions in the nation's watersheds. The committee offers practical guidelines for designing and constructing sustainable compensation wetlands, for assessing the functional endpoints of those wetlands, and for creating the institutional reforms that will identify and secure the desired results.

OPERATIONAL GUIDELINES FOR CREATING OR RESTORING WETLANDS THAT ARE ECOLOGICALLY SELF-SUSTAINING

The compensatory mitigation process has been weakened by insufficient scientific knowledge as well as limitations of wetland regulatory institutions. The committee agreed on 10 guidelines applicable to compensatory mitigation projects. The guidelines suggest that in most cases wetland restoration is preferred to wetland creation. Also, the guidelines address mitigation project setting, site design, landscape setting, the importance of reference wetlands, and long-term management. Special attention must be paid to hydrological and topographical variability, subsurface characteristics, and the hydrogeomorphic and ecological landscape and climate of a site. Systems that are designed to incorporate natural processes will be more likely to ensure long-term sustainability.

1. Consider the hydrogeomorphic and ecological landscape and climate. Whenever possible, locate the mitigation site in a setting of comparable landscape position and hydrogeomorphic class. Do not generate atypical "hydrogeomorphic hybrids"; instead, duplicate the features of reference wetlands or enhance connectivity with natural upland landscape elements (Gwin et al. 1999).

Regulatory agency personnel should provide a landscape setting characterization of both the wetland to be developed and, using comparable descriptors, the proposed mitigation site. Consider conducting a cumulative impact analysis at the landscape level based on templates for wetland development (Bedford 1999). Landscapes have natural patterns that maximize the value and function of individual habitats. For example, isolated wetlands function in ways that are quite different from wetlands adjacent to rivers. A forested wetland island, created in an otherwise grassy or agricultural landscape, will support species that are different from those in a forested wetland in a large forest tract. For wildlife and fisheries enhancement, determine if the wetland site is along ecological corridors such as migratory flyways or spawning runs. Constraints also include landscape factors. Shoreline and coastal wetlands adjacent to heavy wave action have historically high erosion rates or highly erodible soils, and often heavy boat wakes. Placement of wetlands in these locations may require shoreline armoring and other protective engineered structures that are contrary to the mitigation goals and at cross-purposes to the desired functions.

Even though catastrophic events cannot be prevented, a fundamental factor in mitigation plan design should be how well the site will respond to natural disturbances that are likely to occur. Floods, droughts, muskrats, geese, and storms are expected natural disturbances and should be accommodated in mitigation designs rather than feared. Natural ecosystems generally recover rapidly from natural disturbances to which they are adapted. The design should aim to restore a series of natural processes at the mitigation sites to ensure that resilience will have been achieved.

As described in other chapters, regulatory agency personnel often do not have either the time or the education and training to consider important, broader issues such as landscape setting. It is imperative that the Corps, EPA, and other advisory agency personnel receive additional training in landscape ecology and other considerations that are poorly represented in the present mitigation process.

2. Adopt a dynamic landscape perspective. Consider both current and future watershed hydrology and wetland location. Take into account surrounding land use and future plans for the land. Select sites that are, and will continue to be, resistant to disturbance from the surrounding landscape, such as preserving large buffers and connectivity to other wetlands. Build on existing wetland and upland systems. If possible, locate the mitigation site to take advantage of refuges, buffers, green spaces, and other preserved elements of the landscape. Design a system that utilizes natural processes and energies, such as the potential energy of streams as natural subsidies to the system. Flooding rivers and tides transport great quantities of water, nutrients, and organic matter in relatively short time periods, subsidizing the wetlands open to these flows as well as the adjacent rivers, lakes, and estuaries.

3. *Restore or develop naturally variable hydrological conditions.* Promote naturally variable hydrology, with emphasis on enabling fluctuations in water flow and level, and duration and frequency of change, representative of other comparable wetlands in the same landscape setting. Preferably, natural hydrology should be allowed to become reestablished rather than finessed through active engineering devices to mimic a natural hydroperiod. When restoration is not an option, favor the use of passive devices that have a higher likelihood to sustain the desired hydroperiod over the long term. Try to avoid designing a system dependent on water-control structures or other artificial infrastructure that must be maintained in perpetuity in order for wetland hydrology to meet the specified design. In situations where direct (in-kind) replacement is desired, candidate mitigation sites should have the same basic hydrological attributes as the impacted site.

Hydrology should be inspected during flood seasons and heavy rains, and the annual and extreme-event flooding histories of the site should be reviewed as closely as possible. A detailed hydrological study of the site should be undertaken, including a determination of the potential interaction of groundwater with the proposed wetland. Without flooding or saturated soils for at least part of the growing season, a wetland will not develop. Similarly, a site that is too wet will not support the desired biodiversity. The tidal cycle and stages are important to the hydrology of coastal wetlands.

4. *Whenever possible, choose wetland restoration over creation.* Select sites where wetlands previously existed or where nearby wetlands still exist. Restoration of wetlands has been observed to be more feasible and sustainable than creation of wetlands. In restored sites the proper substrate may be present, seed sources may be on-site or nearby, and the appropriate hydrological conditions may exist or may be more easily restored.

The U.S. Army Corps of Engineers (Corps) and Environmental Protection Agency (EPA) Mitigation Memorandum of Agreement states that, "because the likelihood of success is greater and the impacts to potentially valuable uplands are reduced, restoration should be the first option considered" (Fed. Regist. 60(Nov. 28):58605). The Florida Department of

Environmental Regulation (FDER 1991a) recommends an emphasis on restoration first, then enhancement, and, finally, creation as a last resort. Morgan and Roberts (1999) recommend encouraging the use of more restoration and less creation.

5. Avoid overengineered structures in the wetland's design. Design the system for minimal maintenance. Set initial conditions and let the system develop. Natural systems should be planned to accommodate biological systems. The system of plants, animals, microbes, substrate, and water flows should be developed for self-maintenance and self-design. Whenever possible, avoid manipulating wetland processes using approaches that require continual maintenance. Avoid hydraulic control structures and other engineered structures that are vulnerable to chronic failure and require maintenance and replacement. If necessary to design in structures, such as to prevent erosion until the wetland has developed soil stability, do so using natural features, such as large woody debris. Be aware that more specific habitat designs and planting will be required where rare and endangered species are among the specific restoration targets.

Whenever feasible, use natural recruitment sources for more resilient vegetation establishment. Some systems, especially estuarine wetlands, are rapidly colonized, and natural recruitment is often equivalent or superior to plantings (Dawe et al. 2000). Try to take advantage of native seed banks, and use soil and plant material salvage whenever possible. Consider planting mature plants as supplemental rather than required, with the decision depending on early results from natural recruitment and invasive species occurrence. Evaluate on-site and nearby seed banks to ascertain their viability and response to hydrological conditions. When plant introduction is necessary to promote soil stability and prevent invasive species, the vegetation selected must be appropriate to the site rather than forced to fit external pressures for an ancillary purpose (e.g., preferred wildlife food source or habitat).

6. Pay particular attention to appropriate planting elevation, depth, soil type, and seasonal timing. When the introduction of species is necessary, select appropriate genotypes. Genetic differences within species can affect wetland restoration outcomes, as found by Seliskar (1995), who planted cordgrass (*Spartina alterniflora*) from Georgia, Delaware, and Massachusetts into a tidal wetland restoration site in Delaware. Different genotypes displayed differences in stem density, stem height, belowground biomass, rooting depth, decomposition rate, and carbohydrate allocation. Beneath the plantings, there were differences in edaphic chlorophyll and invertebrates.

Many sites are deemed compliant once the vegetation community becomes established, as described in the Coyote Creek case study (Appendix B). If a site is still being irrigated or recently stopped being irrigated, the vegetation might not survive. In other cases, plants that are dependent on surface-water input might not have developed deep root systems. When the surface-water input is stopped, the plants decline and eventually die, leaving the mitigation site in poor condition after the Corps has certified the project as compliant.

7. *Provide appropriately heterogeneous topography.* The need to promote specific hydroperiods to support specific wetland plants and animals means that appropriate elevations and topographic variations must be present in restoration and creation sites. Slight differences in topography (e.g., micro- and mesoscale variations and presence and absence of drainage connections) can alter the timing, frequency, amplitude, and duration of inundation. In the case of some less-studied, restored wetland types, there is little scientific or technical information on natural microtopography (e.g., what causes strings and flarks in patterned fens or how hummocks in fens control local nutrient dynamics and species assemblages and subsurface hydrology are poorly known). In all cases, but especially those with minimal scientific and technical background, the proposed development wetland or appropriate example(s) of the target wetland type should provide a model template for incorporating microtopography.

Plan for elevations that are appropriate to plant and animal communities that are reflected in adjacent or close-by natural systems. In tidal systems, be aware of local variations in tidal flooding regime (e.g., due to freshwater flow and local controls on circulation) that might affect flooding duration and frequency.

8. *Pay attention to subsurface conditions, including soil and sediment geochemistry and physics, groundwater quantity and quality, and infaunal communities.* Inspect and characterize the soils in some detail to determine their permeability, texture, and stratigraphy. Highly permeable soils are not likely to support a wetland unless water inflow rates or water tables are high. Characterize the general chemical structure and variability of soils, surface water, groundwater, and tides. Even if the wetland is being created or restored primarily for wildlife enhancement, chemicals in the soil and water may be significant, either for wetland productivity or bioaccumulation of toxic materials. At a minimum, these should include chemical attributes that control critical geochemical or biological processes, such as pH, redox, nutrients (nitrogen and phosphorus species), organic content, and suspended matter.

9. Consider complications associated with creation or restoration in seriously degraded or disturbed sites. A seriously degraded wetland, surrounded by an extensively developed landscape, may achieve its maximal function only as an impaired system that requires active management to support natural processes and native species (NRC 1992). It should be recognized, however, that the functional performance of some degraded sites may be optimized by mitigation, and these considerations should be included if the goal of the mitigation is water- or sediment-quality improvement, promotion of rare or endangered species, or other objectives best served by locating a wetland in a disturbed landscape position. Disturbance that is intense, unnatural, or rare can promote extensive invasion by exotic species or at least delay the natural rates of redevelopment. Reintroducing natural hydrology with minimal excavation of soils often promotes alternative pathways of wetland development. It is often advantageous to preserve the integrity of native soils and to avoid deep grading of substrates that may destroy natural below-ground processes and facilitate exotic species colonization (Zedler 1996a,b).

10. Conduct early monitoring as part of adaptive management. Develop a thorough monitoring plan as part of an adaptive management program that provides early indication of potential problems and direction for correction actions. The monitoring of wetland structure, processes, and function from the onset of wetland restoration or creation can indicate potential problems. Process monitoring (e.g., water-level fluctuations, sediment accretion and erosion, plant flowering, and bird nesting) is particularly important because it will likely identify the source of a problem and how it can be remedied. Monitoring and control of nonindigenous species should be a part of any effective adaptive management program. Assessment of wetland performance must be integrated with adaptive management. Both require understanding the processes that drive the structure and characteristics of a developing wetland. Simply documenting the structure (vegetation, sediments, fauna, and nutrients) will not provide the knowledge and guidance required to make adaptive "corrections" when adverse conditions are discovered. Although wetland development may take years to decades, process-based monitoring might provide more sensitive early indicators of whether a mitigation site is proceeding along an appropriate trajectory.

WETLAND FUNCTIONAL ASSESSMENT

The goal of no net loss refers to both wetland acres and wetland function, as the functions contribute to the watershed where the wetland is located. Therefore, when setting compensatory mitigation goals, the functions of a wetland proposed for fill need to be precisely characterized and, if possible, quantified, as should the functions of the proposed compensatory mitigation project. Even if the mitigation goal does not seek in-kind replacement of functions, functional assessment provides a foundation for considering the watershed consequences of out-of-kind mitigation. Functional assessment helps determine whether the location and design of a compensation wetland will secure the functions that are emphasized for the watershed.

In practice, mitigation attention often is focused on relatively few of the numerous functions that wetlands can provide—for example, habitat, water-quality improvement, and various hydrological functions (groundwater recharge and floodwater desynchronization). The committee does not believe that a science-based functional assessment should be used to assess or rank all the societal values of a wetland. In some cases, technical assessment and the social values of each function have been merged into one assessment procedure. It is recognized that these functions have human value by the societal, economic, and other services they provide and that the values emphasized should be reflected in the location and design of the compensatory wetland. However, the committee believes there are other points in the process of mitigation planning to consider tradeoff among functions where, based on a systematic functional assessment that evaluates all functions objectively, weighting factors can be introduced into the mitigation planning process to consider the broader perspectives about their relative importance (see Chapter 8 for a discussion of this point).

Complete characterization of a compensatory mitigation site requires an assessment of the level of performance attainable for each wetland function under different site designs. This would include consideration of various natural hydrological, geochemical, and ecological attributes and processes. In addition, functional assessment of prospective compensation sites will help establish the design and the monitoring and assessment procedures for the wetland to be created or restored.

Most wetland scientists argue that science-based, regionally standardized procedures are preferable to best professional judgment in comprehensively evaluating wetland function for both impacted and mitigation sites. As a result, the general absence of a uniform approach to assessing wetlands as multifunctional ecosystems have likely encouraged less complex wetland mitigation designs and rudimentary measures of achieving mitigation goals.

THE FLORISTIC APPROACH

In early wetland mitigation efforts, functional assessment was usually confined to lists or qualitative descriptions. Furthermore, although permit requirements often suggest the need to consider area and function, structural characteristics (usually the amount of vegetation cover) may be used as a criterion to judge whether functional replacement is achieved (e.g., Kentula et al. 1992b). Vegetation structure (e.g., percent cover) is a pervasive example of one structural attribute that is often the default indicator of wetland function. One example of such a singular criterion is the Floristic Quality Assessment developed by Swink and Wilhelm (1979, 1994) for wetlands in the Chicago region and several Midwest states[1] (Andreas and Lichvar 1995; Taft et al. 1997; Herman et al. 1996; Mack et al. 2000). The floristic and similar approaches basically characterize a mitigation site solely on the vegetation present. The assumption underlying this approach is that wetland vegetation is a comprehensive indicator of the hydrological and ecological status of the site, and specific vegetation parameters can be used to indicate the functions of the mitigation site. The reasoning behind this approach is that if the vegetation community is healthy and has "natural" species diversity, the ecological components (e.g., physical, biological, and biochemical) that support the vegetation must be present.

In Swink and Wilhelm's (1979, 1994) application of Floristic Quality Assessment, indicators are based on the site in question; thus, a riparian system would have completely different vegetation parameters than a coastal salt marsh. In the Floristic Quality Assessment, each plant species was assigned a coefficient of conservatism (C) ranging from 0 (ubiquitous species) to 10 (species having narrow habitat tolerances), based on the authors' knowledge of the flora of the Chicago region (Swink and Wilhelm 1979, 1994). Other indicators used in past evaluations include percent canopy and/or ground cover, percent survival of specific indicator species, tree height, and species diversity. In many areas, floristic assessment has been the method of choice because vegetation parameters are easy to measure, provide a dramatic visual indicator of compliance (full canopy, tall trees), and allow resource agencies to write well-defined performance criteria for the mitigation sites.

However, the assumptions and premises of the floristic approach are often unclear or incompletely specified when examining the regional spectrum of wetland types. Low plant diversity is not always characteristic of "inferior" hydrogeological and geochemical settings, and high plant diversity is not necessarily a de facto indicator of the multitude of wetland functions (e.g., NRC 1995). Systematic assessment of more than just floristic quality indicators reduces dependence on such speculative assumptions.

------

[1] D.M. Ladd. The Missouri Floristic Quality Assessment system. The Nature Conservancy, St. Louis, MO, in preparation.

HABITAT EVALUATION PROCEDURES AND THE
HYDROGEOMORPHIC APPROACH

The possible array of procedures now available for functional wetland assessment has grown to the point that there is considerable confusion about what is acceptable or preferable and by which regulator or scientist (e.g., 40 procedures are recognized by Bartoldus (1999); see Appendix H). Most procedures are site-specific, with only a few providing assessments at the wetland system or landscape scale. Many are specifically designed to assess one or a few wetland functions, such as fish and wildlife habitat, and lack any procedures to assess other functions or a comprehensive assessment of all functions. Many limit consideration to wetland functions with societal value. Some were developed to generate scores that are scaled to wetland area, such that functions are explicitly assumed to be multiplicative (which is not always the case). Although most use systematic models, many are based on qualitative and often subjective interpretations rather than measurement of discrete variables or parameters. Some procedures, such as habitat evaluation procedures (HEPs) (USFWS 1980, 1981; Sousa 1985), have become operationally codified in regulatory procedures as either required or recommended elements of wetland assessment. HEP was one of the two functional assessment procedures that Bartoldus (1999) considered applicable in all 50 states. Meanwhile, the lack of a broadly accepted, generalized functional assessment procedure as a universal screening tool has led to hybrids that are designed to meet perceived unique needs, such as that for wetland banking (Stein et al. 2000). Among the 40 procedures evaluated by Bartoldus (1999), only seven have been applied or are being considered to establish credits in mitigation banks.

In the mid-1990s, the Corps and Natural Resources Conservation Service (NRCS) agreed to the formal adoption of the hydrogeomorphic (HGM) approach (see Box 7-1) as a uniform procedure for functional assessment in the Clean Water Act's Section 404 program and the U.S. Department of Agriculture programs (Smith et al. 1995). Because it is exclusively based on wetlands and not social processes and has applicability at both the watershed and the landscape scales, HGM was attractive to wetland scientists. It was seen as particularly applicable to wetland mitigation because target hydrology could be based on the influence of water sources, wetland type, and the relative ease or difficulty of establishing certain hydrological regimes. Another of the recognized strengths of HGM is the assessment of functional performance based on a domain of reference systems that capture the presumed optimum natural function. Reference sites are essential for the precise identification of specific wetland attributes and processes for the mitigation site (e.g., hydrology

---

BOX 7-1
The Hydrogeomorphic Method

HGM classification (Brinson 1993) is a functional classification and as such differs from other wetland classification systems, such as the Cowardin system, which was designed for use in national wetland inventory mapping (Cowardin et al. 1979). The HGM approach classifies a wetland based on its setting in the landscape, its source of water for the wetland, and the dynamics of the water on-site. Setting in the landscape, in the context of HGM, refers to distinctions among, for example, wetlands that occupy depressions, river flood plains, or estuary fringes. Similarly, by water source, the committee means to distinguish among wetlands that receive surface water, as opposed to those that receive primarily precipitation or ground-water. By dynamics of the water on-site, or flow, the committee is interested in distinguishing wetlands that have unidirectional horizontal flow from those that have vertical flow (upwelling) and those that have horizontally bi-directional flow. HGM classification groups wetlands with similar structure and function and emphasizes features of wetlands that are relatively independent of the biogeo-graphical distribution of species and requires recognition of factors external to the wetland (Brinson 1993). Wetlands within a class are assumed to be hydrogeo-morphically and functionally similar and to have functional attributes different from wetlands in other classes. According to the HGM perspective—for example, a slope wetland (i.e., a groundwater-driven wetland) dominated by emergent vege-tation is functionally different from a riverine wetland (i.e., a wetland of the active floodplain) with emergent vegetation.

---

functions in terms of saturation duration, depth, and frequency not only seasonally but also annually). The enhancement of functions (such as control of water levels or flows to enhance vegetation, water quality, or waterfowl habitat) was to be considered outside the domain of reference sites. In addition, fundamental incorporation of reference wetlands meant that assessments were sensitive to regional variations in the functional performance of hydrogeomorphic subclasses. However, in one respect, HGM and similar assessment procedures are still deficient at assessing the effect of wetland mitigation at the landscape scale. Although they may effectively assess the functions of a wetland site in a hydrogeo-morphic, landscape setting, these procedures will not necessarily exam-ine whether the development of a wetland will reduce the functional value of adjacent wetlands or put at risk significant other areas.

HGM AS A FUNCTIONAL ASSESSMENT PROCEDURE

The original Corps-sponsored HGM functional assessment proce-dures have been modified to meet different, often project specific, needs, but most versions have not been accepted by the Corps (Bartoldus 1999). Although many include elements of the HGM concepts and/or terminology, most deviate substantially from the intent, premises, and design of HGM (Brinson 1995; 1996). Most of these do not include complete data sets, particularly, appropriate reference site data. Examples of these derivative procedures include the Washington State Wetland Function Assessment Project (Hruby et al. 1998), Minnesota Routine Assessment Method for Evaluating Wetland Functions (Minnesota Board of Water and Soil Resources 1998), EPA (Bartoldus et al. 1994), Method for Assessment of Wetland Function (MDE method) (Fugro East Inc. 1995), and the Rapid Assessment Procedure (Magee and Hollands 1998).

However, HGM has many useful applications in functional assessment and the mitigation process in general. For instance, the advantages of evaluating mitigation performance using an assessment of hydrological equivalence (Bedford 1996) based on hydrogeomorphic classification (Brinson 1993) are effectively demonstrated by an EPA Wetlands Research Program evaluation of mitigation projects in the Portland, Oregon, metropolitan area (Gwin et al. 1999; Shaffer et al. 1999; Shaffer and Ernst 1999; Magee et al. 1999) (see Box 7-2). HGM classification analysis revealed seven HGM regional classes in the Portland sample wetlands that were defined using the Cowardin system. Analysis showed that the vegetation, soil, and hydrological variables that were measured differed significantly among HGM classes.

Almost all of the mitigation wetlands belonged to HGM classes that were atypical of the region. There were no naturally occurring analogs for the hydrogeomorphic types they represented (Gwin et al. 1999). This explains why the results presented above comparing features of mitigation wetlands and naturally occurring wetlands pointed to differences between the two groups. One would expect differences in hydrogeomorphic features to be reflected in the soils, vegetation, and, especially, the hydrology. These results demonstrate that the diversity in the hydrogeomorphic characteristics of wetland exemplified in the regional HGM classes is related to the diversity of their extant hydrology (Shaffer et al. 1999). Because hydrology is a critical forcing function for other wetland attributes, changes in hydrology can be assumed to have significant effects on a variety of wetland functions.

However, development and testing of HGM at the regional level are inconsistent, uncoordinated, and dependent on needs and funding. Perhaps more important, as exemplified by the Corps's recent development of HGM regional guidebooks for several wetland classes (Ainslie et al. 1999; Brinson et al. 1995), most variables incorporated into the assessment models remain measures of wetland structure rather than processes. This may be the inevitable result of tension between the costs of doing functional assessment in terms of staff time and funds and the available staff and budgets. Compromises may be necessary. Although HGM as a specific functional assessment procedure may not be meeting expectations and may be too costly to implement in all cases, it has put a focus on the need for assessing wetland function at the landscape scale (see Box 7-2).

BOX 7-2
Functional Assessment of Hydrological Equivalence Using HGM in the Portland, Oregon, Metropolitan Region

The Portland metropolitan area, located in northwestern Oregon at the north end of the Willamette Valley, was chosen for study of the functionality of mitigation sites because rapid urbanization and development have placed wetlands in the area at high risk for modification and destruction (Holland et al. 1995). The sites sampled were small (~2 ha) palustrine wetlands ranging from those dominated by emergent marsh to those dominated by open water (Cowardin et al. 1979), that is, the wetland types historically most common in the Willamette Valley (Davis 1995; Guard 1995) and most frequently required as mitigation for permitted losses of freshwater wetlands in the Portland area and the State of Oregon (Kentula et al. 1992a,b). The study wetlands were located in a variety of land-use conditions, including urban, agricultural, and undeveloped. Ninety-six sites (45 naturally occurring and 51 mitigation wetlands) were assessed in terms of morphology, hydrology, soils, and vegetation in the summer of 1993. The mitigation wetlands ranged in age from 1 to 9 years, averaging 5. In addition, the hydrological characteristics of approximately half the sites were monitored through January 1997.

EVALUATION USING STRUCTURAL INDICATORS

Characteristics considered desirable in naturally occurring wetlands are commonly used as permit conditions and design criteria for mitigation wetlands. Therefore, a comparison of naturally occurring wetlands and mitigation wetlands, using variables that have or could be used as permit conditions, illustrates the types of results produced by such an approach. For example, one might conclude that the mitigation wetlands in the Portland study could be called compliant based on the characteristics of the plant community. Within 5 years after construction, most mitigation wetlands would have met a criterion of 80% cover per square meter where emergent vegetation occurred on the site. However, only a small portion of the site was vegetated on many of the mitigation wetlands, because most of the sites were occupied by deep open water. The naturally occurring wetlands and mitigation wetlands both had plant communities composed of about 50% native species. On average, a slightly higher percentage of the species per site was native to the mitigation wetlands (mitigation wetlands = 47%; naturally occurring wetlands = 43%). However, the wetland flora in the area, in general, was degraded by the predominance of exotic species. The species composition of naturally occurring wetlands and mitigation wetlands is different ($p < .0001$) with species richness per

Furthermore, it should be recognized that, as it is currently developed as an assessment tool, HGM is principally a diagnostic method, not a prescriptive "cookbook." In this respect, the HGM models do not specifically lay out design parameters that guarantee the likelihood that hydrology, desired wetland vegetation, and desired animals will be reestab-

---

site higher on mitigation wetlands (p = .0006). However, the species composition of mitigation wetlands less than 3 years in age differed from that of mitigation wetlands more than 3 years in age due to the influx of introduced species, averaging 11 additional species per site as the site aged. So the mitigation wetlands may not maintain native plant species over time, especially in the face of changes occurring with urbanization of the landscape.

In the case of the organic matter content of the soils, naturally occurring wetlands and mitigation wetlands are significantly different. There is less organic matter in the top 5 cm of the soil of mitigation wetlands (p = .0001) and at 15 to 20 cm (p = 0.0551) than in naturally occurring wetlands. There was no substantive relationship between soil organic matter concentration and the age of mitigation wetlands (r^2 = .0232, p = .6003). This suggests that development of a soil organic matter content similar to that of naturally occurring wetlands may not be achieved for a very long time, if ever. Finally, the hydrological characteristics of the mitigation wetlands differed from the naturally occurring wetlands. As mentioned above, mitigation wetlands had more open water than naturally occurring wetlands. On average, 57% of the area of the mitigation wetlands was flooded, while 28% of the area of the naturally occurring wetlands was flooded during the year (p < .001). The predominance of deep open water on mitigation wetlands was indicated by higher mean annual water levels (0.85 m) on mitigation wetlands than on naturally occurring wetlands (0.25 m, p < .001). Hydrological variability also differed between naturally occurring wetlands and mitigation wetlands. The mean difference between the 10th and the 90th percentiles of water levels was 0.60 m for naturally occurring wetlands and 0.32 m for mitigation wetlands (p < .01). The difference between the 10th and the 90th percentiles was used to represent conditions commonly found in the wetlands as it minimizes the effects of extreme storm events.

Given the above analysis, conclusions about the performance of the mitigation wetlands would depend on whether only vegetation characteristics were considered and on how one viewed the predominance of alien plant species. In addition, there is evidence that the conclusion might change with time, especially with time periods longer than the 5-year monitoring requirement often associated with permits. Regardless, these kinds of analyses do not overcome the inherent problems of using structural similarity to infer functional equivalence, let alone determining the effects of permit decisions on the resource as a whole. Faced with this dilemma, it was found that the concept of hydrological equivalence as exemplified in HGM classification brought important insights to the evaluation of mitigation wetlands.

lished or the likelihood that exotics will not invade. HGM provides no analytical structure (e.g., a decision matrix) for inserting information about factors that influence the likelihood that hydrology, desired wetland vegetation, and desired animals will be reestablished or that exotics will not invade. This can only come with explicit application and monitoring using HGM in the design of wetland mitigation projects, and the resulting feedback on the correspondence between HGM indicators and monitored performance. This gap between HGM, and most other functional assessment procedures for that matter and the need for specific scientific and technical guidance for self-sustaining, functional mitigation wetlands remains a major hindrance to effective wetland mitigation.

It is possible that there is no single "best" wetland assessment procedure, because the specific needs vary with the situation, especially if a quick screening technique is needed (Smith 1993). However, in the mitigation process it is essential that there be an ability to relate the structural characteristics of a site to the resulting functions. Only in that way can the compensation site be designed to secure certain functions. The level of the function is calculated relative to levels in reference sites in the same subclass of wetlands within the same watershed or ecoregion. Perhaps functional assessments will evolve to meet this goal. The functional assessment procedure has the following desirable attributes:

• It includes reliable indicators of the important wetland processes (hydrology, sedimentation, and primary production) or a scientifically established structural surrogate of those processes.

• It assesses function over a broad range of performance conditions, such that differences in wetlands can be relatively easily distinguished.

• It is integrative over space and time, and its indicators are not vulnerable to seasonal or other fine-scale temporal or spatial variability.

• It results in a continuous, parametric scale that has not been reduced to a relative rank.

• It assesses all recognized functions so that the assessment encompasses all goals for the mitigation.

RECOMMENDATIONS

1. Dependence on subjective, best professional judgment in assessing wetland function should be replaced by science-based, rapid assessment procedures that incorporate at least the following characteristics:

• Effectively assess goals of wetland mitigation projects.
• Assess all recognized functions.
• Incorporate effects of position in landscape.

- Reliably indicate important wetland processes, or at least scientifically established structural surrogates of those processes.
- Scale assessment results to results from reference sites.
- Are sensitive to changes in performance over a dynamic range.
- Are integrative over space and time.
- Generate parametric and dimensioned units, rather than non-parametric rank.

2. Impact sites should be evaluated using the same functional assessment tools used for the mitigation site.

8

Institutional Reforms for Enhancing Compensatory Mitigation

INTRODUCTION

The committee reviewed wetland restoration and creation projects that were required as a condition of a Clean Water Act (CWA) Section 404 permit. Some sites appear to have met the criteria established for permit compliance and are, or show promise of becoming, functional wetlands in watersheds. However, in some cases, required compensation actions were never initiated or, if initiated, were poorly designed or carelessly implemented. In other cases, the compensation site was placed in a landscape that would not provide the hydrology or associated communities, including uplands, necessary to achieve the desired functions. At some sites, compensation was undertaken and the compliance criteria prescribed by the regulator were being met, but the conditions did not allow for the hydrological variability that is a defining feature of a wetland. At other times the compliance criteria called for the presence of certain plant species without ensuring that site conditions would support them. Meanwhile, at most sites, monitoring was not expected to continue over the longer term, and legal and financial assurances for long-term protection of the site were not present.

Results detailed in previous chapters arise from weaknesses in wetland regulatory institutions. Therefore, on the basis of case studies, the materials provided to the committee, and the available literature, the committee recommends the following goal statement for compensatory mitigation institutions: *Institutions (laws, regulations, and guidance) governing*

fill permitting and compensatory mitigation should promote compensatory miti-
gation sites that meet ecological performance criteria and that result in a matrix
of protected, restored, and created wetlands in the watershed that contribute to
the physical, chemical, and biological integrity of the waters of each watershed.

Wetland management programs should seek to achieve three specific outcomes by ensuring that the following conditions are met:

• Individual compensatory mitigation sites should be designed and constructed to maximize the likelihood that they will make an ongoing ecological contribution to the watershed, and this contribution is specified in advance.

• Compensatory mitigation (i.e., wetlands created or restored to compensate for wetland damage) should be in place concurrent with, and preferably before, permitted activity.

• To ensure the replacement of lost wetlands functions, there should be effective legal and financial assurances for long-term site sustainability of all compensatory wetland projects.

Achieving these results will require modifications to the regulatory program and compensatory mitigation mechanisms described in Chapters 4 and 5. In Chapter 5, permittee-responsible mitigation, on-site, off-site, or at a single-user mitigation bank, was distinguished from third-party mitigation, where some party other than the permit recipient assumes responsibility for the mitigation. This chapter describes institutional reforms governing compensatory mitigation that could move the nation toward the outcomes identified above.

The committee proposes these reforms as a suite of integrated recommendations and urges that they be considered in their entirety and not be selectively implemented. However, for clarity of exposition, the committee first offers an overarching recommendation on the need to move wetland mitigation in the CWA Section 404 program toward a watershed focus and suggests alternative means to move in that direction. Second, because permittee-responsible mitigation will likely continue to be the prevalent form of compensatory mitigation, regulations and guidelines governing this approach should be modified to address its weaknesses. Finally, third-party compensation approaches (mitigation banks, in-lieu fee programs) may offer advantages over permittee-responsible mitigation, especially when compensating for smaller fills, but they also have weaknesses. The committee makes specific suggestions on how to build on the strengths of these compensatory mitigation mechanisms.

Finally, in making all of these recommendations, the committee recognizes that wetland permitting is a decentralized process. In this process, regulators in field offices of federal and state agencies are expected to evaluate a permit applicants' proposal and issue or deny a permit in an expeditious manner. Individual agencies' staffs are required to make an enormous range of decisions in the regulatory process. The Section 404(b)(1) guidelines require a complicated assessment of the practicability of alternatives, expecting regulators to make decisions about the viability of different development proposals. The guidelines require a broad analysis of the extent to which a project, individually or in combination with other past and foreseeable activities, affects a broad array of issues, ranging from water-quality functions to shellfish. Helping a permit recipient design a compensatory wetland project and then assuring that the project is undertaken as designed is yet another task. In all of this, the wetland science is not fully developed, and many of the science and economic issues that must be addressed are site-specific.

In the end, wetland permitting, and the compensatory mitigation required as a part of this process, will need to rely on regulators making informed judgments at many stages in the permitting process. Each of the recommendations is made with an appreciation of the need to inform and support those judgments toward achieving the goals that are described in recommendation 1 in this chapter.

A WATERSHED-BASED APPROACH TO COMPENSATORY MITIGATION

Many call for a watershed approach to wetland management, including for the Section 404 program. The Unified Federal Policy for Ensuring a Watershed Approach to Federal Land and Resource Management defines a watershed approach as follows:

> A framework to guide watershed management that: 1) uses watershed assessments to determine existing and reference conditions; 2) incorporates assessment results into resource management planning; and 3) fosters collaboration with all landowners in the watershed. The framework considers both ground and surface water flow within a hydrologically defined geographical area. (http://cleanwater.gov/ufp/glossary.html)

The 2000 in-lieu fee guidance embraces the watershed approach for in-lieu fee mechanisms, stating,

> Local watershed planning efforts, as a general matter, identify wetland and other aquatic resources that have been degraded and usually have established a prioritization list of restoration needs. In-lieu fee mitigation projects should be planned and developed to address the specific resource needs of a particular watershed" (Fed. Regist. 65(Nov. 7): 66914-66917).

The 1995 mitigation banking guidance (Fed. Regist. 60(Nov. 28):58605-58614) encourages a watershed-based approach as the overall goal of a mitigation bank:

> The overall goal of a mitigation bank is to provide economically efficient and flexible mitigation opportunities, while fully compensating for wetland and other aquatic resource losses in a manner that contributes to the long-term ecological functioning of the watershed within which the bank is to be located. The goal will include the need to replace essential aquatic functions that are anticipated to be lost through authorized activities within the bank's service area. In some cases, banks may also be used to address other resource objectives that have been identified in a watershed management plan or other resource assessment.

Implications of the Watershed Approach

The influence of landscape setting on ecological function has been discussed in several chapters of this report. The committee has argued that the ecological functions of a restored and created wetland acreage in a watershed depend on the design (e.g., size and hydroperiod) of the wetland and on its local setting or context. Also, what may need to be addressed at the watershed scale are the desired wetland functions and how the types and locations of the wetland in the landscape can secure them.

One way to set goals for wetland functions is to seek to replace those lost to the Section 404 permit. Such a compensation goal might imply that the watershed was in some desired condition before the permit was issued, and the compensatory wetland will assure a return to that condition. Exact replacement is also warranted if the particular wetland lost to the permitted activity was the type critical to watershed conditions. At least one of these arguments might lie behind the on-site and in-kind compensation preferences in the 1990 Memorandum of Agreement (MOA) between the U.S. Army Corps of Engineers (Corps) and the EPA (as discussed in Chapter 4). That MOA remains in place and is often reinforced as the new program guidance is issued. For example, the new guidance to govern the in-lieu fee form of third-party mitigation includes a number of sections that continue to emphasize the 1990 MOA. Either of these arguments may be valid, but they need to be analytically defended when the compensation for each permit is being considered.

An alternative approach for determining what is desired in a watershed is to begin with a landscape perspective and seek to emphasize the type and location of compensatory wetlands that are revealed by that perspective. If a watershed approach to compensatory mitigation is taken, as has been recommended by recent guidance, a suite of desired wetland functions may not be secured by strict adherence to a policy of in-kind replacement of the wetland types in the same location as the wetland lost. Stated differently, a preference for in-kind and on-site compensatory wetland would follow from an analytically based watershed assessment to assure that in-kind replacement furthers the watershed goals.

As discussed in Chapter 3, there are many watersheds where existing wetland functions have been degraded, and the mix of wetland types in the watershed is a result of historical development patterns. In such cases, it makes little sense to replicate a degraded system. A watershed approach forces a consideration of this possibility. As a first step in a watershed approach, there is a need to assess what functions would be lost from the permitted activity, and the goal would be to restore the localized functions that had been performed by the impact site (e.g., water quality or storm water).

In some cases, it might be desirable to secure out-of-kind wetland types that have been disproportionately lost from the watershed, if those types would improve watershed functioning. As advocated by Bedford (1999), the mitigation program would achieve greater short- and long-term results by looking at each permitting decision over a broader space and longer time period. Bedford describes this as modifying the boundaries of permit decision-making in time. As a specific example, Magee et al. (1999) found that both natural and mitigation wetlands in Portland, Oregon, had been degraded due to hydroperiod alteration and land-use changes in rapidly urbanizing areas. Mitigation planning that considers the location of projects in relation to larger surface-water and groundwater systems and the extent to which this landscape setting has been or likely will be altered by humans will have a great effect on ecological performance (Bedford 1999). The watershed setting chosen for the compensation wetland should consider the time frame, because ongoing alteration of the landscape could greatly affect projects by affecting groundwater and surface-water patterns.

A watershed perspective provides the context for considering wetland enhancement as mitigation. If the proposed mitigation is to enhance an existing degraded jurisdictional wetland, the result could be a net loss of wetland area in exchange for an increase in wetland functioning. Here, changing the condition of the degraded wetland to some other ecosystem state (enhanced to become a better example of the current type or remodeled to become a more functional/valuable wetland type) at the expense of lost area might be judged a desirable exchange. Whatever the form of mitigation—enhancement, restoration, creation, or preservation—it will rarely be acceptable as mitigation for impacts if the mitigation work is "temporary" in nature, such as simply spraying exotics without addressing the site conditions that would continue to foster re-infestation. Specifically, regulatory agencies should consider each permitting decision over broader geographic areas and longer time periods (i.e., by modifying the boundaries of permit decision-making in time and space).

Another advantage of a watershed perspective is that it clarifies the place of wetland preservation and/or incorporation of upland areas as compensation. Preservation might not appear to offset the permitted loss to the wetland acreage base in the short-term. However, when the goal of a wetland program is viewed from a watershed perspective over a long period, the purpose is to secure a desired matrix of wetland types and locations to achieve the goals of the CWA in the watershed. If, in the future, certain wetlands deemed central to that goal might be compromised, purchase and protection of those wetlands as a part of compensation package might be warranted (Gardner 2000).

Similarly, uplands might be accepted as compensation for filling a wetland. A watershed perspective recognizes that terrestrial connections are especially critical between small wetlands in a regional landscape, and that is recognized in the federal guidance cited earlier. Terrestrial connectivity is essential to the persistence of some wetland species (Semlitsch and Bodie 1998). Historically, when mitigation projects have incorporated upland areas, the focus has been on buffers. The new nationwide permits follow that pattern by allowing the use of buffers as mitigation, even without the inclusion of wetland (see Chapter 4). In 1995, the banking guidance began to encourage the inclusion of uplands in mitigation banks, "to the degree that such features increase the overall ecological functioning of the bank" because

> the presence of upland areas may increase the per-unit value of the aquatic habitat in the bank. If a watershed perspective is taken, some limited acceptance of highly functional uplands [meeting the mitigation obligation] may be given to relatively undisturbed upland areas protected in the bank to reflect the functions inherently provided by such areas (e.g., nutrient and sediment filtration of storm water runoff, wildlife habitat diversity) which directly enhance or maintain the integrity of the aquatic ecosystem and that might otherwise be subject to threat of loss or degradation.

The amount of uplands to be included could be determined by an "appropriate functional assessment methodology . . . to determine the manner and extent, to which such features augment the functions of restored, created or enhanced wetlands and/or other aquatic resources" (Fed. Regist. 60(Nov. 28):58605). This guidance recognizes that wetlands excised from the functions of their surrounding uplands will function at a reduced level.

A watershed perspective will discourage the tendencies to favor fewer large, charismatic compensatory mitigation wetlands that may not yield the important water-quality and habitat functions of many smaller and less "attractive" wetlands (e.g., open-water lakes instead of several small bogs). Making mitigation decisions from a watershed perspective would explicitly recognize the need for and the desired locations of wetlands of all sizes and types and then proactively assure that these sites are protected and restored. A watershed perspective could help to focus on how the water-quality functions might be replaced and would direct attention to the base of the food web. Watershed-scale assessment could consider the long-term connectivity of wetland and upland habitats. The individual projects that would implement a watershed approach would occur on parcels of varying size, so that while mitigation would often be located off-site, it would be located where it would be likely to secure defined watershed goals.

Finally, it should be noted that by focusing on the functions of wetlands within a landscape mosaic, there is no single watershed scale that should be prescribed. Instead, depending on the function of concern, different scales of watershed might be considered in defining the location and type of desired compensatory mitigation (Poiani et al. 2000).

The committee endorses the watershed approach and finds the automatic preference for in-kind and on-site compensatory mitigation of the 1990 MOA to be inconsistent with that approach. The committee is aware of the concern that a watershed approach might weaken the commitment during the permitting process to protect individual wetlands and the functions they provide, with existing wetlands being too readily traded for compensatory wetlands that might not be ecologically functional. However, if recommendations made elsewhere in this report on avoidance and improvements to compensatory mitigation institutions are incorporated into guideline and regulatory revisions, that concern will be addressed. Therefore, the committee recommends that the regulatory agencies consider each permitting decision over broader geographic areas and longer time periods (i.e., by modifying the boundaries of permit decision-making in time and space).

Implementing a Watershed Approach

One concern expressed about the watershed approach is the impracticality of implementation. However, implementing a watershed approach does not mean writing a plan that is expected to guide future permitting decisions. To call for a watershed approach only is to recognize that management of wetland types, functions, and locations requires structured consideration of watershed needs and how wetland types and location serve those needs. A watershed approach means that mitigation decisions are made with a regional perspective, involve multiple agencies, citizens, scientists, and nonprofit organizations, and draw upon multiple funding sources (e.g., permittee-responsible, mitigation banks, and in-lieu fees). A watershed approach means that permitting decisions are integrated with other regulatory programs (e.g., storm-water management or habitat conservation) and nonregulatory programs (e.g., conservation easement programs).

The idealized watershed planning process, described in the 1993 Clinton Administration Wetland Plan (August 24, 1993) suggests far more formality than may be possible or required. The interagency plan states,

> Typically, decisions affecting wetland[s] are made on a project-by-project, permit-by-permit basis. This often precludes the effective consideration of the cumulative effects of piecemeal wetland loss and degradation. It also hampers the ability of State, Tribal, regional, and local governments to integrate wetland conservation objectives into the planning, management, and regulatory tools they use to make decisions regarding development and other natural resource issues. This can often result in inconsistent and inefficient efforts among agencies at all levels of government, and frustration and confusion among the public.

> In contrast, advance planning, particularly comprehensive planning conducted on a watershed basis, offers the opportunity to have strong participation by State, Tribal, and local governments and private citizens in designing and implementing specific solutions to the most pressing environmental problems of that watershed. Advance planning generally involves at least the identification, mapping, and preliminary assessment of relative wetland functions within the planning area. More comprehensive advance planning may identify wetlands that merit a high level of protection and others that may be considered for development, and may also incorporate wetland conservation into overall land use planning at the local level. Advance planning can provide greater predictability and certainty to property owners, developers, project planners, and local governments. (http://www.epa.gov/OWOW/wetland/wwater/wtrshd.html)

Various efforts at structured wetland planning for watersheds have been attempted in many places (e.g., Minnesota, Tennessee, North Carolina, and Tampa Bay, Florida). A recent review of watershed planning for wetlands describes three approaches to wetland planning, where each approach relies on formal analytical processes and primary or secondary data for their execution (White and Shabman 1995). These planning approaches are characterized by the purpose to be served by the planning activity. *Management-oriented wetland planning* has the broadest objective. These plans are expected to replace case-by-case permitting by employing a watershed approach for making advance decisions about all matters related to permitting and sequencing, the required compensation for fills in certain areas, and the location and type of compensation that would be required. In making this determination, both regulatory and nonregulatory programs are coordinated. The above quotation from EPA's Interagency Wetland Plan describes management-oriented wetland planning. The high cost and potential for legal and political disagreements suggest that such broad planning may be too ambitious to be implemented in many watersheds, as has proven to be the case in many instances (White and Shabman 1995). While some broad planning efforts have come to fruition (West Eugene, Oregon), others remain controversial years after their initiation (Hackensack Meadowlands, New Jersey).

Protection-oriented wetland planning has the single objective of discouraging wetland-damaging activities (avoidance) by defining and mapping wetlands by their ecological value in advance of any proposed wetland development project. Such planning may be completed under the advanced identification (ADID) process of the CWA or be part of a general land-use planning process (White and Shabman 1995). These plans would then be used to help define areas that should be avoided in the sequencing process.

Compensation wetland planning identifies watershed needs for types, functions, and general locations of wetlands in the landscape in order to establish restoration priorities for both regulatory and nonregulatory programs. However, the written plan will not include specific locations and designs for the restoration and creation sites. This type of planning might link projects undertaken through both regulatory and nonregulatory programs to secure some desired mosaic of wetlands in the landscape. Such a goal is served by the Southern California Wetland Recovery Project (see Box 8-1). North Carolina's developing statewide wetland restoration plan provides advance planning for choosing specific wetland projects (see Box 8-2 and Appendix B). The North Carolina program's formal plans are expected to guide investments in wetland restoration, although the plans do not identify specific sites. Such large, regional programs can combine the efforts of governmental and mitigation funds to achieve the broad goal of no net loss plus a net gain in wetland area and function.

The committee understands that even the more limited form of compensation planning for wetlands can be costly. Both the California and the North Carolina programs have a permanent staff and supporting resources. Such support will not be available in all areas. Therefore, watershed planning for wetlands will need to proceed without a formal written plan. Instead, reliance on the professional judgment of staff from multiple agencies can set watershed priorities and be the form of compensation wetland planning, given current agency time and resource limitations. At

---

BOX 8-1
Southern California Wetland Recovery Project (SCWRP)

The SCWRP is a partnership of public agencies that work to acquire, restore, and enhance coastal wetland and watersheds between Point Conception and the U.S.-Mexico border. Federal partners are the Corps, EPA, the Fish and Wildlife Service, the National Marine Fisheries Service, and the Natural Resources Conservation Service. State partners are California's Resources Agency, Environmental Protection Agency, Coastal Commission, Department of Fish and Game, State Coastal Conservancy, State Lands Commission, State Water Resources Control Board, and four Regional Water Quality Control Boards (San Diego, Santa Ana, Los Angeles, and Central Coast). Each of these agencies sends top officials to the SCWRP Governing Board. Advisers to the board come from the Wetland Managers Group, the Public Advisory Committee, and the Science Advisory Panel. Five counties (San Diego, Orange, Los Angeles, Ventura, and Santa Barbara) help to identify critical wetland resources and promote education about wetlands and funding of projects. Nine projects totaling over $25 million have already been funded. For 2000 to 2001, the SCWRP has identified 31 projects totaling over $30 million (www.coastalconservancy.ca.gov/scwrp).

---

present, in many emerging fee payment programs the selection of projects for funding is made based on a consensus of professional interagency judgment on watershed needs (Scodari and Shabman 2000). Two institutional reforms could be made to increase the technical quality of these regulator judgments.

BOX 8-2
The North Carolina Wetland Restoration Program

The North Carolina Wetland Restoration Program was created to simplify meeting wetland compensation requirements and to achieve a net gain of wetlands in that state's watersheds. The state is responsible for developing watershed plans to identify areas where restoration actions would be of high priority and of greatest ecological value. Wetland restoration plans for 17 watersheds and their subwatersheds are now finished or are nearing completion. Restoration activities are now under way in a number of watersheds.

To initiate the planning and wetland restoration program, the state provided $6 million to the North Carolina Wetland Restoration Fund, with additional funds to be provided in future years. In addition, the North Carolina Department of Transportation pays $2.5 million each year for a period of 7 years for plan development.

With the fund now in operation, a wetland permit recipient can satisfy compensatory mitigation requirements by paying a fee to the fund. The collected fees are used to repay the wetlands restoration fund (WRF) for wetland restorations that were implemented with the initial state allocation (see Appendix B).

First, the level of scientific expertise in wetland management among the principal staff responsible for the wetland programs must be maintained and allowed to grow as the scientific understanding of watersheds and wetlands continues to advance (see additional discussion later in this chapter). Second, broader-based participation in setting priorities for wetland preservation and restoration projects would be a substitute for complicated and often expensive formal planning efforts and a way to expand limited staff capabilities in any single agency. Absent a formal plan, a watershed approach to compensatory site decisions would be a process that engages community and multiple agency input supported by a panel of wetland experts from the scientific community who are familiar with the watersheds in question. This process could be an addition to an ongoing program, might operate at a state or a substate level, and could be led by federal, state, or local regulatory staff. Such details would vary across the nation.

The Corps itself might take the lead in initiating such watershed processes. Part II of the Corps Standard Operating Procedure (SOP) (USACE 1999a) includes guidelines for incorporating a watershed approach into the regulatory program. It indicates that 0 to 20% of the program's time should be devoted to this issue. The program could take this policy much further if there were a directive to integrate regulatory decisions into a watershed framework. The Corps mitigation policy, also outlined in the 1999 SOP, supports taking just such an approach:

> The aforementioned 1990 EPA/DA MOA establishes a preferred sequence of on-site/in-kind mitigation to off-site/out-of-kind approaches; however, districts should not consider this a required hard and fast policy. Corps field experience has shown ecological value in pursuing practicable and successful mitigation within a broader geographical context. This approach, combined with innovations such as mitigation banks and in-lieu-fee programs, provides proportionately higher ecological gains where the aquatic functions are most needed. The Corps depends on regulators reviewing relevant agency and public comments and applying their best professional judgment in requiring appropriate and practicable mitigation for unavoidable, authorized aquatic resource impacts. The bottom line test for mitigation should be what is best for the overall aquatic environment.

It is not clear that the Corps will take the lead; however, it does appear the agency has the authority to participate fully in watershed-oriented approaches to wetland mitigation. Therefore, the committee encourages the states, with the participation of appropriate federal agencies, to prepare technical plans or to initiate community and interagency consensus processes for setting wetland protection, acquisition, restoration, enhancement, and creation project priorities on an ecoregional (watershed) basis. Recommendations to support such efforts, including the use of functional assessment, development of guidelines, research and continuing professional training of regulatory staff necessary to implement this approach, are provided elsewhere in this chapter.

IMPROVEMENTS IN PERMITTEE-RESPONSIBLE MITIGATION

Permittee-responsible mitigation remains the predominant mitigation mechanism. Improvements in permittee-responsible compensatory mitigation are needed to meet the goals listed earlier in this chapter. Changes will need to be made in what is expected of the applicant and the regulatory agency. These two areas for change are discussed separately. While it is recognized that the recommended changes may increase the costs borne by both the applicant and the agencies, the changes are essential for compensatory mitigation projects to achieve their desired results. It is possible that for smaller fills and certain permittees the costs might increase to the point where third-party compensation systems might be preferred to permittee-responsible mitigation. Cost issues and third-party mitigation are discussed later in this chapter.

Expectations for the Permittee

The committee believes that some party must be held responsible for compliance with a specified mitigation requirement (Chapters 4, 5, and 6). These responsible parties do expect some finite time horizon at the end of which they will have to meet their legal obligations. However, the time frame for ecological development of a wetland project (see Chapter 2) is often longer than the time a permittee would be held responsible for the project. In an effort to mesh these two time imperatives, performance standards (e.g., wetness and certain plants) are often written so that the applicant can demonstrate compliance in a relatively short time period. In addition, the short time frames and clearly measurable standards reduce the burden on the regulatory staff to certify compliance. These are powerful incentives to simplistic wetland design. One result that is especially troubling is the frequent use of criteria that do not recognize the need for appropriate hydrological variability in ecologically functional wetlands. The result has often been sites with continuous open water and stable water levels and overly managed wetland vegetation. Another outcome that follows from the effort to shorten the time required to establish compliance is replacement of bottomland hardwoods with emergent wetlands or open water (Brown and Veneman 1998). Incentives for both the agency and responsible party to favor simplistic wetlands performance criteria would be lessened if legal compliance was based on three obligations:

initiate construction no later than concurrent with the permitted activity; construct and monitor to meet performance requirements; and transfer the long-term responsibility for ownership, monitoring, and adaptive management of the site to a certified wetland management entity.

Initiation of Compensatory Mitigation Projects

Permittees' development projects cannot be delayed until compensation sites meet performance standards. Therefore, for permittee-responsible mitigation there needs to be some practical compromise between the goals of securing functional compensatory wetlands and placing realistic requirements on permit recipients. Toward that end, the first obligation of the permittee should be to initiate the required compensatory project no later than concurrently with the permitted activity. If there is a lag, a financial assurance that would be adequate to initiate, complete, and manage a comparable project at another location should be secured. Such an expectation would assure that permittee-responsible mitigation proceeds in the same expeditious time frame that is comparable to third-party mitigation systems.

Permit Conditions

Linking designs to ecological performance can be extremely difficult, because wetland science and restoration and creation efforts are still developing and must be tailored to individual sites. Therefore, while site designs should reflect current mitigation science and emerging scientific understanding, the initial designs may not always result in the exact wetland properties that were the original intent of the design. However, much can be accomplished within the limits of the current science. The committee believes that enough is understood about wetland hydrology, place in the landscape, soils, and other determinants of wetland structure to specify design requirements that will result in a site that will develop into a wetland and provide for a number of wetland functions. However, as was noted in Chapter 2, some wetland types are more difficult to create or restore than others. In short, we can design sites with a high probability of becoming functional wetlands, but whether particular sites will always result in particular functional outcomes is less certain. Based on the difficulty of restoring or creating wetland type, the permit could require more than an acre of replacement for each acre permitted—a mitigation ratio greater than 1:1. This use of mitigation ratios is already common in setting regulatory requirements.

Permit conditions for legal compliance with the mitigation obligation should recognize this reality. First, the permittee and regulators would work together to design the site according to specified criteria included in the permit. The permittee would then construct by that design and coordinate any changes identified in the field as necessary to meet the performance criteria with the regulator. The mitigation plan review would concentrate on those design factors that will ensure the restoration or implementation of ecological and hydrological processes appropriate to the project. The site should be designed in recognition of hydrological variability and other factors described in other parts of this report. Performance goals in the permit would require that the mitigation project reestablish fundamental wetland processes; attainment of hydrology should be paramount. When societal values are considered in the watershed context, the functions emphasized are likely to be ones that will solve water-quality problems, reduce flooding, and stabilize shorelines. Such goals are usually set in regions where there are existing water-quality problems in the watershed and/or the impact wetlands are of relatively low quality. If the functions to be performed by the mitigation site are not easy to describe, the plan could specify reference sites for comparison with the mitigation site. The degree of similarity to the reference should be similar to the degree of variability among natural sites in the reference set (Brinson and Rheinhardt 1996; White and Walker 1997). Once monitoring shows that specified wetland processes are in place, compliance can be certified.

Monitoring for Performance

The 1990 Corps/EPA mitigation MOA states that, "Monitoring should be directed toward determining whether permit conditions are complied with and whether the purpose intended to be served by the conditions are actually achieved . . . for projects to be permitted involving mitigation with higher levels of scientific uncertainty, such as some forms of compensatory mitigation, long-term monitoring, reporting and potential remedial action should be required." The 1995 banking guidance follows the same basic logic as the 1990 MOA and states, "The period for monitoring will typically be five years; however, it may be necessary to extend this period for projects requiring more time to reach a stable condition (e.g., forested wetlands) or where remedial activities were undertaken." The Corps SOP (USACE 1999b) requires monitoring reports for mitigation banks and "other substantial mitigation," such as in-lieu fees. The banking guidance states, "Annual monitoring reports should be submitted to the authorizing agency(ies), who is responsible for distribution to the other members of the MBRT, in accordance with the terms specified in the banking instrument."

The justification for this approach is that the time to meet different criteria will vary for different attributes and for different projects. Hydrological criteria may be met almost immediately for some restoration projects, while they may take years for creation projects. Vegetative cover and biodiversity criteria may be met within a year in a subtropical climate where topsoil from a donor marsh (often the impact site) is used to "plant" a site versus years when tree seedlings are planted. These are important requirements for ensuring the desired ecological results over time.

Transfer of Long-Term Responsibility

The presumption that once mitigation sites meet their permit criteria they will be self-sustaining in the absence of any management or care is flawed. Once performance criteria have been achieved, some other means to secure long-term monitoring for routine management is needed. Long-term responsibility for a site may require that management actions as adaptations to external forces be taken. Scientists recognize the inherent variability of ecosystems (Pickett and Parker 1994). Also, the likelihood of a major disturbance event at a mitigation site is a matter for concern. A major restoration on the San Diego River complied with permit criteria, but the next flood eliminated the river flow to the site, and the costly transplanted trees were mostly lost to drought. Even if a specific type of wetland has been enhanced, restored, created, or preserved, that type may change subtly or dramatically over time due to such disturbances (Duever et al. 1986; White and Walker 1997). Clearly, mitigation sites have ongoing maintenance and management needs (control of exotics, prescribed fire, and sometimes minor hydrological adjustments); however, individual permittees with no expertise and no long-term interest in a wetland site cannot be expected to manage that site over time.

Long-term management of a site demands that long-term real-estate protection be in place. The 1995 banking guidance states that a bank site should be "protected in perpetuity with appropriate real estate arrangements" and the 2000 in-lieu fee guidance recommends legal arrangements to ensure long-term management. Permittee-responsible mitigation outside single-user mitigation banks has rarely included requirements that the property on which the mitigation is located be legally protected and managed in perpetuity.

For all these reasons, an obligation on the permittee is to transfer the long-term site management and maintenance responsibility, along with a cash endowment for these purposes, to a prescribed management authority (the characteristics for such an authority are discussed elsewhere in this chapter). Such transfer of responsibility has precedent in the Section 404 program. As noted in Chapter 5, the transfer of mitigation responsibility is a defining feature of third-party compensation mechanisms. In third-party compensation, there are potentially two transfers of responsibility. First, the responsibility transfers from the permittee to an entity responsible for the design and construction of the site. Then there may be a second transfer of the constructed site to an organization responsible for long-term maintenance and protection of the site, along with a cash endowment for site management. These organizations would own the site and manage it as a wetland and have access to financial resources to secure that end. The 1995 banking guidance states that bank sponsors need to provide some sort of financial assurances in the event of bank default or failure as well as throughout or beyond its operational life. Florida's administrative rules on mitigation banking require posting of financial assurances for both mitigation implementation and perpetual management (Florida Administrative Code Rule 62-342.400 & 700). The 2000 in-lieu fee guidance recommends financial arrangements to ensure long-term management. This second transfer of responsibility, with a cash endowment, does not occur in permittee-responsible mitigation. Responsibility for long-term maintenance of the site remains with the permittee.

Permit Compliance Conditions for Permittee-Responsible Mitigation

It is impractical to expect permittee-responsible compensatory mitigation to be in place before a permitted fill can proceed. The suggestion here is for concurrent site construction and mandatory use of financial assurances to ensure that construction is by design. Compliance with a compensation requirement for the permittee is defined by the construction according to an agreed-to site plan that will result in a functional wetland. The design is intended to achieve that endpoint, but practically speaking, there may need to be minor, moderate, or sometimes, major modifications to the implemented design. In short, the mitigation provider has the responsibility to get the project on its way to becoming a relatively self-sustaining wetland (i.e., physical, chemical, and biological processes are in place and will lead to a jurisdictional wetland). At this point, the regulators and the long-term site manager have reasonable assurance that the site has developed the processes necessary to require only maintenance. Then, long-term protection and management of the site will be necessary. Therefore, when the permittee responsible for the construction is released from its legal obligation, it would either commit to undertaking the long-term management or make a one-time payment to an appropriate stewardship entity.

These conditions should be placed into the Section 404 permit. Mitigation permits would specify the acreage of jurisdictional wetlands required, implementation and design standards, performance criteria, requirements for monitoring and reporting on performance, a long-term management plan, and documentation and quality-assurance require-ments for the financial assurance and property protection mechanisms needed to implement management of the site in perpetuity. In summary, the committee recommends that the Corps and other responsible regula-tory authorities establish and enforce clear compliance requirements for permittee-responsible compensation to assure that (1) projects are initi-ated no later than concurrent with fill activity; (2) projects are imple-mented and constructed according to established design criteria and use an adaptive management approach specified in the permit; (3) the perfor-mance standards are specified in the permit and attained before permit compliance is achieved; and (4) the permittee provides a stewardship organization with an easement on, or title to, the compensatory wetland site and a cash contribution appropriate for the long-term monitoring, management, and maintenance of the site.

EXPECTATIONS FOR THE REGULATORY AGENCY

A permittee's mitigation compliance requirements are established by the permitting agency. The agency also is responsible for assuring the permit conditions are met. This section includes discussion and recom-mendations directed toward the responsibilities of the regulatory agency.

Recognize Watershed Needs

It is often impossible and, with a watershed perspective, may not be necessary for a compensatory wetland to perform the same wetland func-tions as those of the wetlands at the impact site on a permit-by-permit basis. A watershed perspective demands that the functions expected to be lost because of the cumulative effect of permitting decisions be under-stood, to ensure that the resulting ecosystems within the watershed func-tion at the highest level attainable. On a specific permitting decision, the functions that would be lost must be quantified and a determination must be made on whether there needs to be an exact replacement of those functions and which aspect of the watershed plan would yield the great-est offset for those functions. If a wetland permit is of a significant size or type, a quantitative functional assessment tool would be used (Chapter 7). For smaller or degraded sites, a more rapid assessment, which is re-lated to the local or regional functional assessment method, could be completed.

As has been noted, conducting an assessment does not mean that functional equivalency must be secured between the compensatory wet-land and the impact site. Functional trade-offs might be considered in the context of the needs of the watershed. Such trade-offs should be made the responsibility of the multiagency watershed approach described earlier. Additions to the wetland assessment procedures that recognize the relative value of different functions in a watershed have recently been proposed and might be used to inform those who are engaged in implementing the watershed approach (Wainger et al. 2000).

Therefore, the committee recommends that the Corps and other responsible regulatory authorities use a functional assessment protocol that recognizes the watershed perspective, described in Chapters 3 and 7, to establish permittee compensation requirements. This recommendation applies whether the compensation is made by the permittee or by a third-party.

Recognition of Temporal Lag

Unless the replacement wetlands functions are in place before the permitted impacts occur, there will be some temporal loss of wetland function in the watershed until the replacement wetland is functioning at the same level that the impact site had been. The logic for a compensation ratio greater then 1:1 to compensate for the temporal loss of wetland function (lag time) was developed 10 years ago by King et al. (1993). The authors provided a formula-based system which uses a discount rate for setting the ratios. This approach provides a useful conceptual starting point for determining the lag in offsetting permitted losses of wetland function. The use of the temporal lag concept can be used in either a ratio-based or functional assessment approach to determining the appropriate mitigation requirements. Considering lag time for permittee-responsible mitigation, mitigation banks and in-lieu fees will require a somewhat different approach, based on the timing of the impacts, relative to mitigation performance.

Ratios also have been used for other purposes. For example, ratios have been used to reflect the functional values of the impact site, i.e., the ratio would be higher for a pristine wetland than for a severely degraded wetland; require more acres when preservation is required as a permit condition instead of restoration; reflect scientific uncertainties in replicating certain wetland types; and address time lag. Depending on these various factors, the appropriate ratio may either be higher or lower than 1:1. Because these factors address important issues in assuring adequate compensation of permitted impacts, the committee believes that these are all important in setting mitigation requirements. Looking toward the future, development of quantitatively based rapid assessment methods that could replace ratios might help regulators ensure the functions lost due to permitted impacts are adequately offset (Redmond 2000).

Compliance Inspection and Enforcement

The desire to enforce permit conditions is reflected in the Corps SOPs:

Districts will inspect a relatively high percentage of compensatory miti-gation, to ensure compliance with permit conditions. This includes SPs [standard permits] and GPs [general permits]. This is important because many of the Corps permit decisions require (and presume the success of) compensatory mitigation to offset project impacts. To minimize field visits and the associated expenditure of resources, SPs and GPs with compensatory mitigation requirements should require applicants to pro-vide periodic monitoring reports and certify that the mitigation is in accordance with permit conditions. Districts should review all monitor-ing reports.

Adequate staff and budget must be dedicated to inspection and en-forcement of the permitting and compensatory mitigation elements of the program, but resources are limited. The pending changes to the general permit program may spread resources more thinly, with adverse conse-quences for mitigation compliance and enforcement, unless there is as much as a 15% increase in staff budget (Institute for Water Resources 2000). Although recent court rulings (discussed elsewhere in this report) may restrict the regulatory scope of the federal permitting program and reduce federal staff workloads, the regulatory burden then may fall on limited state resources to fill in gaps created by the court rulings. At present, it appears that the Corps's field staff focuses principally on re-viewing permits and dedicates limited time and attention to inspection and enforcement. This is evidenced by the SOPs issued as guidance for regulatory staff (USACE 1999a). Inspection and enforcement of compen-satory mitigation requirements are not given priority. Although it is re-ported that about 20 to 25% of the annual budget is spent on enforcement of all types (presentation given by John Studt, U.S. Army Corps of Engi-neers, at first meeting of the committee), studies have shown that the Corps does not make sufficient visits to mitigation sites to determine if mitigation projects were constructed as proposed or to evaluate compli-ance of mitigation efforts (Sudol 1996).

The budget blueprint from President George W. Bush calls for more funds for the Section 404 program. Dedicating these funds to more staff resources for inspection of mitigation sites would improve compliance (Allen and Feddema 1996). However, even if such resources are not forth-coming, agencies could make better use of available resources. First, staff specialization might be considered. Some districts now have specialized permitting and mitigation staff; such specialization should be considered for all field units to make better use of limited staff. Second, monitoring and enforcement resources will go farther if structural changes are made in the program. At present (for permittee-responsible compensation), there is a limited incentive for permittee to perform the mitigation in a responsible way. The likelihood of being cited for poorly completed mitigation is low. Penalties can be as much as $25,000 per violation for each day of noncompliance with a mitigation condition. However, significant fines are ineffective if detection of violations is not a priority.

Compliance must be given higher priority (Allen and Feddema 1996; FDER 1991a). The responsible party should conduct an approved monitoring program and provide monitoring results to the appropriate agency, certifying how the conditions of the mitigation site continue to meet the goals of the mitigation project. Supplementing the current self-reporting requirement to be sure there is a common format required for each permittee could increase detection probability. A record-keeping system for the reports would need to be developed. Randomized auditing of the reports would then allow for more efficient use of limited compliance and enforcement resources. The committee recommends that the Corps and other responsible regulatory authorities take actions to improve effectiveness of compliance monitoring before and after project construction. This recommendation applies to both permittee-responsible and third-party mitigation.

Long-Term Stewardship and Management

It was suggested earlier that when the design is completed and performance standards have been achieved, long-term adaptive management responsibility would be shifted to an organization. Whether a wetland is created or restored, the type of wetland, the surrounding watershed condition, and uncertainties in the science all mean that different mitigation projects will require different amounts of time to become functional wetlands. Once these wetlands have attained their permit-specified performance criteria, long-term stewardship is critical to achieving the goals of the CWA. "Long-term stewardship" implies a time frame typically accorded to other publicly valued natural assets, like parks. This time frame emphasizes the importance of developing mitigation wetlands that are self-sustaining so that the long-term costs are not unmanageable.

Appropriate stewardship entities can include a public agency, a non-governmental organization, or a private land manager. The entity assumes responsibility for a portfolio of wetland sites (preserved and compensation) in a watershed or some other defined area. As needed, such organizations need to be identified for each watershed. These organizations would be the repositories for the land encumbrance unless it was already held by a conservation entity that would prefer that the stewardship entity undertake the long-term management responsibilities. In addition, there may need to be a cash payment adequate for monitoring the site or for periodic assessment, as appropriate to the site's self-sustainability. The precise process for certifying such an organization would vary by area. In addition, the amount of payment for the endowment should be appropriately linked to a reasonable expectation of future costs. While there would be some administrative cost for simply making inspections of the site, the need for funds for active management and possible repairs after construction will vary by the condition of the site when it is assumed, the location in the watershed, and other factors. If the site has been constructed to meet the design requirements, the required cash payment might be limited to an annual administrative charge and be based on a sliding scale with the payment in inverse proportion to the self-maintenance capability of the site.

Different mechanisms are available to provide legal and real-estate protection. Donation to a land management agency or qualified land trust and placement of conservation easement or deed restrictions are among the more commonly used mechanisms. Generally, deed restrictions are less desirable because a judge may vacate them. Conservation easements are much stronger mechanisms for this purpose, although each state's laws should be reviewed for potential weaknesses. For example, Florida statutes (§704.06) allow a grantee to give an easement back to the grantor at the grantee's discretion. For this reason, Florida's mitigation banking rule requires that conservation easements be granted to two entities. Deeding of the property's fee to an appropriate conservation land manager, whether public or private, is often the most desirable method of legal protection. However, many of these organizations will only accept lands that are financially endowed and that make strategic additions to their portfolio of landholdings. If donation to such entities is contemplated, it would be wise to work with them during the mitigation planning phase to ensure that they will accept property.

The committee recommends that the Corps, in cooperation with states, encourage the establishment of watershed organizations responsible for tracking, monitoring, and managing all preserved and compensatory wetlands in public ownership or under easement. This recommendation applies for both permittee-responsible and third-party mitigation.

Agency Technical Capacity

Corps regulatory staff should receive continuing training on ecological and hydrological principles necessary to analyze a mitigation design so that there is a reasonable expectation that mitigation projects will meet target functions. In addition, Corps staff could become mitigation specialists who would review mitigation designs. Recommendations obtained by the mitigation specialist would be discussed with the applicant and Corps regulatory staff and ultimately be incorporated into the site location and design requirements of a permit. Once a project is in the ground, it may become evident that aspects of the permitted plan need to be modified. An agency mitigation specialist would have the expertise to troubleshoot the implementation and coordinate adjustments to the plan, the primary goal being that the fundamental ecological and hydrological processes are established.

To move in this direction, the Corps regulatory staff and other responsible agencies must be given opportunity to draw on the ecological and hydrological principles necessary for implementing a watershed approach and developing the site-specific compliance criteria to be placed in permits. Corps and other agency staff must have opportunities for continuing education to ensure that they fully understand the application of these principles in the execution of their permitting and planning responsibilities. Agencies should commit to annual time set aside for each staff member for participation in educational programs, over and above normal regulatory training, including attendance at technical conferences. Therefore, the committee recommends that the Corps commit funds to allow staff participation in professional activities and in technical training programs that include the opportunity to share mitigation experiences across districts.

The committee noted instances where compensatory mitigation was having a positive result in watersheds and other cases that have problems. However, there is insufficient feedback to Corps regulatory staff on whether the performance standards developed for a given project produced the expected results. As a result, the same performance standards are used repeatedly with uncertain results (Streever 1999b). Designing restoration sites to help in learning which approaches work and why (cf. Zedler 2001) can greatly accelerate the learning curve. Unfortunately, there is no mechanism in place to build an experimental design or adaptive management process into mitigation projects in order to learn from these real-world tests of mitigation project design. Therefore, the committee recommends that the Corps establish a research program to study mitigation sites to determine what practices achieve long-term performance for creation, enhancement, and restoration of wetlands.

All of the preceding recommendations increase the likelihood that the mitigation plan will be undertaken as designed and that there will be long-term attention to the site. However, while scientists have developed tools for assessing wetland condition, there is no dependable tool for predicting outcomes of restoration or creation efforts. As an example, hydrogeomorphic (HGM) approach can assist in comparing (*a*) the system that is lost with (*b*) a reference site, or (*c*) the system that is provided through wetland restoration and construction, but HGM cannot predict the ability of a selected restoration or creation site to become similar to *a*. Thus, the committee has proposed that mitigation wetlands be designed and built to meet the performance criteria stated in the permit by following an adaptive management process.

The committee has described how all wetlands types are difficult to restore and create, but a subset of wetlands types is harder to replicate (see Chapter 2). It also notes that there are some basic principles that will increase the likelihood of ecological performance (Chapter 7). To help design the wetland and to set the cash payment for the long-term site maintenance in the face of long-term uncertainty regarding performance, those who write and review permits will benefit from the training and professional development. In addition, to assist permit writers and others in making compensatory mitigation decisions, a reference manual should be developed to help design projects that will most likely achieve mitigation requirements. The manual should be organized around the themes developed in Chapter 7. The committee recommends that the Corps develop such a manual for each region, based in part on the careful enumeration of wetland functions in the 404(b)(1) guidelines and in part on local and national expertise on the difficulty of restoring different wetland types, hydrological conditions, and functions in alternative restoration or creation contexts. Third, these manuals should be updated and improved on a regular basis to reflect the emerging science and the Corps research recommended in recommendation 10 in this chapter.

THIRD-PARTY MITIGATION

A taxonomy of the forms of third-party mitigation was suggested in Chapter 5. A third party might be a commercial mitigation bank that is authorized by a mitigation banking review team (MBRT) process (Chapter 5) to offer wetland credits, measured as acres of wetland type or by functional indices. Private firms are typically the providers of MBRT-certified credits to permittees. The Corps also has the authority to approve compensation offered by third-party mitigation sellers who have not had an MBRT review. The North Carolina program described in Box 8-2 is an example of such a program. Also, the Corps might agree that the compensatory mitigation requirement is met if a fee payment is made to a fee administrator who has a Memorandum of Understanding (MOU) with the Corps. This is defined as an in-lieu fee program in Chapter 5 and comports with the general understanding of such systems in the federal guidance. Finally, the Corps may approve a cash payment to a conservation program on a case-by-case basis as compensatory mitigation. This is termed a cash donation program in Chapter 5.

Private firms began producing wetland credits for sale to permit recipients in the early 1990s (Tabatabai and Brumbaugh 1998). For a comprehensive discussion of the economics of private mitigation sales, see Scodari et al. (1995) and Shabman et al. (1994). As the number of private-sector bankers began to expand, the federal agencies responded by issuing the 1995 mitigation banking guidelines under which an interagency MBRT approves wetland credits for sale to permittees. The MBRT establishes performance criteria that must be achieved by the proposed compensatory wetlands and the number of credits that will be produced by the wetlands and that can be used for compensatory mitigation. Monitoring and repair requirements, performance bonds if credits are sold prior to meeting performance criteria, and assurances for long-term site protection also may be required. Commercial mitigation bankers provide the capital to initiate compensation projects "in advance" of permitted activities, and they provide project development and management expertise. They also readily assume legal responsibility for the mitigation project to comply with permit conditions. For these reasons, credits from commercial mitigation bankers are of potentially high ecological quality, and because of the credit release schedules imposed upon mitigation banks, the credits will be used after some degree of ecological performance has been achieved. These are all important attributes if compensatory mitigation is going to meet the goals listed earlier in this chapter.

Today, commercial mitigation banks are an accepted mitigation option. However, for reasons explained in the literature, growth has been concentrated in a few watersheds (Tabatabai and Brumbaugh 1998), and credit prices may be quite high (Shabman et al. 1998; Rolband et al. 2001). In addition, the MBRT process may call for compensatory wetlands that replace lost wetlands in-kind rather than credits to address watershed priorities.

The goals listed earlier in this chapter allow for few exceptions from compensation requirements for fills of any size or activity. In recent years the Corps has been seeking compensation for fills allowed under general permits. However, the poor ecological performance of permittee-responsible compensation, especially for small fills, suggested a need for third-party mitigation. In fact, many of the first commercial mitigation banks were compensation for losses from general permits. Nonetheless, the absence of commercial mitigation bankers of affordable credits in all watersheds motivated the Corps to rely on cash donations, in-lieu fee payments, and other programs to secure compensatory mitigation (Scodari and Shabman 2000).

An increasing number of the Corps districts have or are developing in-lieu fee agreements (Scodari and Shabman 2000). As described in Chapter 5, in-lieu fee programs are established when a nongovernmental organization or a nonfederal agency is certified by an MOU with the Corps to accept payments from Section 404 permit recipients. This certified fee administrator is responsible for the fees until a decision is made to spend the collected receipts on one or more wetland projects. Because the fee programs are administered by nonprofits and agencies with a wetland protection and restoration mission, there is an expectation by the regulatory agencies that compensatory actions will be undertaken, perform well, and receive long-term stewardship. Banks and, at times, larger permittee-responsible compensation may address watershed goals. Nonetheless, often no formal watershed planning process guides in-lieu fee program expenditures. Instead, a consensus of professional judgments governs the expenditure of the collected fees.

In-lieu fee programs have been subject to criticism (EPA 1999; Gardner 2000). In response, guidance governing in-lieu fee programs was issued in October 2000 by the federal agencies (Fed. Regist. 65(Nov. 7):66914-66917). Also, the General Accounting Office (GAO) is studying the Corps's authorities and uses of fee programs (letter to D. M. Walker, GAO, from Committee on Transportation and Infrastructure, 1999). Two criticisms were that in-lieu fee programs were allowing compensation outside impacted watersheds and that funds were being used for activities other than compensatory wetlands, although it is not clear that these practices are widespread (Scodari and Shabman 2000). The 2000 guidance cautions against these kinds of expenditures (Fed. Regist. 65(Nov. 7):66914-66917). Another criticism is that in-lieu fee programs result in out-of-kind compensation and allow preservation as a compensatory action. In the context of a watershed approach, such decisions may be preferable. However, few programs have a formal watershed plan; instead, best professional judgment is used to ascertain whether a particular restoration or preservation expenditure best serves the watershed (Scodari and Shabman 2000). Because many of these relatively new organizations will be managed by public agencies and/or nongovernmental organizations, there will need to be more attention to cost-based fee setting and accountability rules and procedures to ensure site-level mitigation compliance (Scodari and Shabman 2000; Gardner 2000).

Lag times between the permitted impact and compensation have been a concern for a number of years for all forms of mitigation (King et al. 1993). Lag times result in a temporal loss of wetland function, but more important, it increases the uncertainty that the compensation action will prove ecologically viable. In the fourth recommendation of this chapter, the committee has suggested a way to reduce lag times and increase the certainty of site-level viability for permittee-responsible compensation. However, lag time is further reduced when an MBRT-approved (usually private-sector) mitigation banker finances the initial investment in credit development. Private sector involvement is a source of capital for initiating wetland projects that are, to some extent, developed in advance of permitted activities, a desirable attribute of a compensation system, even though fully functional wetlands are not always in place before credits are purchased and the permitted activity proceeds. In-lieu fee programs have been criticized for not initiating compensatory mitigation in a timely manner, although this criticism cannot be uniformly applied to all operating programs (Scodari and Shabman 2000). Therefore, it appears that all forms of third-party mitigation have some lag time. However, in many cases permittees delay or never initiate compensation projects (Chapter 6), and the time delays with third-party systems should be viewed in that context.

For third-party mitigation there is less uncertainty about long-term outcomes than with permittee-responsible compensatory mitigation. MBRT-certified commercial mitigation banks offer project management expertise, assume responsibility for meeting defined performance criteria, and bring an entrepreneurial desire to seek out improved and lower-cost approaches to securing compensatory mitigation. Once all regulator-approved credits have been sold, the wetland site is either managed by the same third party or transferred to a conservation authority, usually an entity quite similar to those that now enter into MOUs with the Corps to accept fee payments or offer credits that have not been MBRT-approved. Therefore, a common feature of all third-party mitigation is that all compensation sites become the responsibility of a conservation entity with a responsibility for, or organizational mission of, wetland and watershed management. This is a desirable stewardship outcome of all third-party compensatory mitigation systems and was a recommendation the committee made (above) for permittee-responsible mitigation.

The committee understands that the best way to have confidence that compensatory mitigation will serve watershed goals is to have mitigation projects initially designed, implemented, and managed by reliable mitigation experts who are held accountable for certain results. These projects would be of varying wetland types, sizes, and locations to secure priority functions identified by the watershed planning process. Once these results are secured, sites would be transferred to a long-term stewardship entity. Preferably, all of this would occur before the wetlands are used for compensatory mitigation. In addition, the supply of available credits must be large enough and the price of credits must be low enough so that all permits issued can have a compensation requirement that will address the cumulative and secondary consequences of permitted activities. An institutional system to secure these goals may be in reach whenever there is a public funding commitment. The outline of such a system can be described.

A state could create a compensatory wetland fund generally modeled after the North Carolina program (see Box 8-2). Initial capital for the fund could be secured from general revenue sources, such as the federally funded State Revolving Loan Fund (EPA 2000). The funds would be repaid at a later time with fees collected from future permit recipients. The fund, once capitalized, would invest in wetland restoration in watersheds. Wetland projects would meet priorities established in formally developed wetland plans or in plans developed by a consensus of agency regulators and wetland scientists.

Although not the practice in North Carolina, one possibility is that the actual restoration and creation work would be contracted out using a competitive bidding process, drawing in the expertise of the private-sector commercial mitigation bankers. To win contracts, private-sector bidders would need to offer sites that conform to design criteria (see earlier discussion). The winner of each contract award would be paid upon completion of the project after meeting design criteria, although as is now the case with the MBRT process, some small amount of credit may be certified and payment received after the project is initiated. Once design or performance criteria are met, the sites would be transferred to a responsible land management entity for long-term stewardship. This parallels the committee's recommendation for permittee-responsible mitigation. However, note that unlike permittee-responsible mitigation, these compensation wetlands can be constructed before permitted activities proceed. The new fund would take responsibility for quantifying (using a functional assessment protocol) and then selling the credits created by the program to the recipients of permits. The payments required of the permit applicants (the "fee") would be tied to the costs of securing the restoration or creation. As fees are collected, the fund is repaid and new projects can be initiated.

In previous sections of this chapter, the committee offered recommendations to improve permittee-responsible mitigation. Even with those improvements, watershed goals may often be best served by placing compensatory wetlands "off-site." Reliance on third parties for off-site compensation will be necessary. A new institutional mechanism for third-party compensation can be created that draws on the best features of the existing mechanisms. Therefore, the committee recommends that institutional systems be modified to provide third-party compensatory mitigation with all of the following attributes: timely and assured compensation for all permitted activities, watershed integration, and assurances of long-term sustainability and stewardship for the compensatory wetlands.

SUPPORT FOR INCREASED STATE RESPONSIBILITIES

The Clean Water Act was not expressly designed to be a wetland protection act. However, as a result of administrative interpretations (regulations and more informal guidance documents), the CWA has evolved into a principal means for the federal government to protect wetlands. Ambiguous statutory support for a comprehensive wetland protection scheme has proven controversial and has led to litigation. Some judicial decisions have broadened the scope of the CWA Section 404 program (*Natural Resources Defense Council* v. *Callaway*, 392 F. Supp. 685 D.D.C. (1975)), and some have affirmed aspects of the Corps's jurisdiction (*United States* v. *Riverside Bayview Homes, Inc.*, 474 U.S. 121 (1985)). Other cases have called the Corps's authority into question, as was the case in Wilson, in which the Fourth Circuit Court limited the Corps's ability to regulate activities affecting certain isolated waters (*United States* v. *Wilson*, 133 F.3d 251 (4th Cir. 1997)). Similarly, the D.C. Circuit's Court's decision in the *National Mining Association* v. *U.S. Army Corps of Engineers* case invalidated the Corps's so-called Tulloch rule, which sought to regulate incidental discharges associated with excavation activities (*National Mining Association* v. *U.S. Army Corps of Engineers*, 145 F.3d 1399 (D.C. Cir. 1998)). Some issues, such as whether the CWA applies to isolated waters unconnected to traditionally navigable waters, have been ruled on only recently by the U.S. Supreme Court. In the Solid Waste Agency of Northern Cook County (SWANCC) case, the CWA was interpreted narrowly, when the Court held that the Corps exceeded congressional intent when it relied on the presence of migratory birds to assert jurisdiction over isolated waters. These judicial decisions and regulatory responses have contributed to shifting standards of jurisdiction in terms of waters and activities subject to regulation under the CWA. In short, the federal Section 404 program is subject to continuing reinterpretations of the jurisdictional scope of the wetland program, reinterpretations of the activities requiring permits, reinterpretations of the requirements for sequencing, and reinterpretations of the need for compensation if permits are issued. In Chapter 1 and elsewhere in this report, it is suggested that the Section 404 program has increased its effectiveness, although significant improvements can be made, for achieving no net loss of wetland acres and function. Recent court rulings raise the question of whether this momentum can be maintained by national reliance on the Section 404 program alone.

In this setting the committee commends the actions of many states that have expanded their roles and responsibilities for wetland management beyond the review role called for in the Section 401 water-quality certification requirement on Section 404 permits and, in some cases, in statements of coastal zone management consistency. States and regional

---

BOX 8-3
Virginia's New Wetland Permitting Program

Many states have created permitting programs with a scope that exceeds the federal Section 404 program. Recent experience in Virginia demonstrates the need for such programs. Nontidal wetlands in Virginia have not been without some protection. Prior to 1992, a Virginia certification was required for a Section 404 permit. However, as court actions limited the scope of the federal program, the state recognized the urgency of taking legislative action. The Virginia General Assembly passed legislation in 2000 to require a state permit for wetland filling and alteration even if a Section 404 permit is not required. The commonwealth's program will apply to activities of less than 0.5 acres, activities of utility and public service companies, linear transportation projects, activities covered by Corps general permits, and mining activities. In addition, compensatory mitigation will be required under the state program.

---

entities can, and have, taken the lead in expanding permitting authorities in the face of limitations in the federal program (see Box 8-3 for a recent example). The committee recognizes the limits of the Section 404 program as a wetland management tool, but expanded state programs could ensure that wetlands within their territories remain protected regardless of where the federal courts decide to demarcate the boundaries of federal jurisdiction. However, the committee also recognizes that the capacities of states may be limited by statute and by staff and budget support. One possible path for federal agencies and for Congress would be to support expanded state capabilities as a response to the contracting of federal authority. The federal agencies could work with Congress to enhance technical assistance to states that wish to expand their permitting authorities, and to increase funding to states that take such actions or that have programs in place. Therefore, the committee encourages the Corps and the EPA to work with states to expand their permitting and watershed planning programs to fill gaps in the wetlands program.

RECOMMENDATIONS

1. The committee recommends the following goal statement for compensatory mitigation institutions:

Institutions (laws, regulations, and guidance) governing wetland permitting and compensatory mitigation should promote compensatory mitigation sites that meet ecological performance criteria and that result in a matrix of protected, restored, and created wetlands in the watershed that contribute to the physical, chemical, and biological integrity of the waters of each watershed.

Wetland management programs should seek to achieve three specific outcomes by ensuring that the following conditions are met:

1. Individual compensatory mitigation sites should be designed and constructed to maximize the likelihood that they will make an ongoing ecological contribution to the watershed, and this contribution is specified in advance.

2. Compensatory mitigation (i.e., wetlands created or restored to compensate for wetland damage) should be in place concurrent with, and preferably before, a permitted activity.

3. To ensure the replacement of lost wetlands functions, there should be effective legal and financial assurances for long-term site sustainability of all compensatory wetland projects.

2. The committee recommends that the regulatory agencies consider each permitting decision over broader geographic and longer time periods (i.e., by modifying the boundaries of permit decision-making in time and space).

3. The committee encourages states, with the participation of appropriate federal agencies, to prepare technical plans or to initiate interagency consensus processes for setting wetland protection, acquisition, restoration, enhancement, and creation project priorities on an ecoregional (watershed) basis.

4. The committee recommends that the Corps and other responsible regulatory authorities establish and enforce clear compliance requirements for permittee-responsible compensation to assure that (1) projects are initiated no later than concurrent with permitted activity, (2) projects are implemented and constructed according to established design criteria and use an adaptive management approach specified in the permit, (3) the performance standards are specified in the permit and attained before permit compliance is achieved, and (4) the permittee provides a stewardship organization with an easement on, or title to, the compensatory wetland site and a cash contribution appropriate for the long-term monitoring, management, and maintenance of the site. The committee's conclusions reached in Chapter 6 are relevant to the implementation of the recommendation.

5. The committee recommends that the Corps and other responsible regulatory authorities use a functional assessment protocol that recognizes the watershed perspective, described in Chapters 3 and 7, to establish permittee compensation requirements.

6. The committee recommends that the Corps and other responsible regulatory authorities take actions to improve the effectiveness of compliance monitoring before and after project construction.

7. "Long-term stewardship" implies a time frame typically accorded to other publicly valued natural assets, such as parks. This time frame emphasizes the importance of developing mitigation wetlands that are self-sustaining, so that the long-term costs are not unmanageable.

8. The committee recommends that the Corps, in cooperation with states, encourage the establishment of watershed organizations responsible for tracking, monitoring, and managing wetlands in public ownership or under easement.

9. The committee recommends that the Corps and other responsible regulatory authorities commit funds to allow staff participation in professional activities and in technical training programs that include the opportunity to share experiences across districts.

10. The committee recommends that the Corps and other responsible regulatory authorities establish a research program to study mitigation sites to determine what practices achieve long-term performance for creation, enhancement, and restoration of wetland.

11. To assist permit writers and others in making compensatory mitigation decisions, a reference manual should be developed to help design projects that will be most likely to achieve permit requirements. The manual should be organized around the themes developed in Chapter 7. The committee recommends that the Corps develop such a manual for each region, based in part on the careful enumeration of wetland functions in the 404(b)(1) guidelines and in part on local and national expertise on the difficulty of restoring different wetland types, hydrological conditions, and functions in alternative restoration or creation contexts.

12. The committee recommends that institutional systems be modified to provide third-party compensatory mitigation with all of the following attributes: timely and assured compensation for all permitted activities, watershed integration, and assurances of long-term sustainability and stewardship for the compensatory wetlands. The committee encourages the Corps and the EPA to work with the states to expand their permitting and watershed planning programs to fill gaps in the federal wetlands program.

References

Ainslie, W.B., R.D. Smith, B.A. Pruitt, T.H. Roberts, E.J. Sparks, L. West, G.L. Godshalk, and M.V. Miller. 1999. A Regional Guidebook for Assessing the Functions of Low Gradient, Riverine Wetlands in Western Kentucky. Tech. Report WRP-DE-17 Operational Draft. Vicksburg, MS: U.S. Army Corps of Engineers Waterways Experiment Station. [Online]. Available: http://www.wes.army.mil/el/wetlands/wlpubs.html [Dec. 14, 2000].

Allen, A.O., and J.J. Feddema. 1996. Wetland loss and substitution by the Section 404 Permit Program in southern California, USA. Environ. Manage. 20(2):263-274.

Andreas, B.K., and R.W. Lichvar. 1995. Floristic Index for Establishing Assessment Standards: A Case Study for Northern Ohio. Wetlands Research Program Tech. Report WRP-DE-8. Vicksburg, MS: U.S. Army Corps of Engineers Waterways Experiment Station.

Ashworth, S.M. 1997. Comparison between restored and reference sedge meadow wetlands in south-central Wisconsin. Wetlands 17(4):518-527.

Bartoldus, C.C. 1999. A Comprehensive Review of Wetland Assessment Procedures: A Guide for Wetland Practitioners. St. Michaels, MD: Environmental Concern Inc.

Bartoldus, C.C., E.W. Garbisch, and M.L. Kraus. 1994. Evaluation for Planned Wetlands (EPW): A Procedure for Assessing Wetland Functions and a Guide to Functional Design. St. Michaels, MD: Environmental Concern Inc.

Bayley, P.B. 1995. Understanding large river-floodplain ecosystems. BioScience 45(3):153-158.

Bedford, B.L. 1996. The need to define hydrologic equivalence at the landscape scale for freshwater wetland mitigation. Ecol. Applic. 6(1):57-68.

Bedford, B.L. 1999. Cumulative effects on wetland landscapes: Links to wetland restoration in the United States and Southern Canada. Wetlands 19(4):775-788.

Bedford, B.L., and E.M. Preston. 1988. Developing the scientific basis for assessing cumulative effects of wetland loss and degradation on landscape functions: Status, perspectives and prospects. Environ. Manage. 12(5):751-771.

Bedford, B.L., M.R. Walbridge, and A. Aldous. 1999. Patterns in nutrient availability and plant diversity of temperature North American wetlands. Ecology 80(7):2151-2169.

Bell, S.S., M.S. Fonseca, and L.B. Motten. 1997. Linking restoration and landscape ecology. Restor. Ecol. 5(4):318-323.

Berven, K.A. 1990. Factors affecting population fluctuations in larval and adult stages of the wood frog (Rana sylvatica). Ecology 71(4):1599-1608.

Berven, K.A., and D.E. Gill. 1983. Interpreting geographic variation in life-history traits. Amer. Zool. 23:85-97.

Berven, K.A., and T.A. Grudzien. 1990. Dispersal in the wood frog (Rana sylvatica): Implications for genetic population structure. Evolution 44(8):2047-2056.

Beven, K. 1982. On subsurface stormflow: Predictions with simple kinematic theory for saturated and unsaturated flows Hillslope hydrology, permeable soils. Water Resour. Res. 18(6):1627-1633.

Bingham, D.R. 1994. Wetlands for stormwater treatment. Pp. 243-262 in Applied Wetlands Science and Technology, D.M. Kent, ed. Boca Raton, FL: Lewis.

Bishel-Machung, L., R.P. Brooks, S.S. Yates, and K.L. Hoover. 1996. Soil properties of reference wetlands and wetland creation projects in Pennsylvania. Wetlands 16(4):532-541.

Bohlke, J.K., and J.M. Denver. 1995. Combined use of groundwater dating, chemical, and isotopic analyses to resolve the history and fate of nitrate contamination in two agricultural watersheds, Atlantic coastal plain, Maryland. Water Resour. Res. 31(9):2319-2339.

Bornette, G., and C. Amoros. 1996. Disturbance regimes and vegetation dynamics: Role of floods in riverine wetlands. J. Veget. Sci. 7(5):615-622.

Breaux, A., and F. Serefiddin. 1999. Validity of performance criteria and a tentative model of regulatory use in compensatory wetland mitigation permitting. Environ. Manage. 24(3):327-336.

Bren, L.J. 1992. Tree invasion of an intermittent wetland in relation to changes in the flooding frequency of the River Murray, Australia. Aust. J. Ecol. 17(4):395-408.

Bridgham, S.D., C.A. Johnston, J. Pastor, and K. Updegraff. 1995. Potential feedbacks of northern wetlands on climate change. BioScience 45(4):262-274.

Brinson, M.M. 1993. A Hydrogeomorphic Classification for Wetlands. Wetlands Research Program Tech. Report WRP-DE-4. Vicksburg, MS: U.S. Army Corps of Engineers, Waterways Experiment Station. [Online]. Available: http://www.wes.army.mil/el/wetlands/wlpubs.html [Dec. 14, 2000].

Brinson, M.M. 1995. The HGM approach. Natl. Wetlands Newslet. 17(6):7-13.

Brinson, M.M. 1996. Assessing wetland functions using HGM. Natl. Wetlands Newslet. 18(1):10-16.

Brinson, M.M., and R. Rheinhardt. 1996. The role of reference wetlands in assessment and mitigation. Ecol. Applic. 6(1):69-76.

Brinson, M.M., R.D. Rheinhardt, F.R. Hauer, L.C. Lee, W.L. Nutter, R.D. Smith, and D. Whigham. 1995. A Guidebook for Application of Hydrogeomorphic Assessments to Riverine Wetlands. Tech. Rep. WRP-DE-11. Vicksburg, MS: U.S. Army Corps of Engineers, Waterways Experiment Station. [Online]. Available: http://www.wes.army.mil/el/wetlands/pdfs/wrpde11.pdf [April 10, 2001].

Brix, H. 1993. Wastewater treatment in constructed wetland: Systems design, removal processes and treatment performance. Pp. 9-22 in Constructed Wetlands for Water Quality Improvement, G.A. Moshiri, ed. Boca Raton, FL: CRC.

Brown, S.C, and B.L. Bedford. 1997. Restoration of wetland vegetation with transplanted wetland soil: An experimental study. Wetlands 17(3):424-437.

Brown, W., and T.R. Schueler. 1997. National Pollutant Removal Performance Database for Stormwater Best Management Practice. Silver Spring, MD: Center for Watershed Protection.

Brown, S., and P. Veneman. 1998. Compensatory Wetland Mitigation in Massachusetts. Research Bulletin 746. Amherst, MA: Massachusetts Agriculture Experiment Station, University of Massachusetts.

Burke, V.J., and J.W. Gibbons. 1995. Terrestrial buffer zones and wetland conservation: A case study of freshwater turtles in a Carolina bay. Conserv. Biol. 9(6):1365-1369.

Burke, V.J., J.L. Greene, and J.W. Gibbons. 1995. The effect of sample size and study duration on metapopulation estimates for slider turtles (Trachemys scripta). Herpetologica 51(4):451-456.

Carter, V. 1986. An overview of the hydrologic concerns related to wetlands in the United States. Can. J. Bot. 64(2):364-374.

Chescheir, G.M., J.W. Gilliam, R.W. Skaggs, and R.G. Broadhead. 1991. Nutrient and sediment removal in forested wetlands receiving pumped agricultural drainage water. Wetlands 11(1):87-103.

Christensen, N.L., R.B. Burchell, A. Liggett, and E.L. Simms. 1981. The structure and development of pocosin vegetation. Pp. 43-61 in Pocosin Wetlands: An Integrated Analysis of Coastal Plain Freshwater Bogs in North Carolina, C.J. Richardson, M.L. Matthews, and S.A. Anderson, eds. Stroudsburg, PA: Hutchinson Ross Pub.

Clark, J.R., and J. Benforado. 1981. Wetlands of Bottomland Hardwood Forests: Proceedings of a Workshop on Bottomland Hardwood Forest Wetlands of the Southeastern United States, held at Lake Lanier, Georgia, June 1-5, 1980. New York: Elsevier.

Cole, C.A., and R.P. Brooks. 2000a. A comparison of the hydrologic characteristics of natural and created mainstem floodplain wetlands in Pennsylvania. Ecol. Eng. 14(3):221-231.

Cole, C.A., and R.P. Brooks. 2000b. Patterns of wetland hydrology in the Ridge and Valley province, Pennsylvania, USA. Wetlands 20(3):438-447.

Collins, B., and G. Wein. 1995. Seed bank and vegetation of a constructed reservoir. Wetlands 15(4):374-385.

Confer, S.R., and W.A. Niering. 1992. Comparison of created and natural freshwater emergent wetlands in Connecticut (USA). Wetlands Ecol. Manage. 2(3):143-156.

Conner, W.H., R.T. Huffman, W. Kitchen, and panel. 1990. Composition and productivity in bottomland hardwood forest ecosystems: The report of the vegetation workgroup. Pp. 455-479 in Ecological Processes and Cumulative Impacts: Illustrated by Bottomland Hardwood Wetland Ecosystems, J.G. Gosselink, L.C. Lee and T.A. Muir, eds.. Chelsea, MI: Lewis.

Cooper, D.J., L.H. MacDonald, S.K. Wenger, and S.W. Woods. 1998. Hydrologic restoration of a fen in Rocky Mountain National Park, Colorado, USA. Wetlands 18(3):335-345.

Cooper, P.C., and B.C. Findlater. 1990. Constructed Wetlands in Water Pollution Control. Oxford, U.K.: Pergamon.

Corbitt, R.A., and P. Bowen. 1994. Constructed wetlands for wastewater treatment. Pp. 221-242 in Applied Wetlands Science and Technology, D.M. Kent, ed. Boca Raton, FL: Lewis.

Correll, D.L., and D.E. Weller. 1989. Factors limiting processes in freshwater wetlands: An agricultural primary stream riparian forest. Pp. 9-23 in Freshwater Wetlands and Wildlife, R.R. Sharitz and J.W. Gibbons, eds. Washington, DC: U.S. Department of Energy.

Costa, J.E. 1975. Effects of agriculture on erosion and sedimentation in the Piedmont Province, Maryland. Geol. Soc. Am. Bull. 86(9):1281-1286.

Cowardin, L.M., V. Carter, F.C. Golet, and E.T. LaRoe. 1979. Classification of Wetland and Deepwater Habitats of the United States. FWS/OBS-79/31. U.S. Fish and Wildlife Service, Washington, DC. [Online]. Available: http://wetlands.fws.gov/Pubs_Reports/Class_Manual/class_titlepg.htm [Dec. 14, 2000].

Craft, C.B., S.W. Broome, and E.D. Seneca. 1988. Nitrogen, phosphorus and organic carbon pools in natural and transplanted marsh soils. Estuaries 11(4):272-289.

Craft, C.B., E.D. Seneca, and S.W. Broome. 1991. Porewater chemistry of natural and created marsh soils. J. Exp. Mar. Biol. Ecol. 152(2):187-200.

Craft, C.J., J. Reader, J.N. Sacco, and S.W. Broome. 1999. Twenty-five years of ecosystem development of constructed Spartina alterniflora (Loisel) marshes. Ecol. Applic. 9(4):1405-1419.

Curtis, J.T. 1959. The Vegetation of Wisconsin: An Ordination of Plant Communities. Madison: University of Wisconsin Press.

Dahl, T.E. 1990. Wetland Losses in the United States 1780's to 1980's. Washington, DC: U.S. Department of the Interior, Fish and Wildlife Service. [Online]. Available: http://www.npwrc.usgs.gov/resource/OTHRDATA/WETLOSS/WETLOSS.HTM [Dec. 14, 2000].

Dahl, T. E. 2000. Status and Trends of Wetlands in the Conterminous United States 1986 to 1997. Washington, DC: U.S. Department of the Interior, Fish and Wildlife and Service.

Dahl, T.E., and C.E. Johnson. 1991. Wetlands, Status and Trends in the Conterminous United States, mid-1970's to mid-1980's. Washington, DC: U.S. Department of the Interior, Fish and Wildlife Service.

Daniel, 3rd, C.C. 1981. Hydrology, geology, and soils of pocosins: A comparison of natural and altered systems. Pp. 69-108 in Pocosin Wetlands: An Integrated Analysis of Coastal Plain Freshwater Bogs in North Carolina, C.J. Richardson, M.L. Matthews, and S.A. Anderson, eds. Stroudsburg, PA: Hutchinson Ross Pub.

Daniels, R.B., and J.W. Gilliam. 1996. Sediment and chemical load reduction by grass and riparian filters. Soil Sci. Soc. Am. J. 60(1):246-251.

Davis, M.M. 1995. Endemic wetlands of the Willamette Valley, Oregon. Pp. 1-8 in Studies of Plant Establishment Limitations in Wetlands of the Willamette Valley, Oregon. WRP-RE-13. Vicksburg, MS.: U.S. Army Corps of Engineers, Waterways Experiment Station.

Davis, M.L. 1997. Statement of M.L. Davis, Deputy Assistant Secretary of the Army (Civil Works) Before the Transportation and Infrastructure Committee, Subcommittee on Water Resources and Environment. United States House of Representatives. April 29, 1997.

Dawe, N.K., G.E. Bradfield, W.S. Boyd, D.E.C. Tretheway, and A.N. Zolbrod. 2000. Marsh creation in a northern Pacific estuary: Is thirteen years of monitoring vegetation dynamics enough? Conserv. Ecol. 4(2):12. [Online]. Available: http://www.consecol.org/vol4/iss2/art12.

Day, F.P. 1982. Litter decomposition rates in the seasonally flooded Great Dismal Swamp. Ecology 63(3):670-678.

Day, F.P. 1983. Effects of flooding on leaf litter decomposition in microcosms. Oecologia (Berlin) 56(2-3):180-184.

Day, Jr., F.P., S.K. West, and E.G. Tupacz. 1988. The influence of ground-water dynamics in a periodically flooded ecosystem, the Great Dismal Swamp. Wetlands 8(1):1-13.

Dennison, M.S., and J.F. Berry. 1993. Wetlands: Guide to Science, Law, and Technology. Park Ridge, NJ: Noyes.

DeWeese, J. 1994. An Evaluation of Selected Wetland Creation Projects Authorized Through the Corps of Engineers Section 404 Program. Sacramento, CA: U.S. Department of the Interior, Fish and Wildlife Service, Ecological Services, Sacramento Field Office.

Dietrich, W.E., C.J. Wilson, and S.L. Reneau. 1986. Hollows, colluvium, and landslides in soil-mantled landscapes. Pp. 361-388 in Hillslope Processes, A.D. Abrahams, ed. Boston, MA: Allen and Unwin.

Dodd, Jr., C.K. 1992. Biological diversity of a temporary pond herpetofauna in north Florida sandhills. Biodivers. Conserv. 1(3):125-142.

Dodd, Jr., C.K. 1993. Cost of living in an unpredictable environment: The ecology of striped newts, Notophthalmus perstriatus, during a prolonged drought. Copeia 1993(3):605-614.

Dodd, Jr., C.K. 1995. The ecology of a sandhills population of the eastern narrow-mouthed toad, Gastrophryne carolinensis, during a drought. Bull. Fla. Museum Nat. History 38(1-9):11-41.

Dole, J.W. 1965. Spatial relations in natural populations of the Leopard frog, Rana pipiens Shreber, in Northern Michigan. Amer. Midl. Nat. 74(2):464-478.

Douglas, B.C. 1995. Global sea level change: Determination and interpretation. Supplement to Reviews of Geophysics, Vol. 33. U.S. National Report to International Union of Geodesy and Geophysics 1991-1994. American Geophysical Union, Washington, DC. [Online]. Available: http://earth.agu.org/revgeophys/dougla01.

Douglas, B.C. 1997. Global sea level rise: A redetermination. Surv. Geophys. 18(2/3):279-292.

DuBowy, P.J., and R.P. Reaves. 1994. Constructed Wetlands for Animal Waste Management. West Lafayette, IN: Department of Forestry and Natural Resources, Purdue University.

Duever, M.J. 1988. Surface hydrology and plant communities of corkscrew swamp. Pp. 97-118 in Interdisciplinary Approaches to Freshwater Wetlands Research, D.A. Wilcox, ed. East Lansing, MI: Michigan State University Press.

Duever, M.J., J.E. Carlson, J.F. Meeder, L.C. Duever, L.H. Gunderson, L.A. Riopelle, T.R. Alexander, R.L. Myers, and D.P. Spangler. 1986. The Big Cypress National Preserve. Research Report No. 8 of the National Audubon Society. New York: National Audubon Society.

Dunne, T., and R.D. Black. 1970. Partial area contributions to storm runoff production in permeable soils. Water Resour. Res. 6:1296-1311.

Dunne, T., and L.B. Leopold. 1978. Water in Environmental Planning. San Francisco: W.H. Freeman.

Eggers, S. 1992. Compensatory Wetland Mitigation: Some Problems and Suggestions for Corrective Measures. U.S. Army Corps of Engineers, St. Paul District Guidelines, Minnesota.

Elling, A.E., and M.D. Knighton. 1984. Sphagnum moss recovery after harvest in a Minnesota bog. J. Soil Water Conserv. 39(3):209-211.

EPA (U.S. Environmental Protection Agency). 1995. Wetlands Silviculture Site Preparation Guidance and Resolution of Silviculture Issue. [Online]. Available: http://www.epa.gov/OWOW/wetlands/silv.html [Revised August 21, 1997].

EPA (U.S. Environmental Protection Agency). 1999. Draft Interagency Guidance on the Use of In-Lieu Fee Arrangements for Compensatory Mitigation Under Federal Regulatory Programs. May 4, 1999.

EPA (U.S. Environmental Protection Agency). 2000. Clean Water State Revolving Fund. Office of Water, U.S. Environmental Protection Agency. [Online]. Available: http://www.epa.gov/nep/fy2000.htm [June 14, 2001].

Erwin, K.L. 1991. An Evaluation of Wetland Mitigation in the South Florida. Water Management District, Vol. 1. Methodology. West Palm Beach, FL: South Florida Water Management District.

Evans, R.O., J.W. Gilliam, and J.P. Lilly. 1993. Wetlands and Water Quality. Water Quality and Waste Management. AG-473-7. WQWM-115. Raleigh, NC: North Carolina Cooperative Extension Service.

FDER (Florida Department of Environmental Regulation). 1991a. Operational and Compliance Audit of Mitigation in the Wetland Resource Regulation Permitting Process. Report no. AR-249. Office of the Inspector General, Department of Environmental Regulation State of Florida. November 1, 1991.

FDER (Florida Department of Environmental Regulation). 1991b. Report of the Effectiveness of Permitted Mitigation. Florida Department of Environmental Regulation Pursuant to Section 403.918(2)(b), Florida Statutes, Department of Environmental Regulation, State of Florida. March 5, 1991.

FDER (Florida Department of Environmental Regulation). 1992. The Biological Success of Mitigation Efforts at Selected Sites in Central Florida. Biology Section, Division of Technical Services, Department of Environmental Regulation, State of Florida. September 1992.

FDEP (Florida Department of Environmental Protection). 1994. The Biological Success of Created Marshes in Central Florida. Biology Section, Division of Technical Services, Florida Department of Environmental Protection. April 1994.

FDEP (Florida Department of Environmental Protection). 1996. A Biological Study of Mitigation Efforts at Selected Sites in North Florida. Biology Section, Division of Administrative and Technical Services, Florida Department of Environmental Protection. May 1996.

Fenner, T.S. 1991. Cumulative Impacts to San Diego County Wetlands Under Federal and State Regulatory Programs 1985-1989. M.A. Thesis. San Diego State University, San Diego.

Fennessy, S., and J. Roehrs. 1997. A Functional Assessment of Mitigation Wetlands in Ohio: Comparisons With Natural Systems. Columbus: Ohio EPA Division of Surface Water.

Fonseca, M.S., W.J. Kenworthy, and G.W. Thayer. 1998. Guidelines for the Conservation and Restoration of Seagrasses in the United States and Adjacent Waters. NOAA Coastal Ocean Program, Decision Analysis Series No. 12. Silver Spring, MD: U.S. Department of Commerce. [Online]. Available: http://shrimp.bea.nmfs.gov/digital.html.

Fugro East, Inc. 1995. A Method for the Assessment of Wetland Function. Prepared for Maryland Department of the Natural Resources, by Fugro East, Inc., Northborough, MA.

Galatowitsch, S.M., and A.G. van der Valk. 1996. Characteristics of recently restored wetlands in the prairie pothole region. Wetlands 16(1):75-83.

Gallihugh, J.L., and J.D. Rogner. 1998. Wetland Mitigation and 404 Permit Compliance Study, Vol. 1. Report and Appendices A, B, D, E. Vol. 2. Appendix C. Barrington: U.S. Fish and Wildlife Service.

GAO (General Accounting Office) 1988. Wetlands: The Corps of Engineer's Administration of the Section 404 Program. Resources, Community, and Economic Development Division. 88-10. U.S. General Accounting Office, Washington, DC. July 1988.

Gardner, R.C. 1991. Public participation and wetlands regulation. UCLA J. Environ. Law Policy 10(1):1-39.

Gardner, R.C. 1998. Casting aside the Tulloch Rule. Natl. Wetlands Newslet. 20(5):5.

Gardner, R.C. 2000. Money for nothing? The rise of wetland fee mitigation. Va. Environ. Law J. 19(1):2-55.

Gibbs, J.P. 1993. Importance of small wetlands for the persistence of local populations of wetland-associated animals. Wetlands 13(1):25-31.

Gilliam, J.W., J.E. Parsons, and R.L. Mikkelsen. 1996. Water quality benefits of riparian wetlands. Pp. 61-65 in Solutions: A Technical Conference on Water Quality Proceedings, NC State University.

Gilliam, J.W., D.L. Osmond, and R.O. Evans. 1997. Selected Agricultural Best Management Practices to Control Nitrogen in the Neuse River Basin. North Carolina Agricultural Research Service Technical Bulletin 311, North Carolina State University, Raleigh, NC.

Gosselink, J.G., and R.E. Turner. 1978. The role of hydrology in freshwater wetland ecosystems. Pp. 63-78 in Freshwater Wetlands: Ecological Processes and Management Potential, R.E. Good, D.F. Whigham, and R. L. Simpson, eds. New York: Academic Press.

Gosselink, J.G., B.A. Touchet, J.V. Beek, D. Hamilton, and panel. 1990. Bottomland hardwood forest ecosystem hydrology and the influence of human activities: The report of the hydrology workgroup. Pp. 347-387 in Ecological Processes and Cumulative Impacts: Illustrated by Bottomland Hardwood Wetland Ecosystems, J.G. Gosselink, L.C. Lee, and T.A. Muir, eds. Chelsea, MI: Lewis.

Groffman, P.M., E.A. Axelrod, J.L. Lemunyon, and W.M. Sullivan. 1991. Denitrification in grass and forest vegetated filter strips. J. Environ. Qual. 20(3):671-674.

Groffman, P.M., G.C. Hanson, E. Kiviat, and G. Stevens. 1996. Variation in microbial biomass and activity in four different wetland types. Soil Sci. Soc. Am. J. 60(2):622-629.

Guard, J.B. 1995. Wetland Plants of Oregon and Washington. Vancouver, BC: Lone Pine.

Gunderson, L.H., J.R. Stenberg, and A.K. Herndon. 1988. Tolerance of five hardwood species to flooding regimes. Pp. 119-146 in Interdisciplinary Approaches to Freshwater Wetlands Research, D.A. Wilcox, ed. East Lansing, MI: Michigan State University Press.

Gwin, S.E., M.E. Kentula, and E.M. Preston. 1990. Evaluating Design and Verifying Compliance of Wetlands Created Under Section 404 of the Clean Water Act in Oregon. EPA/600/3-90/061. Environmental Research Laboratory, Office of Research and Development, U.S. Environmental Protection Agency, Cornvallis, OR.

Gwin, S.E., M.E. Kentula, and P.W. Shaffer 1999. Evaluating the effects of wetland regulation through hydrogeomorphic classification and landscape profiles. Wetlands 19(3):477-489.

Haas, J.C. 1999. Prevalence of Headwater Wetlands and Hydrologic Flow Paths in a Headwater Wetland. M.S. Thesis. University of Maryland, College Park.

Hammer, D.A., ed. 1989. Constructed Wetlands for Wastewater Treatment: Municipal, Industrial, and Agricultural. Chelsea, MI: Lewis.

Hammer, D.A. 1997. Creating Freshwater Wetlands, 2nd Ed. Boca Raton, FL: CRC Press.

Hardin, E.D., and W.A. Wistendahl. 1983. The effects of floodplain trees on herbaceous vegetation patterns, microtopography and litter. Bull. Torrey Bot. Club 110(1):23-30.

Harper, J.L., J.T. Williams, and G.R. Sagar. 1965. The behavior of seeds in soil. I. The heterogeneity of soil surfaces and its role in determining the establishment of plants from seed. J. Ecol. 53(4)273-286.

Harris, R.R. 1988. Associations between stream valley geomorphology and riparian vegetation as a basis for landscape analysis in the Eastern Sierra Nevada, California, USA. Environ. Manage. 12(2):219-228.

Harvey, H.T., and M.N. Josselyn. 1986. Wetlands restoration and mitigation policies: Comment. Environ. Manage. 10(5):567-569.

Hefner, J.M., B.O. Wilen, T.E. Dahl, and W.E. Frayer. 1994. Southeast Wetlands: Status and Trends, mid-1970's to mid-1980's. Atlanta, GA: U.S. Department of the Interior, Fish and Wildlife Service. [Online]. Available: http://www.nwi.fws.gov/sewet/index.html.

Heimlich, R.E. 1999. U.S. Experiences with Incentive Measures to Promote the Conservation of Wetlands. OECD Working Papers 7(31). Paris, France: Organisation for Economic Co-operation and Development.

Herman, K.D., L.A. Masters, M.R. Penskar, A.A. Reznicek, G.S. Wilhelm, and W.W. Brodowicz. 1996. Floristic Quality Assessment with Wetland Categories and Computer Application Programs for the State of Michigan. Lansing, MI: Michigan Dept. of Natural Resources, Wildlife Division, Natural Heritage Program.

Hill, A.R. 1978. Factors affecting the export of nitrate nitrogen from drainage basins in southern Ontario. Water Res. 12(12):1045-1058.

Hill, A.R. 1996. Nitrate removal in stream riparian zones. J. Environ. Qual. 25(4):743-755.

Hill, A.R., and K.J. Devito. 1997. Hydrologic-chemical interactions in headwater forest wetlands. Pp. 213-230 in Northern Forested Wetlands: Ecology and Managment, C.C. Trettin, M.F. Jurgensen, D.F. Grigal, M.R. Gale, and J.K. Jeglum, eds. Boca Raton, FL: CRC.

Hobbs, R.J., and L.F. Huenneke. 1992. Disturbance, diversity and invasion: Implications for conservation. Conserv. Biol. 6(3):324-337.

Holland, C.C., and M.E. Kentula. 1992. Impacts of section 404 permits requiring compensatory mitigation on wetlands in California. Wetlands Ecol. Manage. 2(3):157-169.

Holland, C.C., J. Honea, S.E. Gwin, and M.E. Kentula. 1995. Wetland degradation and loss in the rapidly urbanizing area of Portland, Oregon. Wetlands 15(4):336-345.

Holman, R.E., and W.S. Childres. 1995. Wetland Restoration and Creation Development of a Handbook Covering Six Coastal Wetland Types. Report No. 289. Raleigh, NC: Water Resources Research Institute of the University of North Carolina.

Hruby, T., T. Granger, K. Brunner, S. Cooke, K. Dublanica, R. Gersib, L. Reinelt, K. Richter, D. Sheldon, A. Wald, and F. Weinmann. 1998. Methods for Assessing Wetland Functions. Vol. I: Riverine and Depressional Wetlands in the Lowlands of Western Washington. Washington State Department of Ecology Publ. 98-106. Olympia, WA: Department of Ecology.

Hunt, R.J. 1996. Do Created Wetlands Replace Wetlands That Are Destroyed? Fact Sheet FS-246-96. Reston, VA: U.S. Department of the Interior, U.S. Geological Survey.

Hunt, R.J., J.F. Walker, and D.P. Krabbenhoft. 1999. Characterizing hydrology and the importance of ground-water discharge in natural and constructed wetlands. Wetlands 19(2):458-472.

Institute for Water Resources. 2000. Cost Analysis for the 1999 Proposal to Issue and Modify Nationwide Permits. Report CEWRC-IWR-P. Alexandria, VA: Institute for Water Resources, U.S. Army Corps of Engineers. January 21, 2000.

IPCC (Intergovernmental Panel of Climate Change). 1990. Climate Change: The IPCC Scientific Assessment, J.T. Houghton, G.J. Jenkins, and J.J. Ephraums, eds. Cambridge: Cambridge University Press.

Jacobs, T.J., and J.W. Gilliam. 1985. Riparian losses of nitrate from agricultural drainage waters. J. Environ. Qual. 14(2):472-478.

Johnson, P.A., D.L. Mock, E.J. Teachout, and A. McMillan. 2000. Washington State Wetland Mitigation Evaluation Study. Phase 1. Compliance. Publ. No. 00-06-016. Olympia, WA: Washington State Department of Ecology.

Johnston, C.A. 1988. Productivity of Wet Soils: Biomass of Cultivated and Natural Vegetation. ORNL/Sub/84-18435/1. Oak Ridge, TN: Oak Ridge National Laboratory.

Johnston, C.A. 1991. Sediment and nutrient retention by freshwater wetlands: Effects on surface water quality. Crit. Rev. Environ. Control 21(5/6):491-565.

Johnston, C.A. 1993. Material fluxes across wetland ecotones in northern landscapes. Ecol. Applic. 3(3):424-440.

Johnston, C.A., N.E. Detenbeck, and G.J. Niemi. 1990. The cumulative effect of wetlands on stream water quality and quantity: A landscape approach. Biogeochemistry 10(2):105-141.

Josselyn, M., J.B. Zedler, and T. Griswold. 1990. Wetland mitigation along the Pacific Coast of the United States. Pp. 3-36 in Wetland Creation and Restoration: The Status of the Science, J.A. Kusler and M.E. Kentula, eds. Washington, DC: Island Press.

Kadlec, R.H., and R.L. Knight. 1996. Treatment Wetlands. Boca Raton, FL: Lewis.

Kentula, M. 1986. Wetland creation and rehabilitation in the Pacific Northwest. Pp. 119-150 in Wetlands Functions, Rehabilitation, and Creation in the Pacific Northwest: The State of Our Understanding, R. Strickland, ed. Olympia, WA: Washington State Department of Ecology.

Kentula, M. 1999. Wetland restoration and creation in: Restoration, Creation, and Recovery of Wetlands. WSP2425. National Water Summary on Wetland Resources. U.S. Geological Survey. [Online]. Available: http://water.usgs.gov/nwsum/WSP2425/restoration.html [Last modified May 3, 1999].

Kentula, M.E., J.C. Sifneos, J.W. Good, M. Rylko, and K. Kunz. 1992a. Trends and patterns in Section 404 permitting requiring compensatory mitigation in Oregon and Washington, USA. Environ. Manage. 16(1):109-119.

Kentula, M.E., R.P. Brooks, S.E. Gwin, C.C. Holland, A.D. Sherman, and J.C. Sifneos. 1992b. An Approach to Improving Decision Making in Wetland Restoration and Creation, A.J. Hairston, ed. EPA/600/R-92/150. Corvallis, OR: U.S. Environmental Protection Agency, Environmental Research Laboratory. August.

King, S.L. 2000. Restoring bottomland hardwood forests within a complex system. Natl. Wetlands Newslet. 22(1):7-8, 11-12.

King, D.M., C.C. Bohlen, and K.J. Adler. 1993. Watershed Management and Wetland Mitigation: A Framework for Determining Compensation Ratios. Solomons, MD: Chesapeake Biological Laboratory.

Kramer, R.A., and L. Shabman. 1993. Wetland drainage in the Mississippi Delta. Land Economics 69(3):249-262.

Kusler, J., and H. Groman. 1986. Mitigation: An introduction. Natl. Wetlands Newslet. 8:2-3.

Kusler, J.A., and M.E. Kentula. 1990. Wetland Creation and Restoration. Washington, DC: Island Press.

LaBaugh, J.W. 1986. Wetland ecosystem studies from a hydrologic perspective. Water Resour. Bull. 22(1):1-10.

Lamoureux, V.S., and D.M. Madison. 1999. Overwintering habitats of radio-implanted green frogs, Rana clamitans. J. Herpetol. 33(3):430-435.

Langis, R., M. Zalejko, and J.B. Zedler. 1991. Nitrogen assessments in a constructed and a natural salt marsh of San Diego Bay. Ecol. Applic. 1(1):40-51.

Lindau, C.W., and L.R. Hossner. 1981. Substrate characterization of an experimental marsh and three natural marshes. Soil Sci. Soc. Am. J. 45(6):1171-1176.

Lindig-Cisneros, R.A., and J.B. Zedler. 2001. Effect of light on Phalaris Arundinaces L. germination. Plant Ecology, in press.

Long, M.M., M. Friley, D. Densmore, and J. DeWeese. 1992. Wetland Losses within Northern California From Projects Authorized Under Nationwide Permit No. 26. U.S. Fish and Wildlife Service, Fish and Wildlife Enhancement, Sacramento Field Office, California. October.

Lowe, G., D. Walker, and B. Hatchitt. 1989. Evaluating manmade wetlands as compensation for the loss of existing wetlands in the St. Johns River Water Management District. Pp. 109-118 in Proceedings of 16th Annual Conference on Wetlands Restoration and Creation, F.J. Webb Jr., ed. Plant City, FL: Hillsborough Community College.

Lowrance, R.R., R.L. Todd, and L.E. Asmussen. 1983. Waterborne nutrient budgets for the riparian zone of an agricultural watershed. Agric. Ecosys. Environ. 10(4):371-384.

Lowrance, R.R., R.L. Todd, and L.E. Asmussen. 1984. Nutrient cycling in an agricultural watershed. 1. Phreatic movement. J. Environ. Qual. 13:22-27.

Mack, J.J., M. Micacchion, L.D. Augusta, and G.R. Sablak. 2000. Vegetation Indices of Biotic Integrity (VIBI) for Wetlands and Calibration of the Ohio Rapid Assessment Method for Wetlands v. 5.0. Final Report, EPA Grant No. CD985276, and Interim Report, EPA Grant No. CD985875, Vol. 1. Wetland Ecology Unit, Division of Surface Water, State of Ohio Environmental Protection Agency, Columbus, OH.

Madison, D.M. 1997. The emigration of radio-implanted spotted salamanders, Ambystoma maculatum. J. Herpetol. 31(4):542-552.

Magee, D.W, and G.G. Hollands. 1998. A Rapid Procedure for Assessing Wetland Functional Capacity Based on Hydrogeomorphic (HGM) Classification. Berne, NY: Association of State Wetland Managers.

Magee, T.K., T.L. Ernst, M.E. Kentula, and K.A. Dwire. 1999. Floristic comparison of freshwater wetlands in an urbanizing environment. Wetlands 19(3):517-534.

Malcom, H.R. 1989. Elements of Stormwater Design. Raleigh, NC: North Carolina State University.

MARA Team (Mid-Atlantic Regional Assessment Team). 2000. Preparing for a Changing Climate: The Potential Consequences of Climate Variability and Change. Mid-Atlantic Overview. Mid-Atlantic Regional Assessment, Pennsylvania. State University. March. [Online]. Available: http://www.essc.psu.edu/mara/results/overview_report/.

Mason, C.O., and D.A. Slocum. 1987. Wetland replication—Does it work? Pp.1183-1197 in Proceedings of the 5th Symposium on Coastal and Ocean Management, May 1987, Vol.1. American Society of Civil Engineers, New York, NY.

Matthews, G.A., and T.J. Minello. 1994. Technology and Success in Restoration, Creation, and Enhancement of Spartina alterniflora Marshes in the United States. Vol. 2. Inventory and Human Resources Directory. NOAA Coastal Ocean Program Decision Analysis Series No. 2. Silver Spring, MD: U.S. Department Commerce, National Oceanic and Atmospheric Administration.

McDonald, C.B., A.N. Ash, and E.S. Kane. 1983. Pocosins: A Changing Wetland Resource. FWS/OBS-83/32. Washington, DC: U.S. Fish and Wildlife Service.

McHugh, J.M. 1989. Hydrologic and Erosional Processes in a Gullied Headwater Basin, Southwestern Wisconsin, U.S.A. M.S. Thesis. University of Illinois at Chicago.

Miao, S.L., and F.H. Sklar. 1998. Biomass and nutrient allocation of sawgrass and cattail along a nutrient gradient in the Florida Everglades. Wetlands Ecol. Manage. 5(4):245-263.

Middleton, B. 1999. Wetland Restoration, Flood Pulsing and Disturbance Dynamics. New York: Wiley.

Minnesota Board of Water and Soil Resources. 1998. Minnesota Routine Assessment Method for Evaluating Wetland Functions (MinRAM). Draft Version 2.0. Minnesota Board of Water and Soil Resources, St. Paul, MN.

Miracle, D.L., C.G. Varnes, and M.G. Cullum. 1998. Assessment of Wetland Vegetation Areas in Northeast Florida. Proceedings of the 1998 ASCE Wetland Engineering River Restoration Conference. ASME, Fairfield, NJ.

Mitsch, W.J., and J.G. Gosselink. 2000. Wetlands, 3rd Ed. New York: Wiley.

Mitsch, W.J., and R.F. Wilson. 1996. Improving the success of wetland creation and restoration with know-how, time, and self-design. Ecol. Applic. 6(1):77-83.

Mitsch, W.J., X. Wu, R.W. Nairn, P.E. Weihe, N. Wang, R. Deal, and C.E. Boucher. 1998. Creating and restoring wetlands. BioScience 48(12):1019-1030.

Mitsch, W.J., J.W. Day Jr., J.W.Gilliam, P.M. Groffman, D.L. Ney, G.W. Randall and N. Wang. 1999. Reducing Nutrient Loads, Especially Nitrate-Nitrogen, to Surface Water, Ground Water, and the Gulf of Mexico: Topic 5, Gulf of Mexico Hypoxia Assessment. NOAA Coastal Ocean Program Decision Analysis Series 19. Silver Spring, MD: NOAA Coastal Ocean Office.

Mockler, A., L. Casey, M. Bowles, N. Gillen, and J. Hansen. 1998. Results of monitoring King County wetland and stream mitigations. [Online]. Available http://www.metrokc.gov.

Montgomery, R.D., and W.E. Dietrich. 1988. Where do channels begin? Nature 336:232-234.

Moore, P.D., and D.J. Bellamy. 1974. Peatlands. New York: Springer-Verlag.

Morgan, K.L., and T.H. Roberts. 1999. An Assessment of Wetland Mitigation in Tennessee. Tennessee Department of Environment and Conservation, Nashville, TN.

Moshiri, G.A. 1993. Constructed Wetlands for Water Quality Improvement. Boca Raton, FL: Lewis.

Moy, L.D., and L.A. Levin. 1991. Are Spartina marshes a replaceable resources? A functional approach to evaluation of marsh creation efforts. Estuaries 14(1):1-16.

Myrick, R.M., and L.B. Leopold. 1963. Hydraulic Geometry of a Small Tidal Estuary. Geological Survey Professional Paper 422-B. Washington, DC: U.S. Government Printing Office.

Neumann, J.E., G. Yohe, R. Nicholls, and M. Manion. 2000. Sea-Level Rise and Global Climate Change: A Review of Impacts to U.S. Coasts. Arlington, VA: PEW Center on Global Climate Change.

Niswander, S.F., and W.J. Mitsch. 1995. Functional analysis of a two-year-old created instream wetland: Hydrology, phosphorus retention, and vegetation survival and growth. Wetlands 15(3):212-225.

Novitzki, R.P. 1985. The effects of lakes and wetlands on floodflows and base flows in selected northern and eastern states. Pp. 143-154 in Proceedings of a Conference—Wetlands of the Chesapeake, H.A. Groman, T.R.Handerson, E.J. Meyers, D.M. Burke, and J.A. Kusler, eds. Washington, DC: Environmental Law Institute.

Novitzki, R.P. 1989. Wetland hydrology. Pp. 47-64 in Wetlands Ecology and Conservation: Emphasis in Pennsylvania, S.K. Majumdar, R.P. Brooks, F.J. Brenner, and R.W. Tiner Jr, eds. Easton: Pennsylvania Academy of Science.

NRC (National Research Council). 1992. Restoration of Aquatic Ecosystems: Science, Technology, and Public Policy. Washington, DC: National Academy Press.

NRC (National Research Council). 1995. Wetlands: Characteristics and Boundaries. Washington, DC: National Academy Press.

O'Connell, M.E. 1999. Processes Affecting the Discharge of Nitrate from a Small Agricultural Watershed Over a Range of Flow Conditions, Maryland Coastal Plain. Ph.D. Thesis. University of Maryland, College Park.

OPPAGA (Office of Program Policy Analysis and Government Accountability). 2000. Policy Review: Wetland Mitigation. Report No. 99-40. Tallahassee, FL: Office of Policy Analysis and Government Accountability.

Owen, C.R., and H.M. Jacobs. 1992. Wetland protection as land-use planning: The impact of Section 404 in Wisconsin, USA. Environ. Manage. 16:345-353.

Panzer, R., D. Stillwaugh, R. Gnaedinger, and G. Derkovitz. 1995. Prevalence of remnant dependence among the prairie- and savanna-inhabiting insects of the Chicago Region. Nat. Areas J. 15(2):101-116.

Pechmann, H.K., D.E. Scott, R.D. Semlitsch, J.P. Caldwell, L.J. Vitt, and J.W. Gibbons. 1991. Declinining amphibian populations: The problem of separating human impact from natural fluctuations. Science 253(5022):892-895.

Peterjohn, W.T., and D.L. Correll. 1984. Nutrient dynamics in an agricultural watershed: Observations on the role of a riparian forest. Ecology 65(5):1466-1475.

Pfeifer, C.E., and E.J. Kaiser. 1995. An Evaluation of Wetlands Permitting and Mitigation Practices in North Carolina. Raleigh, NC: Water Resources Research Institute of the University of North Carolina.

Pickett, S.T.A., and V.T. Parker. 1994. Avoiding the old pitfalls: Opportunities in a new discipline. Restor. Ecol. 2(2):75-79.

Poiani, K.A., B.D. Richter, M.G. Anderson, and H.E. Richter. 2000. Biodiversity conservation at multiple scales: Functional sites, landscapes, and networks. Bioscience 50(2):133-146.

Prestegaard, K.L. 1986. Effects of agricultural sediment on stream channels. Pp.54-62 in The Off-Site Costs of Soil Erosion: Proceedings of a symposium held in May 1985, T.W. Waddell, ed. Washington, DC: Conservation Foundation.

Prestegaard, K.L. 2000. Controls of stream morphology and stratigraphy on hydrological and biogeochemical processes in riparian zones. EOS 81(Suppl. May 9):S255.

Prestegaard, K.L., and A.M. Matherne. 1992. Scale variations in the hydrology of riparian zones, southwestern Wisconsin. EOS 73(Suppl. April 7.):131.

Prestegaard, K. L., A.M. Matherne, and N. Katyl. 1994. Spatial variations in the magnitude of the 1993 floods, Raccoon River basin, Iowa. Geomorphology 10(1/4):169-182.

Prince, H.C. 1997. Wetlands of the American Midwest: A Historical Geography of Changing Attitudes. Chicago: The University of Chicago Press.

Race, M.S. 1985. Critique of present wetlands mitigation policies in the United States based on an analysis of past restoration projects in San Francisco Bay. Environ. Manage. 9(1):71-81.

Race, M.S. 1986. Wetlands restoration and mitigation policies: Reply. Environ. Manage. 10(5):571-572.

Race, M.S., and M.S. Fonseca. 1996. Fixing compensatory mitigation: What will it take? Ecol. Applic. 6(1):94-101.

Reaves, R.P., and M.R. Croteau-Hartman. 1994. Biological aspects of restored and created wetlands. Proceedings of the Indiana Academy of Science 103(3-4):179-194.

Reddy, K.R., and W.H. Smith, eds. 1987. Aquatic Plants for Water Treatment and Resource Recovery. Orlando, FL: Magnolia Pub.

Redmond, A.M. 2000. Dredge and fill regulatory constraints in meeting the ecological goals of restoration projects. Ecol. Eng. 15(3-4):181-189.

Reinartz, J.A., and E.L. Warne. 1993. Development of vegetation in small created wetlands in southeast Wisconsin. Wetlands 13(3):153-164.

Reppert, R. T. 1979. Wetlands Values: Concepts and Methods for Wetlands Evaluation. IWR Research Report 79-R1. Fort Belvoir, VA: U.S. Institute for Water Resources.

Richardson, C.J., D.L. Tilton, J.A. Kadlec, J.P.M. Chamie, and W.A. Wentz. 1978. Nutrient dynamics of northern wetland ecosystems. Pp. 217-241 in Freshwater Wetlands: Ecological Processes and Management Potential, R.E. Good, D.F. Whigham, and R.L. Simpson, eds. New York: Academic Press.

Richardson, C.J., R. Evans, and D. Carr. 1981. Pocosins: An ecosystem in transition. Pp. 3-20 in Pocosin Wetlands, C.J. Richardson, ed. Stroudsburg, PA: Hutchinson Ross.

Riekerk, J., S.A. Jones, L.A. Morris, and D.A. Pratt. 1979. Hydrology and water quality of three small lower coastal plain forested watersheds. Proc. Soil Crop Sci. Soc. Fla. 38:105-111.

Rinaldo, A., S. Fagerazzi, and W.E. Dietrich. 1999. Tidal networks, 2. Watershed delineation and comparative network morphology. Water Resour. Res. 35(12):3905-3918.

Robb, J.T. 2000. Indiana Wetland Compensatory Mitigation Inventory: Final Report. EPA Grant # CD985482-010-0. Indiana Department of Environmental Management, Indianapolis, IN. Prepared for USEPA, Region 5 Acquisition & Assistance Branch, Chicago, IL. May.

Rolband, M.S., A. Redmond, and T. Kelsch. 2001. Wetland Mitigation Banking. Pp. 181-213 in Applied Wetlands Science and Technology, 2nd Ed., D.M. Kent, ed. Boca Raton, FL: Lewis.

Rulifson, R.A. 1991. Finfish utilization of man-initiated and adjacent natural creeks of South Creek Estuary, North Carolina, using multiple gear types. Estuaries 14(4):447-464.

Scherrer, E., T.H. Shear, J.M. Hughes-Oliver, R.O. Evans and T.R. Wentworth. 2001. Analysis of created microtopography in restored forested wetlands in eastern North Carolina. Wetlands. In press.

Schueler, T.R. 1992. Design of Stormwater Wetland Systems: Guidelines for Creating Diverse and Effective Stormwater Wetlands in the Mid-Atlantic Region. Washington, DC: Metropolitan Washington Council of Governments.

Scodari, P., and L. Shabman. 2000. Review and Analysis of In Lieu Fee Mitigation in the Clean Water Act Section 404 Permit Program. Institute for Water Resources, U.S. Army Corps of Engineers. November 2000. [Online]. Available: http://www.iwr.usace.army.mil/iwr/pdf/IWRReport_ILF_Nov00.PDF.

Scodari, P.F., L.A. Shabman, and D. White. 1995. National Wetland Mitigation Banking Study Commercial Wetland Mitigation Credit Markets: Theory and Practice. IWR Report 95-WMB-7. November. Alexandria, VA: Institute for Water Resources. [Online]. Available: http://www.iwr.usace.army.mil/iwr/pdf/95wmb7.pdf.

Seigel, R.A., J.W. Gibbons, and T.K. Lynch. 1995. Temporal changes in reptile populations: Effects of a severe drought on aquatic snakes. Herpetologica 51(4):424-434.

Seliskar, D. 1995. Exploiting plant genotypic diversity for coastal salt marsh creation and restoration. Pp. 407-416 in Biology of Salt-Tolerant Plants, M.A. Khan, and I.A. Ungar, eds. Pakistan: Department of Botany, University of Karachi.

Semlitsch, R.D. 1986. Life history of the northern mole cricket, Neocurtilla hexadactyla Orthoptera Gryllotalpidae, utilizing Carolina-Bay habitats. Ann. Entomol. Soc. Am. 79(1):256-261.

Semlitsch, R.D. 1998. Biological delineation of terrestrial buffer zones for pond-breeding salamanders. Conserv. Biol. 12(5):1113-1119.

Semlitsch, R. 2000. Size does matter: The value of small isolated wetlands. Natl. Wetlands Newslet. 22(1):5-6,13-14.

Semlitsch, R.D., and J.R. Bodie. 1998. Are small isolated wetlands expendable? Conserv. Biol. 12(5):1129-1133.

Semlitsch, R.D., and M.A. McMillan. 1980. Breeding migrations, population size structure, and reproduction of the dwarf salamander, Eurycea quadridigitata, in South Carolina. Brimleyana (3):97-106.

Semlitsch, R.D., D.E. Scott, J.H.K. Pechmann, and J.W. Gibbons. 1996. Structure and dynamics of an amphibian community: Evidence from a 16-year study of a natural pond. Pp. 217-248 in Long-term Studies of Vertebrate Communities, M.L. Cody, and J.A. Smallwood, eds. San Diego: Academic Press.

Shabman, L., P. Scodari, and D. King. 1994. National Wetland Mitigation Banking Study. Expanding Opportunities for Successful Mitigation: The Private Credit Market Alternative. IWR Report 94-WMB-3. January. Alexandria, VA: U.S. Army Corps of Engineers, Water Resources Support Center, Institute for Water Resources.

Shabman, L., K. Stephenson, and P. Scodari. 1998. Wetland credit sales as a strategy for achieving no-net loss: The limitations of regulatory conditions. Wetlands 18(3):471-481.

Shaffer, P.W., and T.L. Ernst. 1999. Distribution of soil organic matter in freshwater emergent/open water wetlands in the Portland, Oregon metropolitan area. Wetlands 19(3):505-516.

Shaffer, P.W., M.E. Kentula, and S.E. Gwin. 1999. Characterization of wetland hydrology using hydrogeomorphic classification. Wetlands 19(3):490-504.

Sharitz, R.R., and J.W. Gibbons. 1982. The Ecology of Southeastern Shrub Bogs (Pocosins) and Carolina Bays: A Community Profile. FWS/OBS-82/04. Washington, DC: U.S. Fish and Wildlife Service.

Sharitz, R.R., R.L. Schneider, and L.C. Lee. 1990. Composition and regeneration of a disturbed river floodplain forest in South Carolina. Pp. 195-218 in Ecological Processes and Cumulative Impacts: Illustrated by Bottomland Hardwood Wetland Ecosystems, J.G. Gosselink, L.C. Lee, and T.A. Muir. Chelsea, MI: Lewis.

Sibbing, J.M. 1997. Mitigation's role in wetland loss. Natl. Wetlands Newslet. 19(1):16-21.

Sifneos, J.C., E.W. Cake, and M.E. Kentula. 1992a. Effects of section 404 permitting on freshwater wetlands in Louisiana, Alabama, and Mississippi. Wetlands 12(1):28-36.

Sifneos, J.C., M.E. Kantula, and P. Price. 1992b. Impacts of section 404 permits requiring compensatory mitigation of freshwater wetlands in Texas and Arkansas. Tex. J. Sci. 44(4):475-485.

Simenstad, C.A., and R.M. Thom. 1992. Restoring wetland habitats in urbanized Pacific Northwest estuaries. Pp. 423-472 in Restoring the Nation's Marine Environment, G.W. Thayer, ed. College Park, MD: University of Maryland.

Simenstad, C.A., and R.M. Thom. 1996. Functional equivalency trajectories of the restored Gog-Le-Hi-Te estuarine wetland. Ecol. Applic. 6(1):38-56.

Skaggs, R.W. 1978. A Water Management Model for Shallow Water Table Soils. Tech. Report No. 134. Raleigh: Water Resources Research Institute of the University of North Carolina.

Smith, R.D. 1993. A Conceptual Framework for Assessing the Functions of Wetlands. Tech. Report WRP-DE-3. Vicksburg, MS: U.S. Army Corps Engineers, Waterways Experiment Station.

Smith, R.D., A. Ammann, C. Bartoldus, and M. Brinson. 1995. An Approach for Assessing Wetland Functions Using Hydrogeomorphic Classification, Reference Wetlands, and Functional Indices. Wetland Research Program Tech. Report WRP-DE-9. Vicksburg, MS: U.S. Army Corps Engineers, Waterways Experiment Station. [Online]. Available: http://www.wes.army.mil/el/wetlands/wlpubs.html.

Snodgrass, J.W., J.W. Ackerman, A.L. Bryan Jr., and J. Burger. 1999. Influence of hydroperiod, isolation, and heterospecifics on the distribution of aquatic salamanders (Siren and Amphiuma) among depression wetlands. Copeia 1:107-113.

Snodgrass, J.W., M.J. Komoroski, A.L. Bryan, and J. Burger. 2000. Relationship among isolated wetland size, hydroperiod, and amphibian species richness: Implications for wetland regulations. Conserv. Biol. 14(2):414-419.

Sousa, P.J. 1985. Habitat Suitability Index Models: Red-Spotted Newt. Biological Report 82(10.111). Washington, DC: U.S. Fish and Wildlife Service, U.S. Department of the Interior.

Stein, E.D., F. Tabatabai, and R.F. Ambrose. 2000. Wetland mitigation banking: A framework for crediting and debiting. Environ. Manage. 26:233-250.

Storm, L., and J. Stellini. 1994. Interagency Follow-Through Investigation of Compensatory Wetland Mitigation Sites: Joint Agency Staff Report. EPA 910-R-94-006. Seattle, WA: U.S. Environmental Protection Agency.

Strand, M.N. 1997. Wetlands Deskbook, 2nd Ed. Washington, DC: Environmental Law Institute.

Streever, W.J. 1999a. Performance standards for wetland creation and restoration under Section 404. Natl. Wetlands Newslet. 21(3):10-12, 13.

Streever, W.J. 1999b. Examples of Performance Standards for Wetland Creation and Restoration in Section 404 Permits and an Approach to Developing Performance Standards. Tech. Notes WRP WG-RS-3.3. U.S. Army Engineer Research and Development Center, Vicksburg, MS. January.

Stromberg, J.C., R. Tiller, and B. Richter. 1996. Effects of groundwater decline on riparian vegetation of semiarid regions: The San Pedro, Arizona. Ecol. Applic. 6(1):113-131.

Sudol, M.F. 1996. Success of Riparian Mitigation as Compensation for Impacts Due to Permits Issued Through Section 404 of the Clean Water Act in Orange County, California. Ph.D. Dissertation. University of California, Los Angeles.

Suhayda, J.N. 1997. Modeling impacts of Louisiana barrier islands on wetland hydrology. J. Coastal Res. 13(2):686-693.

Swink, F., and G. Wilhelm. 1979. Plants of the Chicago Region: A Checklist of the Vascular Flora of the Chicago Region, With Keys, Notes on Local Distribution, Ecology, and Taxonomy, and a System for Evaluation of Plant Communities. Lisle, IL: Morton Arboretum.

Swink, F., and G. Wilhelm. 1994. Plants of the Chicago Region. Indianapolis: Indiana Academy of Science.

Tabatabai, F., and R. Brumbaugh. 1998. National Wetland Mitigation Banking Study. The Early Mitigation Banks: A Follow-Up Review. IWR Report 98-WMB-Working Paper. Institute for Water Resources Report, Water Resources Support Center, U.S. Army Corps of Engineers, Alexandria, VA.

Taft, J.B., G.S. Wilhelm, D.M. Ladd, and L.A. Masters. 1997. Floristic quality assessment for vegetation in Illinois, a method for assessing vegetation integrity. Erigenia 15:3-24.

Thullen, J.S., and D.R. Eberts. 1995. Effect of temperature, stratification, scarification, and seed origin on the germination of Scirpus acutus Muhl. seeds for use in constructed wetlands. Wetlands 15(3):298-304.

Titus, J.H. 1990. Microtopography and woody plant regeneration in a hardwood floodplain swamp in Florida. Bull. Torrey Bot. Club 117(4):429-437.

Titus, J.G. 1999. Can we save our coastal wetlands from a rising sea? Natl. Wetlands Newslet. 21(2):5-6,14.

Torok, L.S., S. Lockwood, and D. Fanz. 1996. Review and comparison of wetland impacts and mitigation requirements between New Jersey, USA, Freshwater Wetlands Protection Act and Section 404 of the Clean Water Act. Environ. Manage. 20(5):741-52.

Tweedy, K.L. 1998. Hydrologic Characterization and Modeling of Two Prior Converted Wetlands in Eastern North Carolina. M.S. Thesis. North Carolina State University.

Updegraff, K., J. Pastor, S.D. Bridgham, and C.A. Johnston. 1995. Environmental and substrate controls over carbon and nitrogen mineralization in northern wetlands. Ecol. Applic. 5(1):151-163.

USACE (U.S. Army Corps of Engineers). 1978a. Preliminary Guide to Wetlands of Peninsular Florida. Tech. Report Y-78-2. Vicksburg, MS: U.S. Army Corps of Engineers, Waterways Experiment Station.

USACE (U.S. Army Corps of Engineers). 1978b. Preliminary Guide to Wetlands of Puerto Rico. Tech. Report Y-78-3. Vicksburg, MS: U.S. Army Corps of Engineers, Waterways Experiment Station.

USACE (U.S. Army Corps of Engineers). 1978c. Preliminary Guide to Wetlands of West Coast States. Tech. Report Y-78-4. Vicksburg, MS: U.S. Army Corps of Engineers, Waterways Experiment Station.

USACE (U.S. Army Corps of Engineers). 1978d. Preliminary Guide to Wetlands of Gulf Coastal Plain. Tech. Report Y-78-5. Vicksburg, MS: U.S. Army Corps of Engineers, Waterways Experiment Station.

USACE (U.S. Army Corps of Engineers). 1978e. Preliminary Guide to Wetlands of the Interior United States. Tech. Report Y-78-6. Vicksburg, MS: U.S. Army Corps of Engineers, Waterways Experiment Station.

USACE (U.S. Army Corps of Engineers). 1987. Wetlands Delineation Manual. Tech. Report Y-87-1. Vicksburg, MS: U.S. Army Corps of Engineers, Waterways Experiment Station.

USACE/EPA (U.S. Army Corps of Engineers and U.S. Environmental Protection Agency). 1990. Memorandum of Agreement Between the Environmental Protection Agency and Department of the Army Concerning the Determination of Mitigation Under the Clean Water Act Section 404 (b)(1) Guidelines. February 8, 1990.

USACE (U.S. Army Corps of Engineers). 1999a. Army Corps of Engineers Standard Operating Procedures for the Regulatory Program. October 15, 1999. [Online]. Available: http://www.nwp.usace.army.mil/op/g/notices/Reg_Stan_SOP.pdf.

USACE (U.S. Army Corps of Engineers). 1999b. Memorandum for Commanders, Major Subordinate Commands and District Commands, from Russell L. Fuhrman, Major General, USA Director of Civil Works. Subject: Program Consistency and Reporting on the U.S. Army Corps of Engineers Regulatory Program. April 8, 1999.

USACE (U.S. Army Corps of Engineers). 2000. Mitigation Guidelines and Monitoring Requirements. Special Public Notice 97-00312-SDM. Los Angeles District, U.S. Army Corps of Engineers, Los Angeles, CA.

USDA (U.S. Department of Agriculture). 1982. National List of Scientific Plant Names. Vol. I. List of Plant Names. Vol. II. Synonymy. SCS-TP-159. Washington, DC: Soil Conservation Service.

USDA (U.S. Department of Agriculture). 1985. Hydric Soils of United States. Washington, DC: Soil Conservation Service.

USDA (U.S. Department of Agriculture). 1987. Hydric Soils of United States, 2nd Ed. Washington, DC: Soil Conservation Service.

USDA (U.S. Department of Agriculture). 1991. Hydric Soils of United States, 3rd Ed. Misc. Pub. No. 1491. Washington, DC: Soil Conservation Service.

USDA (U.S. Department of Agriculture). 1997. Chesapeake Bay Riparian Handbook: A Guide for Establishing and Maintaining Riparian Forest Buffers. NA-TT-02-97. U.S. Department of Agriculture, Forest Service, Northeastern Area, State and Private Forestry, Washington, DC.

USFWS (U.S. Fish and Wildlife Service). 1980. Habitat Evaluation Procedure (HEP). Ecological Service Manual ESM 102. Fort Collins, CO: U.S. Fish and Wildlife Service.

USFWS (U.S. Fish and Wildlife Service). 1981. Standards for the Development of Habitat Suitability Index Models. ESM 103. Washington, DC: Division of Ecological Services, U.S. Fish and Wildlife Service, Department of the Interior.

USFWS (U.S. Fish and Wildlife Service). 1983. Natural and Modified Pocosins: Literature Synthesis and Management Options. FWS/ OBS-83-04. Washington, DC: Department of the Interior.

van Duren, I.C., R.J. Strykstra, A.P. Grootjans, G.N.J. ter Heerdt, and D.M. Pegtel. 1998. A multidisciplinary evaluation of restoration measures in a degraded Cirsio-Molinietum fen meadow. Appl. Veg. Sci. 1(1):115-130.

Vymazal, J. 1995. Algae and Element Cycling in Wetlands. Boca Raton, FL: Lewis.

Wainger, L.A., D.M. King, and J.S. Boyd. 2000. Expanding Wetland Assessment Procedures: Landscape Indicators of Relative Wetland Value Indicators, with Illustrations for Scoring Mitigation Trades. University of Maryland, Center for Environmental Science.

Want, W.L. 1994. Law of Wetlands Regulation. Deerfield, IL: Clark Boardman Callaghan.

Welsch, D.J. 1991. Riparian Forest Buffers: Function and Design for Protection and Enhancement of Water Resources. USDA Forest Service Pub. NA-PR-07-91. Radnor, PA: U.S. Department of Agriculture, Forest Service, Northeastern Area, State and Private Forestry, Forest Resources Management.

White, D., and L. Shabman. 1995. National Wetland Mitigation Banking Study Watershed-Based Wetlands Planning: A Case Study Report. IWR Report 95-WMB-8. December. Alexandria, VA: Institute for Water Resources. [Online]. Available: http://www.wrc-iwr.usace.army.mil/iwr/pdf/95wmb8.pdf [April 10, 2001].

White, P.S., and J.L. Walker. 1997. Approximating nature's variation: Selecting and using reference information in restoration ecology. Restor. Ecol. 5(4):338-349.

Wilcox, D.A. 1988. The necessity of interdisciplinary research in wetland ecology: The Cowles Bog example. Pp. 1-9 in Interdisciplinary Approaches to Freshwater Wetlands Research, D.A. Wilcox, ed. East Lansing: Michigan State University Press.

Wilcox, D.A., S.I. Apfelbaum, and R.D. Hiebert. 1984. Cattail invasion of sedge meadows following hydrologic disturbance in the Cowles Bog Wetland Complex, Indiana Dunes National Lakeshore. Wetlands 4(1):115-128.

Williams, G.P. 1977. Washington, D.C.'s Vanishing Springs and Waterways. Geological Survey Circular 752. Washington, DC: U.S. Dept. of the Interior, Geological Survey.

Wilson, R.F., and W.J. Mitsch. 1996. Functional assessment of five wetlands constructed to mitigate wetland loss in Ohio, USA. Wetlands 16(4):436-451.

Winston, R. 1996. Design of an urban, groundwater dominated wetland. Wetlands 16(4):524-531.

Winter, T.C. 1986. Effect of groundwater recharge on configuration of the water-table beneath sand dunes and on seepage in lakes in the sandhills of Nebraska. J. Hydrol. 86(3-4):221-237.

Winter, T.C. and M.-K. Woo. 1990. Hydrology of lakes and wetlands. Pp. 159-187 in The Geology of North America. Surface water hydrology, Vol. 0-1, M.G. Wolman, and H.C. Riggs, eds. Boulder, CO: The Geological Society of America.

Woodwell, G.M. 1956. Phytosociology of Coastal Plain Wetlands of the Carolinas. M.A. Thesis. Duke University.

Zedler, J.B. 1990. A Manual for Assessing Restored and Natural Coastal Wetlands with Examples from Southern California. T-CSGCP-021. La Jolla: California Sea Grant College.

Zedler, J.B. 1996a. Ecological issues in wetland mitigation: An introduction to the forum. Ecol. Applic. 6(1):33-37.

Zedler, J.B. 1996b. Coastal mitigation in southern California: The need for a regional restoration strategy. Ecol. Applic. 6(1):84-93.

Zedler, J. 1998. Replacing endangered species habitat: The acid test of wetland ecology. Pp. 364-379 in Conservation Biology for the Coming Age, 2nd Ed., P.L. Fiedler and P.M. Kareiva, eds. New York: Chapman and Hall.

Zedler, J.B. 1999. The ecological restoration spectrum. Pp. 301-318 in An International Perspective on Wetland Rehabilitation, W. Streever, ed. Dordrecht, The Netherlands: Kluwer.

Zedler, J.B, ed. 2001. Handbook for Restoring Tidal Wetlands. Boca Raton, FL: CRC.

Zedler, J.B., and J.C. Callaway. 1999. Tracking wetland restoration: Do mitigation sites follow desired trajectories? Restor. Ecol. 7(1):69-73.

Zedler, J.B., and R. Langis. 1991. Authenticity: Comparisons of constructed and natural salt marshes of San Diego Bay. Restor. Manage. Notes 9(1):21-25.

Zedler, P.H., C.K. Frazier, and C. Black. 1993. Habitat creation as a strategy in ecosystem preservation: An example from vernal pools in San Diego County. Pp. 239-247 in Interface Between Ecology and Land Development in California, J.E. Keeley, ed. Los Angeles, CA: Southern California Academy of Sciences.

Zentner, J. 1988. Wetland projects of the California State Coastal Conservancy: An assessment. Coast. Manage. 16(1):47-67.

Appendixes

Appendix A

Survey of Studies: Comparison of Mitigation and Natural Wetlands

TABLE A-1 Survey of Studies: Comparison of Mitigation and Natural Wetlands

Region	Time Period	# Sites	Scope
Massachusetts	1983 to 1994	114	Vegetation (% cover) Size Hydrology If project was built
Portland, Oregon	1987 to 1993	95	Freshwater emergent and open-water wetlands, soil organic matter (SOM), hydrology
Orange County, California	1979 to 1993	70	Vegetation, hydrology

Findings	Reference
79.9% mitigated for impacts <5000 ft^2 54.4% noncompliant	Brown and Veneman (1998)
70.1% involved impacts to forested wetlands 61.4% designed to produce scrub/shrub 38.6% actually produced no wetlands 36.8% actually produced open wet meadows	
Plant communities at replication sites differed significantly from wetlands they were designed to replace. Similarity did not increase between new and 12-year-old projects.	
Compliance but not similarity between replicated and impacted plant communities increased with greater completeness of the replication plan and Order of Conditions.	
Mean SOM concentrations were higher in naturally occurring wetlands (NOWs) than in mitigating wetlands.	Shaffer and Ernst (1999)
No significant change in SOM concentration in soils in mitigating wetlands (MWs) sampled.	
For a subset of wetlands measured for hydrology, there was a significant negative relationship between SOM and the extent of inundation by standing water.	
Success of mitigation, in terms of SOM, could be improved by better project design and better management of soils during project construction.	
Thirty of the 70 (43%) met all of their permit conditions and were considered successful; these projects comprised 195 ac.	Sudol (1996)
Six sites (9%) comprising 52 ac did not meet any of their permit conditions and were considered failures.	
Mitigation in Orange County has been unsuccessful. There has been a net loss of wetland and riparian habitat.	

continued

TABLE A-1 Continued

Region	Time Period	# Sites	Scope
Portland, Oregon	1993	45 natural 51 mitigation 1-11 years; mean 5 years	Small (≤ 2 ha) Freshwater, palustrine wetlands in rapidly urbanizing area Plant species richness (presence/absence) and composition of natural and mitigation wetlands Relationships between floristic characteristics and variables describing land use, site conditions, and mitigation activities
Susquehannah River watershed, Pennsylvania	1993	20 reference; 44 created	Soil organic matter, matrix chroma, bulk density, total nitrogen, pH
Iowa, Minnesota, South Dakota	1989 to 1991	62	Restored prairie potholes Basin morphometry, hydrology, and vegetation zone development

Findings	Reference
Overall species richness was high (365 plant taxa), but >50% of species on both natural and mitigation wetlands were introduced.	Magee et al. (1999)
Wetlands surrounded by agricultural and commercial/ industrial/transportation corridor uses had more introduced species per site than those surrounded by undeveloped land.	
Wetlands in the urbanizing study area are floristically degraded.	
Current wetland management practices are replacing natural marsh and wet meadow systems with ponds, changing composition of plant species assemblages.	
Compared to reference wetlands, wetland creation projects contained less SOM at 5 cm and unlike reference sites, SOM content was uniform between 5 and 20 cm. Created wetlands contained less silt at 5 cm and more sand and less clay at 20 cm. Wetland creation projects had higher pH, bulk density, and matrix chroma and lower total nitrogen.	Bishel-Machung et al. (1996)
No relationship was found between time elapsed since construction and soil organic matter content in wetland creation projects.	
Earthen dams installed on 73% restoration sites.	Galatowitsch and van der Valk (1996)
About 60% of basins had predicted hydrology or held water longer than predicted.	
Twenty percent were hydrological failures and either never flooded or had significant structural problems.	
Most had developed emergent and submersed aquatic vegetation zones, but only a few had developed wet prairie and sedge meadow vegetation zones.	

continued

TABLE A-1 Continued

Region	Time Period	# Sites	Scope
Florida	1990	40 projects	Surface hydrology, vegetation
Central Florida	1993	10 natural, 10 created	Dipterans in freshwater herbaceous wetlands
Galveston Bay, Texas	Fall 1990, spring 1991	10 created, 5 natural	Densities of nekton and infauna in salt marshes

Findings	Reference
Forty of 195 permitted projects had undertaken some type of mitigation activity. Only four of those 40 projects met all the stated permit goals. Twenty-four of the 40 projects contained success criteria, but for 23 (57.5%) of the projects the success criteria were inappropriate.	Erwin (1991)
Of the 1,058 acres required by permit to be created for all 40 projects, only 530.6 (50%) acres had actually been constructed.	
Location and persistence were not in the criteria. Twenty-three (57.5%) of the 40 projects were located where surrounding existing or future land uses may prevent the wetlands from providing their intended functional values. Only three projects had a long-term management plan.	
Twenty-five (62.5%) of the projects had hydrological problems.	
32 (80%) projects were colonized by undesirable plant species. Permits for 22 projects required removal of problematic plants, but attempts to control them were undertaken in only 13 (59%) projects. Postconstruction monitoring was required for 39 projects, but adequate monitoring had been undertaken in only 15 (38%).	
No convincing evidence of differences in natural and created wetland dipteran communities.	Streever et al. (1996)
Densities of daggerblade grass shrimp were not significantly different among marshes, but the size of these shrimp was significantly smaller than in natural marshes.	Minello and Webb (1997)
Densities of the marsh grass shrimp and of three commercially important crustaceans were significantly lower in created marshes than in natural marshes.	
Fish densities in vegetation were significantly lower in created marshes than in natural marshes.	
Natural and created marshes did not differ in species richness of nekton.	
Marsh elevation and tidal flooding are key characteristics affecting use by nekton and should be considered in marsh construction projects.	

continued

TABLE A-1 Continued

Region	Time Period	# Sites	Scope
Ohio	1994 to 1995	5 replacement	Hydrology, soils, vegetation, wildife, water quality
Texas		10	Invertebrates, fish
North Carolina		6	Sediment/soil, invertebrates
North Carolina		5	Sediment/soil
North Carolina	Marshes established 1971 to 1974 and monitored for 25 years	2 constructed marshes 2 natural marshes	Above-ground biomass, soil, benthic infauna, carbon, total nitrogen
South Carolina		2	Sediment/soil, plants, invertebrates, fish
Texas		3	Plants, invertebrates, fish
California		2	Sediment/soil, plants, fish, topography
North Carolina		*Multiple*	Plants, invertebrates

Findings	Reference
Eighty percent were in compliance with legal requirements and demonstrated medium-to-high ecosystem success	Wilson and Mitsch (1996)
	Minello and Webb (1997)
	Sacco et al. (1994)
	Craft et al. (1988)
Constructed marshes: macrophyte community developed quickly and within 5 to 10 years, above-ground biomass and MOM were equivalent or exceeded corresponding values in the natural marshes.	Craft et al. (1999)
After 15-25 years, benthic infauna and species richness were greater in the natural marshes.	
Soil bulk density decreased and organic carbon and total nitrogen increased over time in constructed marshes.	
Nitrogen accumulation was much higher in constructed marshes than in natural marshes.	
Different ecological attributes develop at different rates, with primary producers achieving equivalence during the first 5 years, followed by the benthic infauna community 5-10 years later.	
	LaSalle et al. (1991)
	Minello and Zimmerman (1992)
	Haltiner et al. (1997)
	Seneca et al. (1976)

REFERENCES

Bishel-Machung, L., R.P. Brooks, S.S. Yates, and K.L. Hoover. 1996. Soil properties of reference wetlands and wetland creation projects in Pennsylvania. Wetlands, 16(4):532-541.

Brown, S., and P. Veneman. 1998. Compensatory wetland mitigation in Massachusetts. Research Bulletin Number 746. Amherst, MA: Massachusetts Agriculture Experiment Station, University of Massachusetts.

Craft, C., J. Reader, J.N. Sacco, and S. Broome. 1999. Twenty-five years of ecosystem development of constructed Spartina alterniflora (Loisel) marshes. Ecol. Applic. 9(4):1405-1419.

Craft, C.B., S.W. Broome, and E.D. Seneca. 1988. Nitrogen, phosphorus and organic carbon pools in natural and transplanted marsh soils. Estuaries 11(4):272-289.

Erwin, K.L. 1991. An Evaluation of Wetland Mitigation in the South Florida. Water Management District, Vol. 1. Methodology. West Palm Beach, FL: South Florida Water Management District.

Galatowitsch, S.M., and A.G. van der Valk. 1996. Characteristics of Recently Restored Wetlands in the Prairie Pothole Region. Wetlands 16(1):75-83.

Haltiner, J., J.B. Zedler, K.E. Boyer, G.D. Williams, and J.C. Callaway. 1997. Influence of physical processes on the design, functioning and evolution of restored tidal wetlands in California (USA). Wetlands Ecol. Manage. 4(2):73-91.

LaSalle, W.M., M.C. Landin, and J.G. Sims. 1991. Evaluation of the flora and fauna of a Spartina alterniflora marsh established on dredged material in Winhay Bay, South Carolina. Wetlands 11(2):191-208.

Magee, T.K., T.L. Ernst, M.E. Kentula, M.E. and K.A. Dwire. 1999. Floristic comparison of freshwater wetlands in an urbanizing environment. Wetlands 19(3):517-534.

Minello, T.J. and J.W. Webb, Jr. 1997. Use of natural and created Spartina alterniflora salt marshes by fishery species and other aquatic fauna in Gavelston Bay, Texas, USA. Mar. Ecol Prog. Ser. 151(1/3):165-179.

Minello, T.J., and R.J. Zimmerman. 1992. Utilization of natural and transplanted Texas salt marshes by fish and decapod crustaceans. Mar. Ecol. Prog. Ser. 90(3):273-285.

Sacco, J.N., E.D. Seneca, and T.R. Wentworth. 1994. Infaunal community development of artificially established salt marshes in North Carolina. Estuaries 17(2):489-500.

Seneca, E.D., L.M. Stroud, U. Blum, and G.R. Noggle. 1976. An Analysis of the Effects of the Brunswick Nuclear Power Plant on the Productivity of Spartina alterniflora (smooth cordgrass) in the Dutchman Creek, Oak Island, Snow's Marsh, and Walden Creek Marshes, Brunswick County, North Carolina, 1975-1976. 3rd Annual Report to Carolina Power and Light. Raleigh, NC: Carolina Power and Light Com.

Shaffer, P.W., and T. Ernst. 1999. Distribution of soil organic matter in freshwater emergent open water wetlands in the Portland, Oregon metropolitan area. Wetlands 19(3):505-516.

Streever, W.J., K.M. Portier, and T.L. Crisman. 1996. A comparison of dipterans from ten created and ten natural wetlands. Wetlands 16(4):416-428.

Sudol, M.F. 1996. Success of Riparian Mitigation as Compensation for Impacts Due to Permits Issued Through Section 404 of the Clean Water Act in Orange County, California. Ph.D. Dissertation. University of California, Los Angeles.

Wilson, R.F., and W.J. Mitsch. 1996. Functional assessment of five wetlands constructed to mitigate wetland losses in Ohio. Wetlands 16(4):436-451.

Appendix B

Case Studies

EVERGLADES NATIONAL PARK

The Hole-in-the-Donut (HID) is a 2,509 ha (6,200-ac) mitigation bank (Florida's first) consisting of abandoned agricultural land that is surrounded by Everglades National Park land with its native vegetation (Doren 1997, ENP 1998). Recently acquired by the park, this tract is being rid of its monotypic exotic vegetation through a massive and intensive eradication program that involves removal of the exotic trees, their roots, and the soil, then allowing the native wetland vegetation to reestablish itself.

Everglades National Park was established in 1947; it is the largest national park in the United States at 0.6 million ha (1.5 million ac). Often called a "river of grass" for its shallow, slow-moving surface water, the park supports many native species and thousands of highly valued birds. Unfortunately, it also supports 217 nonindigenous plant species, which continually threaten to displace the natives (Doren 1997). The previously farmed areas of HID now support a monotypic stand of Brazilian pepper (*Schinus terebinthifolius*) that is virtually impenetrable by humans. Found only in a small area in 1975, the pepper trees now enjoy an extensive distribution, in part due to wide dispersal of their berries by fruit-eating birds.

Prior to rock plowing the natural limestone rock for agriculture, the HID area was naturally dominated by a mosaic of marl prairie and related vegetation types. Two disturbances are thought to be responsible for the ability of *Schinus terebinthifolius* to establish and expand following agriculture: (1) abandonment and exposure of an artificial soil and (2) reduced fire frequency (see Figure B-1). The rapid growth and high productivity of this invasive tree are likely due to changes in the soil brought about by rock plowing—namely, increased nutrient release and increased soil depth that allows seedlings to withstand 2-5 months of drought each year (Doren 1997).

Several attempts were made to remove *Schinus* in the 1970s and 1980s (Doren 1997). These involved herbicides, disking, bulldozing, burning, mowing, and planting and seeding of natives, hardwoods, and pines. All failed. One treatment in 1974 and another in 1983 involved the removal of rock-plowed soil down to the level of hard porous limestone substrate. The promising results led to larger soil-removal treatments 1989; soil was partially removed on 6 ha and completely removed on 18 ha. The site with partial soil removal was recolonized by *Schinus*, but the area without soil was not (Doren 1997).

The present effort to remove *Schinus* from about 2,529 ha (6,250 ac = about 10 miles^2) grew out of the success of the earlier trial with total soil removal. The restoration target is a muhly grass-sawgrass prairie over 90% of the area and upland hammocks covering about 10% of the area. The hammocks, or mounds, would support pineland and hardwoods. The current plan (ENP 1998) proposes to remove about 5,000,000 yd^3 of material over 20 years. Trees are first bulldozed and then shredded and composted. Sediment is trucked to fill old limestone quarries and borrow pits. Once the nearby borrow pits are filled, the spoils will be mounded in place to create upland hammocks and mowed to prevent *Schinus* dominance. Trees will then be planted on the mounds to restore 40 to 60 ha (100-150 ac) of pineland and hardwood forest. It is expected that one mound would be about 7 m (20 ft) high with 3:1 slopes, covering about 10 ha (25 ac). One mound could accommodate the spoils from about 7% of

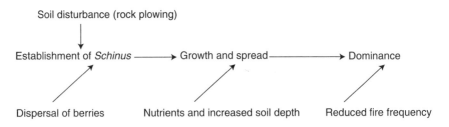

FIGURE B-1 Conceptual model of factors facilitating the invasion of *Schinus terebinthifolius*.

the HID, drawing materials from $3\frac{1}{2}$ to 4 years' remediation period. Ultimately, there would be 5-12 mounds covering 60 to 80 ha (150-200 ac) or 2 to 3% of the project area. The scale of the project is also indicated by the amount of on-site trucking, which is calculated to involve 26 trucks driving over 750,000 km (458,640 miles) per year.

The current marl prairie restoration program is conducted with mitigation funds that result from a cooperative program involving Everglades National Park, Dade County, the Florida Department of Environmental Protection, and the U.S. Army Corps of Engineers (Doren 1997). A mitigation "credit" consists of 0.4 ha (1 ac) of restoration, costing $30,000 in 1999 (M. Norland, National Park Service (NPS), personal communication). Two-thirds of the mitigation money acquired goes to restoration, and one-third goes to additional land purchase.

The scientific basis for the program is the hypothesis that soil removal will eliminate the conditions responsible for the presence of *Schinus*. A scientific advisory panel was established in 1996 to call for and review research proposals concerning this large-scale problem with invasive vegetation, to include studies of the current program and alternative approaches to eradication; the panel was subsequently disbanded by National Park Service administrators.

References

Doren, R.F. 1997. Restoration Research Themes and Hypotheses for Hole-in-the-Donut (HID). South Florida Natural Resources Center, Everglades National Park. [Online]. Available: http://everglades.fiu.edu/hid/ [June 28, 2001].

ENP (Everglades National Park). 1998. Environmental Assessment: Hole-in-the-Donut Soil Disposal. Everglades National Park, Florida. U.S. Department of the Interior, National Park Service. Denver, CO: Denver Service Center, National Park Service. 177 pp.

COYOTE CREEK MITIGATION SITE

Synopsis

The Coyote Creek mitigation site was installed in 1993 to partially satisfy permit requirements pursuant to Section 404 of the CWA. The site was designed to provide off-site mitigation for impacts to nine creeks in Santa Clara County that were impacted by the construction of State Route 85 in San Jose, California. The mitigation goal is to develop 24.4 ac of stratified native riparian habitat adjacent to Coyote Creek, similar to riparian habitats found along other creeks (used as model sites) in Santa Clara County, California.

To achieve the mitigation goal, the ground surface elevation of an agricultural field adjacent to Coyote Creek was lowered by removing topsoil. A levee adjacent to the field was breeched, and a diversion channel was excavated through the agricultural field to create the mitigation area. The mitigation area was then extensively planted with tree, shrub, and herbaceous riparian plant species to create four plant associations, including streamside, floodplain, valley oak forest, and slope communities.

The monitoring plan for the mitigation site called for measurement of various site parameters over a 15-year period to track the success of the site and its overall status. The two parameters used to measure success included percent survival of planted species (short-term success criteria) and establishment of a trend toward percent cover of tree and shrub species similar to that of a mature riparian community (long-term success criteria). In addition, other parameters to be monitored at the site included species composition, plant vigor and health, plant height, basal area, natural reproduction, species diversity, root development, hydrology, and photo documentation. While all parameters were to be measured, only percent survival (short-term success criteria) and a trend toward mature percent cover (long-term success criteria) were used to judge the overall success or failure of the mitigation site. The short-term success criteria called for the survival of 60 to 90% of planted riparian species measured over a 5-year period. The long-term success criteria called for establishment of a trend toward 75% cover for tree species and 45% cover for shrub species at the site (not specified for a particular plant association) measured over a 10-year period (annually until year 5 and biennially until year 10).

Because of contracting difficulties, monitoring reports for years 2 through 4 were submitted to the Corps together during year 4. Monitoring reports for years 2 through 4 indicated that the short-term success criteria were met by year 3 (1996) and that no further monitoring of percent survival would be performed. The monitoring reports submitted for year 6 (1999) indicated that the site was showing a trend toward satisfaction of the long-term success criteria. Other parameters measured at the site but not used to judge the site's overall success or failure indicated a trend toward successful establishment of riparian vegetation, as determined by an increase in the mean tree basal area and tree height. No information in the monitoring reports indicated the status of the site relative to species composition, plant vigor and health, natural reproduction, species diversity, or root development.

Hydrological evaluation of the site through 1996 indicated that flow frequency was maintained according to predicted models. Recent hydrological evaluations (1992 to 2000), however, indicate that the diversion channel has been dry since the spring of 1997 because of sediment deposits at the channel inlet as a result of high winter flows and overgrowth of giant reed (*Arundo donax*) on the deposited sediment. Year 2000 monitoring is scheduled to be conducted on a biennial basis, and the inlet remains blocked, precluding flows from entering the diversion channel.

Introduction

The California Department of Transportation applied for a Department of the Army permit in 1991 to construct State Route 85, a new 18-mile-long 6-lane freeway in San Jose, California. The freeway was designed to link two existing highway corridors, Interstate 280 to the north and U.S. 101 to the south. The construction of State Route 85 resulted in impacts to nine creeks regulated by the Corps pursuant to Section 404 of the CWA. The impacts included construction of bridges over the creeks, construction of storm drain outfalls, realignment of creek channels, installation of hardscape erosion control measures such as rip rap and channel lining, placing fill into the creeks to facilitate widening, culverting flows beneath new sections of road, and installation of flood control facilities such as berms, flood walls, and check dams. The overall project resulted in the placement of approximately 7,600 yd^3 of fill within Corps jurisdiction and impacted a total of 9 ac of riparian habitat.

Coyote Creek Mitigation

The Coyote Creek mitigation site is as an off-site mitigation area. It was designed to provide riparian woodland habitat common to California's Santa Clara Valley, as partial replacement for riparian habitat impacts on nine creeks along the proposed State Route 85 corridor. The Coyote Creek mitigation site is located near U.S. 101 in San Jose, California.

The mitigation project required extensive reworking of the site's topography and hydrology and installation and establishment of large numbers of native riparian plant species. The mitigation site was designed to provide 24.4 ac of riparian habitat adjacent to Coyote Creek. Monitoring for the project was designed to assess the development of riparian habitat from the time of grading and plant installation until the project met or exceeded all success criteria or by mutual agreement between the California Department of Transportation and resource agencies (H. T. Harvey and Associates 1992).

In 1993, the mitigation site was graded and soil removed (approximately 10 to 15 ft deep) in an effort to bring the final grade closer to the groundwater table. In addition, a meandering 2,300-ft-long, 9-ft-wide, 2-ft-deep channel was constructed through the center of the site (USACOE 1991, Public Notice 18998S92). The channel was designed to carry water diverted from Coyote Creek through two breaches in the adjacent levee. Once the diversion channel was created, the site was vegetated with four plant associations, including streamside, floodplain, valley oak forest, and slope communities. The streamside association extended 25 ft on each side of the channel and included willows and associated understory species. The floodplain association included an overstory of cottonwoods, a shrub layer of willows and blackberry, and an understory consisting of an herbaceous seed mix. The valley oak forest association included an overstory of valley oaks and an understory consisting of an herbaceous seed mix. The slope association included an overstory of sycamore, California walnut, and buckeye; a shrub layer of elderberry, toyon, coyote brush; and an understory consisting of an herbaceous seed mix.

Mitigation Monitoring Parameters and Success Criteria

The mitigation monitoring plan established short- and long-term success criteria, including percent survival and percent cover (H. T. Harvey and Associates 1992). Various other site evaluation parameters also were monitored; however, these parameters provided information about the site and did not constitute specific criteria on which the success of the site would be judged. These parameters included species composition, plant vigor and health, plant height, plant basal area, natural reproduction, species diversity, root development, photo documentation, and hydrological evaluation (frequency and extent of flow in the created channel and functioning of the created channel itself).

Short-Term Success Criterion

Percent survival formed the basis for establishment of the short-term success criterion. This criterion applies to survival of plants installed at the site and was broken down into two categories, including overall survival and cumulative survival. Overall survival includes that of original woody plants and plants installed to replace dead or dying original plantings. This criterion specifies successful accomplishment of this monitoring parameter when 80% survival of woody plants in the slope and valley oak associations is achieved and when 90% survival of woody plants in the floodplain and streamside plant associations is achieved. This parameter was designed to be measured for a period of 5 years after initial plant installation. Cumulative survival calculates the survival of original plantings only. This criterion specifies successful accomplishment of this monitoring parameter when 60% survival (original plantings only) of woody plants in the slope and valley oak associations is achieved and when 70% survival of woody plants in the floodplain and streamside associations is achieved. This parameter was designed to be measured for a period of 5 years after initial plant installation.

Long-Term Success Criterion

Percent cover formed the basis for establishment of long-term success criterion. This criterion calls for achievement of a *steady trend* toward reaching the *ultimate goal* of 75% cover for trees and 45% cover for shrubs. The mitigation monitoring report indicated that after the fifth year, the percent cover by trees and shrubs would be monitored as the prime indicator of increasing habitat values. This parameter was designed to be measured for a 10-year period with yearly monitoring occurring through year 5 and biennial monitoring occurring thereafter until year 10.

Duration of Monitoring

The mitigation monitoring plan called for a 15-year monitoring period in which plant survival and species composition would be measured for a total of six consecutive years starting with year 0 (H. T. Harvey and Associates 1992). Plant vigor and health, plant height, basal area, natural reproduction and species diversity were to be monitored yearly through year 5 and every other year through year 10. Root development was to be monitored in years 3, 4, and 5, while hydrological monitoring and photo documentation of the site were to be carried out yearly for 15 consecutive years, including year 0.

Site Installation and Postinstallation Site Review

Grading, installation of water control structures, and planting took place at the Coyote Creek mitigation site from May through September 1993. A total of 10,484 container plants of riparian species were installed between May and June. Results from monitoring visits in September 1993 (year 0) indicated that insects, dust, drought, and browse damage were the most common plant damage factors, although it was noted that none were significantly impeding restoration efforts (H. T. Harvey and Associates 1993). Actions implemented to respond to the September 1993 monitoring results included weed control, mulch application, installation of foliage protectors, erosion control remedial measures, installation of groundwater monitoring systems, and insect/rodent abatement.

Mitigation Monitoring and Site Development

Due to funding and contracting issues associated with the California Department of Transportation, mitigation monitoring reports for 1994,

1995, and 1996 were not submitted until the spring of 1997. The results of the monitoring reports submitted for 1994 and 1995 indicated that the success criterion for vegetation was met in both years; however, results were inconclusive for success of the hydrological criterion, and the Department of Transportation recommended reevaluation of the hydrological criteria in conjunction with the Corps.

Monitoring conducted at the site in 1996 indicated that it continued to improve after cessation of irrigation in 1995. The 1996 monitoring report noted that the overall and cumulative success criteria goals had been met. As a result, subsequent monitoring reports would no longer document short-term success criterion parameters. Hydrologic monitoring was not performed.

Management recommendations for 1996 included the need to evaluate the system for collecting hydrological data, as well as performance of weed control along the perimeter of the site to reduce potential fire hazard, increasing monitoring for erosion control needs during the wet season and installation of remedial erosion control measures, and weed control throughout the site particularly for giant reed and water primrose (*Ludwigia peploides*), which were noted to cover approximately 60% of the site.

The 1997 monitoring report specified that, since the short-term success criterion had been achieved, measurements for this criterion were no longer taken at the site (California Department of Transportation 1997). The 1997 monitoring report showed that measurements taken at the site for percent cover (long-term success criterion) indicated that the site was continuing to show a steady trend toward reaching the ultimate goal of 75% cover for trees and 45% cover for shrubs (California Department of Transportation 1997). For the hydrological evaluation, the report noted that equipment installed on Coyote Creek to monitor flow frequency and the extent of flooding had not functioned since 1995 and was washed away during high flows in 1997. The report recommended that field observation with documented field notes and photographs be used to monitor flow frequency and stability of the diversion channel.

The monitoring report also noted that the diversion channel did not receive water from Coyote Creek throughout the monitoring period from spring through fall of 1997 (California Department of Transportation 1997). Field surveys to determine the cause of reduced flows revealed that sediment deposits had blocked the inlet to the diversion channel (California Department of Transportation 1998). In addition, in the general area of the weir inlet into the diversion channel, there was a massive growth of giant reed. The giant reed appears to have stabilized deposited sediments and assisted in blocking flow into the diversion channel. The monitoring report documents that the disruption of flows into the diversion channel was predicted in the hydrological model developed for the site (California Department of Transportation 1998; H.T. Harvey and Associates 1990).

No monitoring was conducted in 1998 due to institution of biennial monitoring as per mitigation plan (H.T. Harvey and Associates 1992).

The year 1999 marked the start of biennial monitoring. Only the long-term success criterion was monitored since the short-term success criterion had been met in earlier monitoring years. Results of monitoring in 1999 indicated that the site was continuing to show a steady trend toward reaching the ultimate goal of 75% cover for trees and 45% cover for shrubs. Other parameters previously monitored, including species composition, plant vigor and health, basal area, natural reproduction, species diversity, and root development, were not monitored. Further, it was recommended in the monitoring report that only the percent cover criterion be evaluated in future reports since the "other vegetative parameters do not contribute information to the trend for percent cover which is the long-term success criterion for the site" (California Department of Transportation 2000). It was also recommended that the length of sampling be reduced such that all monitoring would be concluded by year 10 (2003) since the long-term success criterion (percent cover) only specifies monitoring through year 10 and it appears that the long-term success criterion will be met. The current schedule calls for monitoring on a biennial basis through 2008.

Visual hydrological evaluations in 1999 showed that the diversion channel had been successfully maintained "in that the extent and frequency of flow are comparable to the model" (California Department of Transportation 2000). The hydrology in the diversion channel was not considered successful in that the diversion channel appeared to have been dry since the spring of 1997. The monitoring cited another agency, the Santa Clara Valley Water District, as having mitigation plans involving Coyote Creek and that part of the mitigation involved removing the giant reed from the creek and dredging the inlet to restore flow to the diversion channel. These activities are expected to restore flow from Coyote Creek into the diversion channel. To date, the problem of flows being blocked at the diversion channel inlet has not been corrected.

References

California Department of Transportation. 1997. Annual Monitoring Report Army Corps Permit 18998S92.

California Department of Transportation. 1998. Annual Monitoring Report Route 85-Coyote Creek Mitigation Site at Bernal Road and Highway.

California Department of Transportation. 2000. Caltrans Route 85-Coyote Creek Mitigation Monitoring Report for 1999.

H.T. Harvey & Associates. 1990. Route 85-Coyote Creek Mitigation Conceptual Revegetation Plan, 2nd Rev. Prepared for the California Department of Transportation, Oakland, CA. October 18, 1990.

H.T. Harvey & Associates. 1992. Final Approved Route 85-Coyote Creek Mitigation Project Site Performance Monitoring Plan. File 449-09. H.T. Harvey & Associates.

H.T. Harvey & Associates. 1993. Route 85-Coyote Creek Mitigation Site 1993 Annual Monitoring Report. Project No. 449-15. H.T. Harvey & Associates.

NORTH CAROLINA WETLANDS RESTORATION PROGRAM

The North Carolina Wetlands Restoration Program (NCWRP) was established by the North Carolina General Assembly in 1996. The purpose of the NCWRP is to restore, enhance, preserve, and create wetlands, streams, and riparian areas throughout the state's 17 major river basins. The goals of the program are to restore functions and values lost through historic, current, and future wetland impacts; to achieve a net increase in wetland acres, functions, and values in all of North Carolina's major river basins; to provide a consistent and simplified approach to address mitigation requirements associated with permits or authorizations issued by the U.S. Army Corp of Engineers (Corps); and to increase the ecological effectiveness of required wetland mitigation and promote a comprehensive approach to the protection of natural resources.

The NCWRP established that all "compensatory mitigation" in North Carolina required as a condition of a Section 404 permit or authorization issued by the Corps be coordinated by the North Carolina Department of Environment and Natural Resources (DENR) consistent with basinwide plans for wetland restoration and rules developed by the North Carolina Environmental Management Commission (EMC). All compensatory wetland mitigation, whether performed by DENR or by permit applicants, shall be consistent with basinwide restoration plans. The emphasis of mitigation is expressly on replacing targeted functions in the same river basin (but not necessarily on in-kind or on-site mitigation) unless it can be demonstrated that restoration of other areas outside the impacted river basin would be more beneficial to the overall purposes of the wetlands restoration program.

Development and implementation of basinwide wetland and riparian restoration plans for each of the state's 17 river basins was a statutory mandate of the program. A key component of the basinwide approach is development of local watershed plans (LWPs) to protect and enhance water quality, flood prevention, fisheries, wildlife habitat, and recreational opportunities in each of the 17 river basins. LWPs are developed cooperatively with representatives of local governments, nonprofit organizations, and local communities. They provide an opportunity for local stakeholders, including residents, community groups, businesses, and industry, to play a role in shaping the future of their watershed. The NCWRP then utilizes the Basinwide Wetlands and Riparian Restoration Plans to target and prioritize degraded wetland and riparian areas, which, if restored, could contribute significantly to the goal of protecting and enhancing watershed functions.

The rational for focusing NCWRP restoration resources in priority watersheds was based on two assumptions. First, it was assumed that, although most watersheds in the state could benefit from wetlands and riparian area restoration, restoration may be more effective, efficient, and feasible in certain watersheds. Second, some watersheds need restoration sooner than others in order to either preserve their threatened natural resources or improve their degraded status before it becomes too late to make a difference. Prioritizing watersheds based on restoration feasibility and the critical nature of restoration needs helps to ensure that resources are used in the most efficient manner to maximize achievement of program goals.

An applicant may satisfy compensatory wetland mitigation requirements by the following actions, provided those actions are consistent with the basinwide restoration plans and also meet or exceed the requirements of the Corps: payment of a fee established by DENR into the Wetlands Restoration Fund (WRF); donation of land to the Wetlands Restoration Program (WRP) or to other public or private nonprofit conservation organizations as approved by DENR; participation in a private wetland mitigation bank; and preparing and implementing a wetland restoration plan.

The WRF was established as a nonreverting fund within DENR and was seeded with a $6 million appropriation from the Clean Water Management Trust Fund and $2.5 million annually from the North Carolina Department of Transportation for a period of 7 years for the development of LWPs. The WRF provides a repository for monetary contributions and donations or dedications of interests in real property to promote projects for the restoration, enhancement, preservation, or creation of wetlands and riparian areas and for payments made in lieu of compensatory mitigation. Funds expended from the WRF for any purpose must be in accordance with the basinwide plan and contribute directly to the acquisition, perpetual maintenance, enhancement, restoration, or creation of wetlands and riparian areas, including the cost of restoration planning, long-term monitoring, and maintenance of restored areas. Monetary fees to the WRF are established by the EMC on a standardized schedule on a per-acre basis based on ecological functions and values of wetlands permitted to be lost.

The DENR must report each year by November 1 to the Environmental Review Commission regarding its progress in implementing the WRP and its use of monies in the WRF. The report must document statewide wetland losses and gains and compensatory mitigation. The report must also provide an accounting of receipts and disbursements of the WRF, an analysis of the per-acre cost of wetlands restoration, and a cost comparison on a per-acre basis between the NCWRP and private mitigation banks.

The Basinwide Wetlands and Riparian Restoration Plan contains the following information: a statement of the restoration goals for each river basin; a map of each priority subbasin showing water-quality information, watershed boundaries, and land cover by type (agricultural, forested, or developed); a narrative overview of the river basin, including general information on existing water-quality-related problems; summary information on natural resources; descriptions of each priority subbasin; and data on wetland impacts.

The NCWRP completed all 17 basinwide wetlands and riparian restoration plans in 1998. Update plans are scheduled to be reviewed and revised in accordance with the Division of Water Quality's 5-year revision schedule for basinwide water-quality plans. LWPs are currently being developed. Since execution of the first contract in 1998, the NCWRP has 37 restoration projects in different stages of development that together will restore an estimated 61 ac of wetlands and 99,637 linear feet of stream and 219 ac of streamside buffers. The NCWRP is in its infancy, and implementation of projects is relatively recent. Little success of restoration targeted wetland functions has yet to be quantified on any project; however, development of basinwide wetland and riparian restoration plans, local watershed plans, and implementation of restoration projects is proceeding on schedule.

Appendix C

Analyses of Soil, Plant, and Animal Communities for Mitigation Sites Compared with Reference Sites

Trajectories for restoration in various mitigation studies are shown in comparison to conditions in reference marshes. The ">" and "<" signs mean that the equilibrium takes more time or less time, respectively, than the age of the mitigation site (years) when the survey was conducted or the data were modeled to project an age.

TABLE C-1 Analysis of Soil, Plant, and Animal Communities for Mitigation Sites Compared with Reference Sites

Component	Location	Sites	Years	Source/Notes
1. Soils				
Organic matter/% carbon	California	1	>22, if ever	Zedler and Callaway (1999); salt marsh
	Louisiana	30	>20	Turner et al. (1994); backfilled marsh
	North Carolina	7	>17	Sacco et al. (1994); planted salt marsh
	Illinois	2	>7	Mitsch and Flanagan (1996); fresh marsh
	Oregon	1	>5	Gwin et al. (1990); fresh marshes
	Pennsylvania	44	?	Bishel-Machung et al. (1996); fresh marsh
	Metanalysis	19	>>10	Streever (2000); coastal dredged sites
	South Carolina	2	>3	LaSalle et al. (1991); salt marsh
Macroorganic concentration	North Carolina	1	>3	Moy and Levin (1991); planted salt marsh from uplands
	North Carolina	5-7	>25	Craft et al. (1988); Craft (2000)
Carbon, nitrogen, and	California	1	40+	Zedler and Callaway (1999); salt marsh
phosphorus concentration	North Carolina	5	>30	Craft et al. (1988); salt marsh
	Illinois	2	>7	Mitsch and Flanagan (1996); fresh marsh
Exchangeable ions	South Carolina	2	>3	LaSalle et al. (1991); salt marsh
Grain size	North Carolina	1	>3	Moy and Levin (1991); planted salt marsh from uplands
Nutrient cycling	Texas	3	>17	Montagna (1993)
Sulfide and nitrogen	California	1	>15	Zedler (1990)
Average of six soil indices	California	4	>5	Zedler and Langis (1991); salt marsh
Nutrient exchange	North Carolina	1	>5	Craft et al. (1991); salt marsh
2. Plants - trees	Ohio	10	>50	Niswander and Mitsch (1995); riparian wetland, simulation model
Plant cover	Atlantic and Gulf	68	5-7	Matthews and Minello (1994); literature review
	Louisiana	30	>20	Turner et al. (1994); backfilled marsh
	Oregon	1	3	Frenkel and Morlan (1990); hay farm restoration
Height and biomass	California	4	>5	Zedler and Langis (1991); salt marsh
	California	1	>11, if ever	Zedler and Callaway (1999); salt marsh
	North Carolina	1	5	Broome et al. (1986)

Species richness/cover	Connecticut	5	>10	Confer and Niering (1992); freshwater marshes
	Illinois	2	>7	Mitsch and Flanagan (1996); fresh marsh
Native plant species	Portland, Oregon	51	>5	Magee et al. (1999); fresh marshes
Ratio above- to below-ground biomass	North Carolina	1	10	Broome et al. (1986)
Below-ground biomass	Texas	14	10	Shafer and Streever (2000); salt marsh
Three taxa	Metanalysis	12-14	>10	Streever (2000); coastal dredge sites
3. Fish and fisheries				
Finfish number, biomass	California	1	>2	Chamberlain and Barnhart (1993); salt marsh
Finfish number	California	1	5	Zedler (1990)
Fish species number	Florida	21	10	Roberts (1991)
Fish and shrimp	North Carolina	Review	<3	Fonseca et al. (1990); sea grasses
Marsh resident fish biomass	North Carolina	3	>3	Minello (2000)
Marsh fisheries species	Texas		>6	Rulifson (1991); salt marsh
Biomass and number of finfish and shrimp	Florida	1	>2	Moy and Levin (1991); planted salt marsh from uplands
4. Marsh invertebrates				
Marsh infauna number	North Carolina	1	>3	Moy and Levin (1991); planted salt marsh from re-graded uplands
Marsh infauna biomass	California	4	>5	Zedler and Langis (1991); salt marsh
Marsh infauna biomass	North Carolina	1	2	Cammen (1976); salt marsh
Marsh infauna biomass	North Carolina	1	1	Cammen (1976); salt marsh
Marsh infauna biomass	North Carolina	7	>25	Craft (2000)
Marsh infauna species, and species proportions	North Carolina	7	<17	Sacco et al. (1994); planted salt marsh
Marsh infauna species number and biomass	California	4	>5	Zedler and Langis (1991); salt marsh
	North Carolina	7	>17	Sacco et al. (1994); fresh marsh

continued

TABLE C-1 Continued

Component	Location	Sites	Years	Source/Notes
Marsh infauna biomass, and number	Florida	1	>2	Vose and Bell (1994); salt marsh impoundment
Larval dipterans	Florida	10	<11	Streever et al. (1996); fresh marsh
Macrobenthos	South Carolina	2	4-8	LaSalle et al. (1991); salt marsh
Epibenthos	Washington	1	>5	Simenstad and Thom (1996); salt marsh
Fish abundance	Metanalysis	11	5?	Streever (2000)
Total crustaceans abundance	Metanalysis	9	>12	Streever (2000)
5. Birds				
Endangered species	California	1	>15	Zedler (1990)
Bird species number	Florida	21	<10	Roberts (1991)
Waterfowl	Iowa	30	<3	van Rees-Siewart and Dinsmore (1996); fresh marsh
Natural assemblages	Iowa	30	>5	van Rees-Siewart and Dinsmore (1996); fresh marsh
	Portland, Oregon	51	>5	Magee et al. (1999); fresh marsh
	Metanalysis	NA	?	Streever (2000); dredged marshes
	Texas	7	>13	Melvin and Webb (1998); dredged marshes

References

Bishel-Machung, L., R.P. Brooks, S.S. Yates, and K.L. Hoover. 1996. Soil properties of reference wetlands and wetland creation projects in Pennsylvania. Wetlands 16(4):532-541.

Broome, S.W., E.D. Seneca, and W.W. Woodhouse, Jr. 1986. Long-term growth and development of transplants of the salt-marsh grass Spartina alterniflora. Estuaries 9:63-74.

Cammen, L.M. 1976. Abundance and production of macroinvertebrates from natural and artificially established salt marshes in North Carolina. Amer. Midl. Nat. 96(2):487-493.

Chamberlain, R.H., and R.A. Barnhart. 1993. Early use by fish of a mitigation salt Marsh, Humbolt Bay, California. Estuaries 16(4):769-783.

Confer, S.R., and W.A. Niering. 1992. Comparison of created and natural freshwater emergent wetlands in Connecticut (USA). Wetlands Ecol. Manage. 2(3):143-156.

Craft, C. 2000. Co-development of wetland soils and benthic invertebrate communities following salt marsh creation. Wetlands Ecol. Manage. 8(2/3):197-207.

Craft, C.B., S.W. Broome, and E.D. Seneca. 1988. Nitrogen, phosphorus and organic carbon pools in natural and transplanted marsh soils. Estuaries 11(4):272-289.

Craft, C.B., E.D. Seneca, and S.W. Broome. 1991. Porewater chemistry of natural and created marsh soils. J. Exp. Mar. Biol. Ecol. 152(2):187-200.

Fonseca, M.S., W.J. Kenworth, D.R. Colby, K.A. Rittmaster, and G.W. Thayer. 1990. Comparisons of fauna among natural and transplanted eelgrass Zostera marina meadows: criteria for mitigation. Mar. Ecol. Prog. Ser. 65 (3):251-264.

Frenkel, R.W., and J.C. Morlan. 1990. Restoration of the Salmon River Salt Marshes: Retrospect and Prospect. Corvallis, OR: Department of Geosciences, Oregon State University.

Gwin, S.E., M.E. Kentula, and E.M. Preston. 1990. Evaluating Design and Verifying Compliance of Wetlands Created Under Section 404 of the Clean Water Act in Oregon. EPA/600/3-90/061. Environmental Research Laboratory, Office of Research and Development, U.S. Environmental Protection Agency, Cornvallis, OR. 122 pp.

LaSalle, W.M., M.C. Landin, and J.G. Sims. 1991. Evaluation of the flora and fauna of Spartina alternifora marsh established on dredged material in Winhay Bay, South Carolina. Wetlands 11(2):191-208.

Magee, T.K., T.L. Ernst, M.E. Kentula, and K.A. Dwire 1999. Floristic comparison of freshwater wetlands in an urbanizing environment. Wetlands 19(3):517-534.

Matthews, G.A., and T.J. Minello. 1994. Technology and Success in Restoration, Creation, and Enhancement of Spartina alterniflora Marshes in the United States. Vol. 2. Inventory and Human Resources Directory. NOAA Coastal Ocean Program Decision Analysis Series No. 2. Silver Spring, MD: U. S. Dept. Commerce, National Oceanic and Atmospheric Administration.

Melvin, S.L., and J.W. Webb. 1998. Differences in the avian communities of natural and created Spartina alterniflora salt marshes. Wetlands 18(1):59-69.

Minello, T.J. 2000. Temporal development of salt marsh value for nekton and epifauna: utilization of dredged material marshes in Galveston Bay, Texas, USA. Wetlands Ecol. Manage. 8(5):327-341.

Minello, T.J., J.R. Zimmerman, and E.F. Klima. 1987. Creation of fishery habitat in estuaries. Pp. 106-120 in Beneficial Uses of Dredged Materials, Proceedings of First Interagency Workshop, 7-9 October 1986, Pensacola, Florida, M.C. Landin and H.K. Smith, eds. Tech. Report D-87-1. Vicksburg, MS: U.S. Army Corps of Engineers.

Mitsch, W.J., and N. Flanagan. 1996. Comparison of Structure and Function of Constructed Deepwater Marshes with Reference Freshwater Marshes. A study at the Des Plaines River Wetland Demonstration Project in northeastern Illinois. RF Project No. 729179, The Ohio State University Research Foundation, Columbus, OH.

Montagna, P.A. 1993. Comparison of Ecosystem Structure and Function of Created and Natural Seagrass Habitats in Laguna Madre, Texas. Final Report. Tech. Report No. TR/93-007. Port Aransas, TX: University of Texas at Austin, Marine Science Institute.

Moy, L.D., and L.A. Levin. 1991. Are Spartina marshes a replaceable resources? A functional approach to evaluation of marsh creation efforts. Estuaries 14(1):1-16.

Niswander, S.F., and W.J. Mitsch. 1995. Functional analysis of a two-year-old created in-stream wetland: hydrology, phosphorus retention, and vegetation survival and growth. Wetlands 15(3):212-225.

Roberts, T.H. 1991. Habitat Value of Man-Made Coastal Marshes in Florida. Technical Report WRP-RE-2. Vicksburg, MS: U.S. Army Engineer Waterways Experiment Station.

Rulifson, R.A. 1991. Finfish utilization of man-initiated and adjacent natural creeks of South Creek Estuary, North Carolina using multiple gear types. Estuaries 14(4):447-464.

Sacco, J.N., E.D. Seneca, and T.R. Wentworth 1994. Infaunal community development of artificially established salt marshes in North Carolina. Estuaries 17(2):489-500.

Shafer, D.J., and W.J. Streever 2000. A comparison of 28 natural and dredged material salt marshes in Texas with an emphasis on geomorphological variables. Wetlands Ecol. Manage. 8(5):353-366.

Simenstad, C.A., and R.M. Thom. 1996. Functional equivalency trajectories of the restored Gog-Le-Hi-Te estuarine wetland. Ecol. Applic. 6(1):38-57.

Streever, W.J. 2000. Spartina alterniflora marshes on dredged material: a critical review of the ongoing debate over success. Wetlands Ecol. Manage. 8(5):295-316.

Streever, W.J., K.M. Portier, and T.L. Crisman. 1996. A comparison of Dipterans from ten created and ten natural wetlands. Wetlands 16(4):416-428.

Turner, R.E., J.M. Lee, and C. Neill. 1994. Backfilling canals to restore wetland: empirical results in coastal Louisiana. Wetlands Ecol. Manage. 3(1):63-78.

van Rees-Siewert, K.L., and J.J. Dinsmore. 1996. Influence of wetland age on bird use of restored wetlands in Iowa. Wetlands 16(4):577-582.

Vose, F.E., and S.S. Bell. 1994. Resident fishes and macrobenthos in mangrove-rimmed habitats: evaluation of habitat restoration by hydrologic modification. Estuaries 17(3):585-596.

Zedler, J.B. 1990. A Manual for Assessing Restored and Natural Coastal Wetlands with Examples from Southern California. Report. No. T-CSGCP-021. La Jolla: California Sea Grant College.

Zedler, J.B., and J.C. Callaway. 1999. Tracking wetland restoration: do mitigation sites follow desired trajectories? Restor. Ecol. 7(1):69-73.

Zedler, J.B., and R. Langis. 1991. Authenticity: comparisons of constructed and natural salt marshes of San Diego Bay. Restor. Manage. Notes 9(1):21-25.

Appendix D

California Department of Fish and Game, South Coast Region; Guidelines for Wetland Mitigation

Contributed by William E. Tippets,
Habitat Conservation Supervisor, California Department of
Fish and Game, South Coast Region

These ratios should be considered as general guidelines for mitigation for impacts to streams and associated habitat.

1:1	Low-value habitat	E.g., isolated freshwater marsh, unvegetated streams.
2:1	Medium-value habitat	E.g., disturbed mulefat scrub, highly disturbed willow riparian.
3:1	High-value habitat	E.g., willow riparian, possibly with some exotics, rare/unique habitats.
5:1	Endangered species habitat	E.g., mature willow riparian with least Bell's vireo.
5:1	Impacts beyond permitted in the SAA/violations	This can vary, depending on the quality, temporal loss, location, etc., but should have a compensatory factor in addition to the above guidelines of 1:1 to 5:1.

Other considerations:

It is important to consider "no net loss to wetlands." Streams should be considered under this no-net-loss policy to ensure that adequate creation is represented (rule of thumb, a minimum of 1:1 of their mitigation for permanent impacts should include creation). Creation, restoration, and/or enhancement could make up the balance of the mitigation measures. Preservation is usually looked at as a recommended avoidance measure, but preservation and protection of significant wetlands can be part of the entire project's measures to be considered. Instances involving lower-quality habitat impacts may be mitigated by nonnative exotic plant removal.

Freshwater marsh restores more successfully than a multilayered willow riparian community, which may have a significant temporal loss component in its mitigation requirements.

Temporary impacts should preferably be restored on-site and should account for mitigation for the temporal loss.

The above ratios consider acreage of impact. Individual tree ratios/ requirements can be incorporated as part of the plan to ensure sufficient mitigation. Also, guidelines for impacts to individual mature oak and sycamore trees are mitigated based on the size of tree impacted, at appropriate planting centers and with appropriate native understory. This may require the applicant to obtain additional land beyond that required for the habitat acreage requirement as described above.

Appendix E

Examples of Performance Standards for Wetland Creation and Restoration in Section 404 Permits and an Approach to Developing Performance Standards

WRP Technical Note WG-RS-3.3
January 1999

Examples of Performance Standards for Wetland Creation and Restoration in Section 404 Permits and an Approach to Developing Performance Standards

PURPOSE: This technical note accomplishes the following: a) defines performance standards for wetland creation and restoration, b) provides 20 example performance standards for wetland creation and restoration projects required by Section 404 permits, c) summarizes seven sets of performance standard guidelines used by Corps of Engineers Districts and one set of guidelines under development, and d) outlines an approach to developing new performance standards or revising existing performance standards.

PERFORMANCE STANDARDS DEFINED: Under Section 404 of the Clean Water Act of 1977, wetland creation and restoration can be required as compensatory mitigation for unavoidable wetland loss. Performance standards, in the context of this technical note, are **observable or measurable attributes** that can be used to determine if a compensatory mitigation project meets its objectives. Performance standards are frequently called "success criteria" but may also be known by other names, such as "success standards" or "release criteria."

Individual Section 404 permits provide both general and special conditions regarding permitted activities. **General conditions** include standardized information relevant to all permitted projects, such as time limits for completion of permitted activities, requirements to report historic or archaeological remains found in the course of permitted activities, and requirements to allow inspection of permitted projects by U.S. Army Corps of Engineers representatives. **Special conditions** include additional information pertinent to specific projects or regions, such as refueling procedures for equipment, safety requirements, sediment control requirements, and seasonal timing of permitted activities. In permits that require restoration or creation of wetlands as compensatory mitigation, performance standards should be included as special conditions.[1]

WHY PERFORMANCE STANDARDS ARE IMPORTANT: Performance standards allow the Corps of Engineers to determine if the objectives of compensatory mitigation required by a Section

---

[1] "Army regulations authorize mitigation requirements to be added as special conditions to an Army permit. . ."–Memorandum of Agreement between the Environmental Protection Agency and the Department of the Army Concerning the Determination of Mitigation Under the Clean Water Act Section 404(b)(1) Guidelines, 1990.

WRP Technical Note WG-RS-3.3
January 1999

404 permit have been successfully fulfilled. Performance standards should generally reflect Corps of Engineers guidelines calling for a minimum of "one for one functional replacement"[1] of wetlands unavoidably impacted by permitted activities. Performance standards also facilitate enforcement actions for projects that fail to comply with Section 404 permit conditions.

PERFORMANCE STANDARDS AND FUNCTIONAL REPLACEMENT: In recent years, a large literature has developed that offers post hoc assessment of compensatory mitigation wetlands. Most post hoc studies compare created or restored wetlands to nearby natural reference wetlands on the basis of a number of attributes, such as vegetation community composition, benthic invertebrate community composition, and water quality. This literature suggests that many wetlands created and restored as compensatory mitigation do not replace the structure and functions of lost natural wetlands. Although many authors have offered opinions regarding the cause of poor structural and functional replacement, few authors have attempted to relate performance standards required by permits with results of post hoc studies comparing compensatory mitigation wetlands and natural reference wetlands. There is a clear need for studies designed to link performance standards required by permits with the ability of created or restored wetlands to replace lost wetland structure and functions.

EXAMPLES FROM PERMITS: Table 1 summarizes performance standards from Section 404 permits and mitigation plans referenced by permits. Examples were compiled by reviewing permit files available at Corps of Engineers District offices and requesting copies of permit files from District offices. Over 300 permits were reviewed to compile examples for Table 1; however, the table represents selected examples rather than a comprehensive summary of Section 404 permit performance standards.

Many permits that required compensatory mitigation did not include performance standards. In some permits, items designated as "performance standards" or "success criteria" did not meet the definition of performance standards used in this technical note; for example, instructions regarding planting techniques were frequently called performance standards. No attempt was made to comprehensively review or representatively sample all Section 404 permits, so no conclusions can be drawn regarding the number of permits issued without performance standards.

Table 1 shows that there are no universally used performance standards for compensatory mitigation. Even within Districts, performance standards may vary from permit to permit. The absence of universal performance standards probably reflects the ongoing evolution of the Section 404 regulatory process as well as differences in regional or site-specific ecological conditions and regional needs.

At least seven distinct approaches can be identified from the examples in Table 1. Most examples combine two or more of these approaches. These approaches include:

---

[1] As per the Memorandum of Agreement between the Environmental Protection Agency and the Department of the Army Concerning the Determination of Mitigation Under the Clean Water Act, Section 404(b)(1) Guidelines, 1990.

Table 1. Summary of Performance Standards from Selected Section 404 Permits Requiring Compensatory Mitigation[1]

Example Number	Performance Standards	Time Frame	Location/ Type/ Year	Size
1	50% survival of planted trees, including replanting efforts, after two growing seasons	3 years, after which natural regeneration is relied upon	Mississippi/ bottomland hardwoods/ 1997	Restoration of 2.17 acres
2	75% survival of planted *Juncus roemerianus*; 4,800 plants per acre after 3 growing seasons	3-year minimum, with 75% survival for 2 years following any replanting	Alabama/ salt marsh/ 1985	Creation of 40 acres
3	75% site survival, defined as [(number of "planting cells" with "species survival" over 35% ÷ total number of planting cells) x 100]; species survival is the [(number of surviving plants in each "planting cell" ÷ number of plants originally planted in the "planting cell") x 100]; the "planting cell" is a discrete cluster of plants as illustrated on the planting or landscaping plan, or, if planting is not in discrete clusters, the cell is the entire site; after 3 years, site will be 80% vegetated with hydrophytic vegetation having an indicator status of FAC or wetter, excluding *Typha* spp. and *Myriophyllum spicatum*, and with less than 5% cover by 28 noxious or invasive species (noxious and exotic species are listed in permit)	3 years following completion of construction	Massachusetts/ cranberry bog and shrub swamp/ 1998	Creation of 2.8 acres and enhancement of 1.1 acres
4	85% of the site vegetated by the planted species and/or naturally regenerated vegetation approved by regulatory agencies	5-year endpoint	Maryland/ forested wetland/ 1996	Restoration of 850 linear feet of stream banks
5	80% wetland vegetation cover in herbaceous wetlands and 80% survival of planted stock in scrub-shrub wetlands, as measured using an approved method	Not specified	Idaho/ herbaceous and scrub-shrub wetlands/ 1995	Creation of 8 acres
6	Sustain 85% or greater cover by obligate and/or facultative wetland plant species; less than 10% cover by nuisance plant species; "proper hydrological condition"	5 years, with requirement for contingency plan after 3 years if performance standards are not achieved and requirement for ongoing monitoring after 5 years if performance standards are not met	Florida/ forested and herbaceous wetlands/ 1991	Creation of 11.8 acres forested wetlands and 10.1 acres herbaceous wetlands
7	85% areal cover by planted herbaceous species and 75% areal cover by planted woody species; specifically prohibits open water ponds	2 years, with provision for replanting if areal cover requirements are not achieved	Maryland/ forested and emergent freshwater wetland/ 1990	Creation of 5.09 acres palustrine forested wetlands and 0.66 acre palustrine emergent and scrub-shrub wetlands

[1] Projects were selected to offer examples of a range of performance standards required by Section 404 permits. Abbreviations FAC, FACW, and OBL and the terms "facultative" and "obligate" refer to the *National List of Plant Species that Occur in Wetlands*. Throughout this table, performance standards were paraphrased directly from permit files; no attempt was made to clarify language used in permit files.

(Sheet 1 of 4)

WRP Technical Note WG-RS-3.3
January 1999

Table 1. (Continued)				
Example Number	Performance Standards	Time Frame	Location/ Type/ Year	Size
8	Hydrology must meet wetland definition of 1987 Corps of Engineers Wetland Manual, with saturation to the surface of the soil for 12.5% (31 days) of the growing season; at least 50% of woody vegetation must be FAC or wetter, with woody vegetation stem counts of 400 per acre or canopy cover of 30% or greater by woody vegetation; at least 50% of all herbaceous vegetation must be FAC or wetter with aerial cover of at least 50% in emergent wetland areas (exclusive of "shrub/scrub or sapling/forest vegetation")	5 years	Virginia/ forested wetland/ 1995	Restoration of 8.5 acres and creation of 1.7 acres on-site; restoration of 17.2 acres off-site
9	Herbaceous zones will have 80% cover with 50% or more cover by species listed as FAC or wetter, with plants rooted for at least 12 months, with plants showing natural reproduction, and with no species other than sawgrass constituting more than 30% cover; forested zones to have a minimum density of 400 live trees per acre with natural reproduction and at least 50% cover by species listed FAC or wetter with no one species contributing greater than 30% of the species represented; cattail, primrose willow, Brazilian pepper, punk trees, Australian pine, and other exotic vegetation limited to 10% or less of total cover; muck layer in "Area C" must average at least 6 in. in depth at the end of 25 years; all conditions must be met without intervention in the form or irrigation, planting, or plant removal for 3 consecutive years in herbaceous wetlands and 5 consecutive years in forested wetlands	At least 3 years for herbaceous wetlands, at least 5 years for forested wetlands, and up to 25 years for development of muck	Florida/ herbaceous and forested wetlands/ 1998	Creation of 1,441 acres herbaceous wetlands, 145 acres forested wetlands, 40 acres "deep muck pockets," and 68 acres open water
10	Emergent and aquatic bed portions of mitigation site not to be inundated with salt or brackish water; less than 10% cover by invasive species during any monitoring event; staged vegetation requirements as follows: Year 1: 100% survival of planted stock, 50% cover in emergent areas Year 2: 80% survival by planted stock, 20% cover by native shrub species, 70% cover in emergent areas Year 3: 70% survival and 40% cover by native shrub species, 80% cover in emergent areas Year 5: 60% cover by native shrub species, 100% cover in emergent areas	5 years	Washington/ emergent, scrub-shrub, and forested wetland/ 1998	Enhancement of 1.12 acres scrub-shrub wetlands and 3 acres emergent wetlands; creation of 0.46 acre scrub-shrub, 0.4 acre forested, and 4.42 acres emergent wetlands
11	80% survival of planted stock each year; at least 50% native perennials by end of year 5; staged vegetation percent cover requirements for wet-mesic meadow / shallow marsh / "no planting zone" (used to experimentally assess natural recruitment) as follows: Year 1: 15% / 10% / no requirement Year 2: 30% / 20% / 20% Year 3: 45% / 30% / 30% Year 4: 60% / 40% / 40% Year 5: 75% / 50% / 50%	8 years	Illinois/ emergent wetland/ 1995	Enhancement of 1.47 acres and creation of 30.68 acres wet-mesic meadow and shallow marsh
12	Less than 5% cover by nuisance and exotic plant species; planted and non-nuisance wetland plant species to have areal cover of 50% in first year, 70% in second year, and 80% in third year, with provisions for remedial planting to meet percentage requirements	5 years, with requirement for ongoing monitoring if percentage requirements are not met	Florida/ freshwater marsh and wet prairie/ 1990	Creation of 10 acres freshwater marsh and wet prairie with additional enhancement and preservation of cypress domes and other wetlands
				(Sheet 2 of 4)

WRP Technical Note WG-RS-3.3
January 1999

Table 1. (Continued)

Example Number	Performance Standards	Time Frame	Location/ Type/ Year	Size
13	Permanently vegetated stand over 85% of disturbed area after first growing season (replacement of dead plants required); documentation of saturated soil; documentation of tidal hydrology; no *Phragmites* infestation; documentation of "animal use" for portion of site	5 years	New Jersey/ salt marsh/ 1990	Creation of 4.2 acres *Spartina alterniflora* marsh and 24 acres open water and intertidal wetland
14	Must meet the regulatory definition of wetlands, and water within the mitigation area should function "as the intended type of water of the United States"	Indefinite (active until performance standards are met and verified by Corps of Engineers)	Texas/ emergent and open water/ 1997	Creation and preservation of 54 acres emergent wetland and 145 acres open water
15	Must meet the regulatory definition of wetlands; specified portions of the mitigation area must meet the definitions of palustrine forested, palustrine scrub-shrub, and palustrine emergent wetland types as per the document *Classification of Wetlands and Deepwater Habitats of the United States*; cover by hydrophytic plants ("those with a regional indicator status of FAC, FAC+, FACW+/-, or OBL"); vegetation not to consist of more than 10% areal cover by any combination of *Phragmites australis* (common reed) or *Lythrum salicaria* (purple loosestrife); all performance standards must be met for 3 consecutive years	5 years, to be extended as necessary to fulfil the requirement of meeting all performance standards for 3 consecutive years	New York/ scrub-shrub, and emergent palustrine wetlands/ 1998	Creation of 12.9 acres and enhancement of 12.13 acres
16	No rills or gullies greater than 12 in. deep; no single plant species from the seeding mixture may constitute more than 50% of species found in the site; two or more native species present; vegetative cover equal to 75% of test plot cover (test plots are plots established at numerous locations to determine viability of plant community development)	5 years, with provisions for early release	Alaska/ emergent wetlands/ 1998	Restoration of up to 261 acres, as needed to restore impacts from gold mining
17	No less than 33% of natural stem densities found in adjacent areas	1 year	Alaska/ emergent wetlands/ 1997	Restoration of up to 19 acres, as needed to restore "exposed earthworks" resulting from construction
18	Areal cover in 90% of planted area equivalent to natural reference marsh; benthic invertebrates and fish with 75% similarity to natural reference marsh, and fish with 75% biomass of fish in natural reference marsh; upper soil horizon with 1% organic matter by dry weight	5 years, after which additional mitigation acreage is required	Alabama/ salt marsh/ 1988	Creation of 25.3 acres
19	Vernal Pool Habitat Suitability Index (VPFI) ≥ 0.55 with 60% of pools > 0.7 [VPFI = a $\div$ (a + b), where a = number of species the pool and the "vernal pool species list" share, and b = number of species in the pool not on the "vernal pool species list;" the list includes those species typically found in the region's vernal pools]; hydrology assessed as suitable on the basis of presence of wetland plants	4 years, with requests for extensions to be given favorable consideration	California/ vernal pools/ 1996	Creation of 27 acres

(Sheet 3 of 4)

WRP Technical Note WG-RS-3.3
January 1999

Table 1. (Concluded)				
Example Number	**Performance Standards**	**Time Frame**	**Location/ Type/ Year**	**Size**
20	The combined relative cover of targeted exotic species, including *Senecio mikanioides* (German ivy) and *Vinca major* (periwinkle), will be less than 5% after 5 years; visual observations of inundation, soil saturation within 12 in. of the soil surface, water marks, drift lines, sediment deposits, and drainage patterns will indicate that the site is as wet or wetter than a nearby reference site; over time, there will be an increase in the numbers and kinds of riparian obligate bird species relative to the numbers and kinds of generalist bird species; 0.23 stems of woody vegetation m^{-2} unless deviation from this density appears to be caused by natural phenomena, the results of which are also apparent at a reference site; 75% cover by native riparian scrub species including herbaceous and shrub strata; evidence of natural seedling recruitment; within 5 years, the mitigation wetland must show conditions similar to pre-impact conditions at the site to be impacted by permitted activities on the basis of narrative descriptions that characterize 14 variables described in the sixth draft *Model for the Santa Margarita River Watershed*— these variables, which are part of a hydrogeomorphic (HGM) approach to functional assessment of wetlands, include 1) V_{contig}, for contiguous vegetation cover, 2) V_{subin}, for subsurface flow into wetland, 3) V_{topo}, for topographic complexity, 4) V_{organ}, for soil organic matter, 5) V_{trees}, for abundance of trees, 6) V_{offsap}, for off-channel saplings, 7) $V_{offshrub}$, for off-channel shrubs, 8) V_{ratio}, for ratio of native to non-native vegetation, 9) V_{offcwd}, for off-channel coarse woody debris, 10) V_{offfwd}, for off-channel fine woody debris, 11) V_{decay}, for stage of decay of coarse wood, 12) $V_{offlitter}$, for off-channel leaf litter, 13) $V_{agedist}$, for stand age distribution, and 14) V_{arundo}, for presence of *Arundo donax* (requirements to meet variables are staged over 5 years to recognize improved function with time but only the 5-year requirements are presented here)	5 years	California/ floodplain wetland/ 1997	Restoration of 8.9 acres
				(Sheet 4 of 4)

a. Requirements for survival of planted stock (examples 1-3, 5, and 10-11).

b. Requirements for plant density or percent cover by plants (examples 2-13, 16-18, and 20).

c. Requirements that are staged over time so that different performance standards must be met as the wetland matures (examples 10-12 and 20).

d. Requirements that specifically reference documents developed for the purpose of wetland delineation, such as the 1987 *Corps of Engineers Wetlands Delineation Manual* (the "87 Manual") and U.S. Fish and Wildlife Service lists of wetland indicator status for plant species (examples 3, 6, 8, 9, 14, and 15).

e. Use of indices to compress large amounts of information (examples 3 and 18-20).

WRP Technical Note WG-RS-3.3
January 1999

f. Reliance on natural reference wetlands (sometimes called "control" wetlands) or other sites as a benchmark (examples 16-20).

g. Requirements specifically limiting occurrence of exotic and nuisance plant species (examples 3, 6, 9, 10, 12, 13, 15, and 20).

All examples explicitly consider vegetation. Some examples explicitly consider vertebrate and invertebrate abundances and diversity, soil characteristics, and hydrological conditions.

SUMMARIES OF PERFORMANCE STANDARD GUIDELINES: Performance standard guidelines were compiled from permitting guidelines provided by Corps of Engineers District offices. Summaries presented here focus on the portion of permitting guidelines dealing with performance standards. All but one example, the *Washington State Department of Transportation* guidelines, are in use at Army Corps of Engineers District offices. The *Washington State Department of Transportation* guidelines, which are still in draft form, are part of an effort undertaken by a committee of wetland professionals, including employees of the Army Corps of Engineers.

All seven of the approaches to performance standards described from examples in Table 1 also appear in performance standard guidelines. Several of the summarized guidelines elaborate on definitions of terms, such as "objective" and "performance standard." Similarly, several of the summarized guidelines elaborate on the need for unambiguous language within permits, including both the language used to describe performance standards and the language used to describe required methods for monitoring performance standards. Several guidelines also recognize a need for flexibility when writing performance standards.

- **St. Paul District's 1992 Guidelines.** *Compensatory Wetland Mitigation: Some Problems and Suggestions for Corrective Measures*, by Steve Eggers, was published by the U.S. Army Engineer District, St. Paul, in February 1992. This document, based in part on field inspections of 30 compensatory mitigation wetlands in Minnesota and Wisconsin, offers guidance on goals, design, construction, long-term protection, and monitoring, as well as performance standards. The report notes that "Lack of specific requirements for measuring the success of compensatory mitigation was one of the most notable deficiencies of past permits." The report also notes that up to 50 years may be necessary to determine success of some systems, but that this is not feasible for most projects, and that fair evaluation of performance standards for herbaceous wetlands may require less time than evaluation of performance standards for shrub or forested wetlands. Comparison to a reference wetland is advocated as a means of determining success of compensatory mitigation wetlands, as is use of performance standards with predetermined levels of vegetation cover, such as "80 percent survival of planted shrubs after 3 years, or 75 percent of the mitigation site must be vegetated by the end of the second growing season."

- **New England District's Guidelines.** The New England District's guidelines regarding Section 404 permit special conditions are given in an undated document entitled *New England District Staff Guidance for Mitigation Special Conditions*. The document includes suggestions regarding topics such as plant species that should be excluded from areas around

WRP Technical Note WG-RS-3.3
January 1999

compensatory mitigation sites and the use of conservation covenants. Several performance standards are listed, including the following:

a. Three-quarters of all cells at a site should have at least 35 percent survival of planted stock. ("Cells" and "survival" are defined in example 3 of Table 1.)

b. Areal cover of 80 percent, excluding open water areas, by noninvasive hydrophytes should occur by a specific date. Purple loosestrife (*Lythrum slicaria*), cattails (*Typha latifolia, Typha angustifolia,* and *Typha glauca*), common reed (*Phragmites australis*), and reed canary grass (*Phalaris arundinacea*) are listed as invasive species.

c. No unstablilized slopes should be present.

This document is periodically reviewed and revised based on experience and "lessons learned."

• **Norfolk District's 1995 Guidelines.** Norfolk District has a document dated 16 November 1995, entitled *Branch Guidance for Wetlands Compensation Permit Conditions and Performance Criteria,* that covers topics such as required information for site design plans, performance bonds, and requirements for hydrological data assessment before planting. The document stresses the need for flexibility: "This guidance is intended to be flexible; it is the decision of project managers and their supervisors whether any condition is appropriate for a particular wetland construction project." Point 6 of the document lists performance standards, or "performance criteria." These performance standards include:

a. Hydrology must meet the criteria for a wetland as per the *Corps of Engineers Wetlands Delineation Manual,* with growing season specified. The number of days with saturation to the soil surface should also be specified in order to allow some control over the wetland type that would develop on a site.

b. At least 50 percent of all plants must be facultative or wetter.

c. For woody vegetation, stem counts of 400 per acre must be achieved until canopy cover is 30 percent or greater.

d. In areas of emergent herbaceous vegetation, areal cover must be at least 50 percent.

• **Baltimore District's 1994 Guidelines.** Baltimore District's 1994 guidelines, entitled *Maryland Compensatory Mitigation Guidance,* were developed by the Interagency Mitigation Task Force, with representatives from eight state and Federal agencies. Guidelines include information about topics such as replacement ratios, site selection, monitoring reports, sampling methods, and performance standards. Different performance standards are given for tidal emergent wetlands, non-tidal emergent wetlands, non-tidal scrub-shrub wetlands, and non-tidal forested wetlands. For example, tidal emergent wetland performance standards include:

WRP Technical Note WG-RS-3.3
January 1999

a. Forty-five percent cover by emergent wetland species with a minimum stem density of 43,650 living stems per acre by the second growing season.

b. Seventy percent cover by emergent wetland species with a minimum stem density of 43,650 living stems per acre by the third growing season.

c. Eighty-five percent cover by emergent wetland species with a minimum stem density of 43,650 living stems per acre by the fifth growing season.

d. For regularly flooded compensatory mitigation wetlands (intended to support plant species such as *Spartina alterniflora, Scirpus robustus,* and *Peltandra virginica*), tides must alternately flood and expose the land surface at least once each day, while for irregularly flooded compensatory mitigation wetlands (intended to support species such as *Spartina patens, Iva frutescens, Juncus roemerianus,* and *Typha angustifolia*), tides should flood the land surface less often than once daily.

Emphasis on vegetation is justified because "sites without sufficient plant biomass support low populations of fish and wildlife and provide insignificant water quality functions. . . [and] techniques to measure vegetation are accomplished economically and require minimum training and equipment."

• **Seattle District's 1994 Guidelines for Freshwater Wetlands.** Seattle District's *Guidelines for Developing Freshwater Wetlands Mitigation Plans and Proposals,* dated March 1994, resulted from collaboration of six federal and state agencies. Guidelines include information on ecological assessment of impacted sites, wetland delineation, mitigation sequencing, monitoring, goals and objectives, and performance standards. The document clearly links objectives and performance standards by defining performance standards as "the measurable values of specific variables that establish when objectives have been met" and by stating that specific performance standards will depend on project objectives. Variables that might be considered for use as performance standards include dissolved oxygen, nutrient levels in water, survival rates of planted vegetation, species diversity, water flows, and water depths. The document also offers several specific examples that show how performance standards could be linked to objectives, two of which are transcribed verbatim here:

Objective c. The vegetated portions around the open water will have 3 acres each of emergent, scrub-shrub, and forested vegetation classes.

> **Performance Standard #1:** *The emergent vegetation will cover at least 3 acres of the wetland after five years, and the cover of native emergent species will be at least 80% in these 3 acres as measured by belt transects. The standard deviation of the mean cover value in the sampling quadrats will be less than 1/4 of the mean value (i.e. SD < (1/4 x 0.8); therefore SD < 0.2).*

> **Performance Standard #2:** *The scrub/shrub vegetation will cover at least 3 acres after five years with an 80% cover of native scrub shrub species in this area as*

WRP Technical Note WG-RS-3.3
January 1999

*measured by belt transects. The standard deviation of the mean cover value will
be less than 1/4 of the mean.*

Performance Standard #3: *The forest vegetation will cover at least 3 acres after
20 years with a canopy cover of at least 40% of native species in these 3 acres.*

Objective d. The area of open water will provide habitat for at least two species of
amphibians within five years.

Performance Standard: *The use of the wetland by two species [of] amphibians will
be documented by live trapping, and/or observation of egg masses during the
breeding season.*

- **Los Angeles District's Proposed Guidelines for Riparian Habitat.** Los Angeles District's
document *Special Public Notice; Proposed Riparian Habitat Mitigation and Monitoring
Guidelines,* distributed for comment between 15 August and 15 September 1997, includes
information on topics such as sequencing, site selection, identification of riparian habitat,
and compliance assurance. In part "e" of a section on mitigation design and planning, performance standards (called "success criteria" in this document) are briefly discussed, as transcribed verbatim below:

  e. Propose realistic success criteria based on the purpose of the mitigation, design of the
  site, and the variables and functions found in the HGM. Develop initial HGM scores
  for the mitigation site after the proposed grading based solely on physical characteristics. Estimate performance curves and time to establish partial and full success of the
  site based on HGM score. The Corps will be intimately involved with this aspect of
  the plan.

  HGM refers to the hydrogeomorphic approach to wetland assessment. The decision to
  use HGM in performance standards resulted from studies suggesting that compensatory
  wetlands could meet performance standards required by earlier guidelines even though
  they "were unsuccessful at restoration or creation of fully functional, riparian habitat."
  The philosophy behind the HGM approach is described in Smith et al. (1995).

- **Chicago District Mitigation Guidelines.** The *Chicago District Mitigation Guidelines and
  Requirements,* dated 30 April 1998, describes issues such as site selection, mitigation ratios,
  long-term management requirements, and enforcement. A section on performance standards
  for compensatory mitigation focuses on vegetation but also suggests that applicants should
  propose performance standards for other functions, such as improvement of water quality
  and provision of wildlife habitat. Use of existing measures, such as the Index of Biological
  Integrity, is encouraged. Vegetation performance standards include the following:

  a. The mean coefficient of conservatism must be greater than or equal to 3.5. Coefficient
  of conservatism values for plant species found in the Chicago District are designated
  in Swink and Wilhelm (1994). These values indicate the degree to which a plant species is representative of an undisturbed native community; a value of 0 is assigned to

WRP Technical Note WG-RS-3.3
January 1999

plants that occur almost exclusively in altered habitats, such as highway verges, while a value of 10 is assigned to plants that occur almost exclusively in remnant undisturbed habitats, such as some fens. Coefficients are not assigned for introduced species.

b. The native floristic quality index, described in Swink and Wilhelm (1994), must be greater than or equal to 20. The native floristic quality index is computed as $I = CN^{1/2}$, where I is the index value, C is the mean coefficient of conservatism value, and N is the number of native species.

c. The mean wetness coefficient (based on regional wetland indicator status) must indicate the presence of a wetland.

d. After 5 years, no area greater than 0.5 m^2 will be devoid of vegetation in areas intended to be vegetated, except in areas with emergent and aquatic communities.

e. After 5 years, the three most dominant species in wetland communities cannot be non-native or weedy. Non-native and weedy species include *Typha* spp., *Phragmites australis, Poa compressa, Poa pratensis, Lythrum salicaria, Salix interior, Echinochloa crusgalli,* and *Phalaris arundinacea.*

Performance standards are staged over time in that there are requirements for annual increases in native mean coefficient of conservatism values and native floristic quality index values.

- **Washington State Department of Transportation.** State, Federal, and private sector wetland professionals in Washington have been working together since May 1997 to "bring more clarity to the issues surrounding the use of success standards in wetland mitigation." A working draft of their suggestions has been published on the World Wide Web.[1] This document suggests that appropriate development of performance standards requires consideration of regulatory requirements, wetland functions, wetland construction methods, wetland monitoring methods, and expected or achievable quantitative values for monitored wetland attributes. Also, this document suggests that attempts to develop universally applicable performance standards are not appropriate because every project is unique. A number of terms are defined as part of this document, including "goal," "objective," "performance objective," and "success standard (or performance standard)." A goal is a broad statement about a project's intended outcomes, objectives are more specific statements about intended outcomes, performance objectives are the subset of objectives that will be considered in evaluating the project, and performance standards are observable or measurable attributes linked to performance objectives. For example, a goal might be restoration of 10 acres of scrub-shrub wetland. Objectives might include provision of floodflow attenuation and storage, food chain support, habitat for fish and amphibians, and water quality improvement. One performance objective related to the water quality improvement objective might be sediment retention.

---

[1] http://www.sws.org/regional/pacificNW/98meeting/Ossinger2.html#fnO

WRP Technical Note WG-RS-3.3
January 1999

The performance standard linked to this performance objective could be 90 percent cover by herbaceous vegetation, which, according to the technical literature, acts to some degree as a surrogate measure of sediment retention. Suggested potential performance standards include herbaceous plant cover, woody plant cover, survival of planted species, cover by invasive plant species, plant species diversity, slope, aquatic invertebrate diversity, presence of specific aquatic invertebrate taxa, presence of specific hydrological conditions, presence of specific soil conditions, and site use by specific wildlife taxa. Despite the long list of potential performance standards offered in this document, the authors recommend restraint in applying these and other standards: "DON'T GET CARRIED AWAY! Remember the purpose of stating performance objectives and success standards: you want to evaluate the success of your project. Usually it takes only a few performance objectives to adequately do this."

DEVELOPING OR REVISING PERFORMANCE STANDARDS: To streamline the Section 404 permitting process, regulatory staff should be provided with performance standard guidelines or templates listing minimum performance standards for various wetland types. While guidelines could help regulators prepare performance standards for permit special conditions, templates could be inserted directly into permit special conditions and be altered as needed to fit specific situations.

Ideally, performance standards should a) refer to practicably measurable or observable attributes that reflect compensatory mitigation objectives, and b) lead to compensatory mitigation that replaces the structure and functions of wetlands lost as the result of permitted activities. When research results linking performance standards with successful replacement of lost wetland structure and functions are not available, development or revision of performance standards relies on the opinions of wetland professionals involved with the regulatory process. The 12-step plan outlined below offers one means of generating performance standard guidelines or templates based on a consensus opinion of wetland professionals, including regulatory staff, scientists, and others.

A 12-step Plan

Step 1. Staff identifies the region for which performance standards are to be developed, recognizing that community needs and expectations–particularly in the sense of what might be considered "practicable"–will vary from region to region, as will ecological conditions. In some cases, the region will be defined by District boundaries.

Step 2. Staff identifies wetland types for which performance standards are to be developed. In general, it will be difficult or impossible to develop performance standards that could be applied to all wetland types.

Step 3. Staff identifies workshop participants and a coordinator. Workshop participants should include experienced Corps regulatory staff, representatives from other government agencies, and at least one person with extensive knowledge of wetland restoration research; consultants and others might also be invited to participate. The coordinator will be responsible for facilitating two workshop sessions, reviewing relevant documentation, and writing and revising performance standard guidelines or developing a template. Workshop coordinators should plan to devote 80 hr or more to development of guidelines or templates. Other workshop participants should plan on a 4- to 8-hr commitment.

WRP Technical Note WG-RS-3.3
January 1999

Step 4. Coordinator gathers and reviews relevant documentation, including selected permits issued in the region and reports from studies designed to assess regional mitigation success. HGM model variables and functional capacity indexes may be useful as performance standards, so relevant models should be reviewed along with other information.

Step 5. Coordinator gathers information about practices outside of the region that might be of interest to workshop participants. (This technical note summarizes some of this information.)

Step 6. Coordinator summarizes information gathered in steps 4 and 5 for participants in the first of two workshops. The coordinator's presentation should be limited to existing information; it should not suggest new or improved performance standards. Workshop participants offer opinions regarding important issues and potential new or improved performance standards.

Step 7. Based on opinions of workshop participants and other information, coordinator drafts performance standard guidelines or templates. In general, performance standard guidelines should be no more than one or two pages in length, and templates listing minimum performance standards may be as short as one page.

Step 8. Coordinator presents draft guidelines or templates to workshop participants who discuss them in an open forum in the second of two workshops.

Step 9. Coordinator revises draft guidelines or templates based on participants' comments.

Step 10. Regulatory supervisors review revised draft guidelines or templates.

Step 11. Coordinator finalizes draft guidelines or templates to the satisfaction of regulatory supervisors.

Step 12. Guidelines or templates are distributed for use by regulatory staff.

By bringing together regulators, scientists, and other stakeholders, the 12-step plan ensures that the best available professional knowledge will be considered while practical issues will not be ignored. However, effectiveness of guidelines or templates developed from the 12-step plan should be periodically reviewed. Ideally, the review process should include collection of data that relate achievement of performance standards to replacement of lost wetland structure and functions.

POINT OF CONTACT: For additional information, contact Dr. Bill Streever (601-634-2942, *streevw@ex1.wes.army.mil*). This technical note should be cited as follows:

> Streever, B. (1999). "Examples of performance standards for wetland creation and restoration in Section 404 permits and an approach to developing performance standards." *WRP Technical Notes Collection* (TN WRP WG-RS-3.3). U.S. Army Engineer Research and Development Center, Vicksburg, MS. *www.wes.army.mil/el/wrp*

WRP Technical Note WG-RS-3.3
January 1999

REFERENCES

Smith, R.D., Ammann, A., Bartoldus, C., and Brinson, M. M. (1995). "An approach for assessing wetland functions using hydrogeomorphic classification, reference wetlands, and functional indices," Technical Report WRP-DE-9, U.S. Army Engineer Waterways Experiment Station, Vicksburg, MS, 1-72.

Swink, F., and Wilhelm, G. (1994). *Plants of the Chicago Region*. Indiana Academy of Science, Indianapolis, IN.

Appendix F

Memorandum for Commanders, Major Subordinate Commands, and District Commands, April 8, 1999

DEPARTMENT OF THE ARMY
U.S. Army Corps of Engineers
WASHINGTON, D.C. 20314-1000

REPLY TO
ATTENTION OF:

CECW-OR

0 8 APR 1999

MEMORANDUM FOR COMMANDERS, MAJOR SUBORDINATE COMMANDS AND
DISTRICT COMMANDS

SUBJECT: Program Consistency and Reporting on the U.S. Army Corps of Engineers
Regulatory Program

1. The U.S. Army Corps of Engineers Regulatory Program is a vital part of the overall Civil
Works Mission. It is also a very visible aspect of the Corps interaction with the public on a
day-to-day basis. Consistency of administration of the Corps Regulatory Program is essential.
There are differences geographically in the extent and functions and values of the waters of the
United States we regulate, and there are different public views on the importance of regulating
these waters. However, we must not allow these regional differences to dictate the
administrative process that we subject the regulated and general public to as we carry out this
important and visible mission.

2. The result of your evaluations of permit applications will certainly reflect regional
differences, but the process each district uses should be as consistent as possible nationwide. In
addition to the need to be fair and consistent to applicants nationwide, with the limited resources
available for program implementation, we must ensure that districts are using essentially the
same procedures in fairness to the regulatory personnel in every district. We cannot afford to
have some districts doing substantially more evaluation than other districts on similar types of
projects. We recognize that the regulations, policy, and guidance provide substantial latitude in
the procedures used. On the one hand, this promotes fair and reasonable decision making on
very diverse permit applications. On the other hand, it creates the potential for dramatically
different program administration nationwide. Some of the flexibility in the program guidance
involves administrative procedures that can result in substantial expenditure of program
resources.

3. Because of the need to administer the Corps Regulatory Program as consistently as possible,
I am providing the enclosed Standard Operating Procedures (SOP) for implementation 30 days
from the date of this memorandum. The 30-day period will provide time for each Major
Subordinate Command (MSC) and district to review and understand the SOP. Should you have
any questions or comments on the SOP please contact my Regulatory Branch, CECW-OR. Each
district will be expected to execute its Regulatory Program consistent with this SOP. In addition,
MSC oversight visits will assess how well each district is complying with this SOP, and
decisions on annual resource allocation, as well as requests for additional funds will be partially
based on how well the district is complying with the SOP. The SOP does not cover every aspect
of the Corps Regulatory Program, it focuses on certain elements of the program that involve
substantial resource allocation and differences in program implementation nationwide.

CECW-OR
SUBJECT: Program Consistency and Reporting on the U.S. Army Corps of Engineers Regulatory Program

Moreover, the SOP does not provide new policy guidance, it simply focuses on where we believe that districts should be within the "zone of discretion" that exists in the Corps Regulatory Program guidance. Finally, the SOP will be a living document. As my staff and the MSC regulatory team identify needed changes, those changes will be made.

4. The SOP consists of three distinct parts. Part I provides focus on specific policy and program administration issues where we have noted distinct and unacceptable differences among the districts nationwide. Part II provides a corporate prioritization of the various types of activities that each district engages in as it administers the program. We have listed these activities in terms of what is above the line and what is below the line. All districts will be actively involved in all activities that are listed as above the line. Activities that are listed as below the line are those that may or may not be accomplished, and will be accomplished to varying degrees, but only after the above the line activities are fully executed. The MSCs will factor into annual budget allocations the extent to which each district is involved in below the line activities. MSCs should review district activities and consider reprogramming below the line funds to other districts that are not sufficiently funded for above the line activities. Districts should normally expend 25 percent or less of its allocation on below the line activities.

5. Part III of the SOP includes new definitions of the reporting requirements for the Regulatory Program. It is essential that HQUSACE have accurate information on Regulatory Program performance. Information on performance will be factored into annual resource allocations, and will be used by HQUSACE in our defense of our annual budget requests to Congress. We believe that the method of reporting time for making permit decisions is too flexible, and does not accurately reflect the time an applicant must wait for a final decision. The revised definitions in Part III of the SOP eliminate ALL stopping of the clock on permit evaluation time. We will now simply keep information on when the application was received, when it was considered complete, and when the district made the permit decision. Districts must also focus more on what reasons delayed the decision. In this way we can work to understand and reduce delays in decision making by the Corps. I am not faulting any district for how they have reported performance in the past, the guidance allowed too much flexibility. From now on, however, I want each district to ensure that reporting is absolutely accurate under the revised, and tightened, definitions in Part III of the SOP.

6. I recognize that by tightening the definitions for reporting many districts will be either amber or red. The reasons for being amber or red include too much workload, too many delays by other Federal or State agencies, and extensive delays by the applicants themselves. We need to better

CECW-OR
SUBJECT: Program Consistency and Reporting on the U.S. Army Corps of Engineers Regulatory Program

identify the reasons that districts are amber or red and react to those reasons appropriately. I do expect every district commander to become actively involved in determining the sources of delay in permit decision making. District commanders must also recognize that your district regulatory staff may be administering the program as efficiently as possible, and simply cannot make the final decisions within the reporting requirements because of external delays or excessive workload. I will be working with the division commanders to identify changes in process and changes in resource allocation to meet the various districts' needs. Some districts may have situations driven by external delay that will result in long term amber or red indicators. We will work to understand and document the reasons for these situations and work to remedy the problems, including pursuing/allocating additional resources, if resource limitations are deemed to be a contributing factor. The bottom line is that districts should identify any internal efficiencies that can improve performance and implement them, and identify the external sources of delay and accurately report them to HQUSACE.

7. This memorandum and the enclosed SOP are critical to fair and reasonable implementation of the Corps Regulatory Program, and to our efforts to ensure that the program is properly resourced in the future. Many recent changes have dramatically increased the workload and complexity of the program, and we need to document accurately what level of service we are able to provide with the current resource allocation to the program.

FOR THE COMMANDER:

Encl

RUSSELL L. FUHRMAN
Major General, USA
Director of Civil Works

DISTRIBUTION:
COMMANDER, MISSISSIPPI VALLEY DIVISION
COMMANDER, NORTH ATLANTIC DIVISION
COMMANDER, NORTHWESTERN DIVISION
COMMANDER, MISSOURI RIVER REGL. HQDS
COMMANDER, GREAT LAKES & OHIO RIVER DIVISION
COMMANDER, GREAT LAKES REGL. HQDS
COMMANDER, PACIFIC OCEAN DIVISION
COMMANDER, SOUTH ATLANTIC DIVISION
COMMANDER, SOUTH PACIFIC DIVISION

CECW-OR
SUBJECT: Program Consistency and Reporting on the U.S. Army Corps of Engineers
Regulatory Program

DISTRIBUTION: (CONT)
COMMANDER, SOUTHWESTERN DIVISION
COMMANDER, MEMPHIS DISTRICT
COMMANDER, NEW ORLEANS DISTRICT
COMMANDER, ROCK ISLAND DISTRICT
COMMANDER, ST. LOUIS DISTRICT
COMMANDER, ST. PAUL DISTRICT
COMMANDER, VICKSBURG DISTRICT
COMMANDER, BALTIMORE DISTRICT
COMMANDER, NEW ENGLAND DISTRICT
COMMANDER, NEW YORK DISTRICT
COMMANDER, NORFOLK DISTRICT
COMMANDER, PHILADELPHIA DISTRICT
COMMANDER, KANSAS CITY DISTRICT
COMMANDER, OMAHA DISTRICT
COMMANDER, PORTLAND DISTRICT
COMMANDER, SEATTLE DISTRICT
COMMANDER, WALLA WALLA DISTRICT
COMMANDER, BUFFALO DISTRICT
COMMANDER, CHICAGO DISTRICT
COMMANDER, DETROIT DISTRICT
COMMANDER, HUNTINGTON DISTRICT
COMMANDER, LOUISVILLE DISTRICT
COMMANDER, NASHVILLE DISTRICT
COMMANDER, PITTSBURGH DISTRICT
COMMANDER, ALASKA DISTRICT
COMMANDER, HONOLULU DISTRICT
COMMANDER, CHARLESTON DISTRICT
COMMANDER, JACKSONVILLE DISTRICT
COMMANDER, MOBILE DISTRICT
COMMANDER, SAVANNAH DISTRICT
COMMANDER, WILMINGTON DISTRICT
COMMANDER, ALBUQUERQUE DISTRICT
COMMANDER, LOS ANGELES DISTRICT
COMMANDER, SACRAMENTO DISTRICT
COMMANDER, SAN FRANCISCO DISTRICT
COMMANDER, FORT WORTH DISTRICT
COMMANDER, GALVESTON DISTRICT
COMMANDER, LITTLE ROCK DISTRICT
COMMANDER, TULSA DISTRICT

Appendix G

Army Corps of Engineers Standard Operating Procedures for the Regulatory Program

INTRODUCTION

These Standard Operating Procedures (SOP) are comprised of three parts. The first part, "Policies and Procedures for Processing Department of the Army Permit Applications," highlights critical portions of the U.S. Army Corps of Engineers implementing regulations to be used in reviewing permit applications. The second part, "Regulatory Program Priorities," provides "above the line" and "below the line" guidance for districts to prioritize their Regulatory Program administration efforts. The third part, "Revised Quarterly Permit Data System (QPDS) Definitions," provides clarification of definitions for inputting information into QPDS. The SOP has three purposes. First, by highlighting important existing procedures and policy, the SOP serves to facilitate consistent program implementation. Second, prioritizing Regulatory Program efforts helps to facilitate efficient national program direction to best achieve the three goals of the Regulatory Program: (1) Protect the environment; (2) Make reasonable decisions; and, (3) Enhance Regulatory Program efficiency. Third, clarification of the QPDS definitions is intended to unequivocally narrow the current gaps in their interpretation to provide an accurate data base that is essential to the analysis of workload, performance, and therefore, resource needs. When applied in conjunction with effective communication on budget needs and good workload indicators, the SOP serves to assure the most equitable distribution of funds proportionate to the district's respective workloads. The Corps must strive to implement its Regulatory Program as consistently as possible across the country in fairness to the regulated public and individual districts in protection of the aquatic environment.

STANDARD OPERATING PROCEDURES

PART I INDEX

1. Scope of Analysis
2. Jurisdiction
3. Wetland Delineations
4. Forms of Permits
5. Discretionary Authority
6. Pre-Application Meetings
7. Complete Application
8. Project Purpose
9. Preparing Public Notices
10. Internal Coordination
11. Permit Evaluation/Public Hearings
12. Appropriate Level of Analysis-404(b)(1) Guidelines
13. The Public Interest Determination
14. Section 401 Certification and Coastal Zone Management
15. Endangered Species Act
16. Documentations-EA/SOF/Guidelines Compliance
17. Conditioning Permits
18. Compensatory Mitigation
19. Duration of Permits
20. Permit Modifications and Time Extensions
21. Enforcement/Compliance
22. File Maintenance
23. Reporting

PART I

Policies and Procedures for Processing Department of the Army Permit Applications

Part I highlights existing policies and procedures to be used in re-viewing applications for Department of the Army (DA) permits under Section 404 of the Clean Water Act (CWA), Section 10 of the Rivers and Harbors Act (RHA) of 1899, and Section 103 of the Marine Protection, Research, and Sanctuaries Act of 1972. It is not intended to be comprehensive or to replace the implementing regulations for the U.S. Army Corps of Engineers Regulatory Program (33 CFR 320-330) or other official policy guidance contained in Memoranda of Agreement (MOA), Regulatory Guidance Letters (RGLs), etc. Part I simply highlights critical policies and procedures that are major factors in administering a consistent program nationwide. These critical policies and procedures, however, are those that the Major Subordinate Commands and CECW-OR will evaluate in our future consistency reviews. In addition, consistency with Part I will be a factor in responding to districts' requests for additional resources.

1. Scope of Analysis.

- *Corps determines scope*
- *33 CFR 325 Appendix B,C*

Scope of analysis has two distinct elements, determining the Corps Federal action area and how the Corps will evaluate indirect, or secondary, adverse environmental effects. The Corps determines its action area under 33 CFR 325 Appendix B and C. Generally, the action area includes all waters of the United States, as well as any additional area of non-waters where the Corps determines there is adequate Federal control and responsibility to include it in the action area. The action area always includes upland areas in the immediate vicinity of the waters of the United States where the regulated activity occurs. For example, the action areas for a road access to uplands for a residential development is the road crossing of waters of the United States and the upland area in the immediate vicinity of the road crossing. In a similar case where there is not only the road crossing, but also considerable additional impacts to waters within the residential development (interior road crossings, house fills, stormwater control berms/dams, etc.), then the Corps action area is the whole residential development. The Corps analyzes all adverse environmental effects within the action area. In addition, the Corps is responsible for analyzing the direct and indirect impacts of its permit decisions, within the action area once the scope of analysis (which defines the corresponding action area) is properly determined. Direct impacts are those that happen in direct response to the permitted activity (e.g., the direct impact of dam construction is the loss of habitat in the dam footprint). Indirect impacts, on the other hand, are those removed in time and/or distance in relation to the permitted activity (e.g., the indirect impact of dam construction is the inundation of the area behind the dam). Another example would be habitat and/or fisheries impacts downstream of the dam associated with hydroperiod changes. Both direct and indirect impacts must be evaluated within site-specific and cumulative impact contexts. It is appropriate for the Corps to evaluate these impacts and render final permit and mitigation decisions based on its evaluation. It is not appropriate, however, for the Corps to consider indirect impacts that are beyond the action area in its regulatory decisionmaking, where those impacts would have occurred regardless of the Corps decision on the permit (e.g., habitat fragmentation, increases in traffic and noise could be judged as these types of impacts).

2. Jurisdiction.

- *33 CFR Parts 323, 328, and 329*
- *Corps determines exemptions (if no special case)*
- *Exemptions do not allow waters conversions*

Not every area that looks like a wetland or other waters of the United States is jurisdictional, and not all activities in jurisdictional waters are subject to regulation. Guidance on jurisdiction is found in the preamble to the 1986 consolidated regulation, 33 CFR Part 328 and in Parts 323.3, 328.3, and 329. Part 329 only addresses Section 10 navigable waters. These regulations provide guidance on jurisdictional determinations, as well as, on areas that are not regulated and on activities that are exempt from regulation. The preamble to 33 CFR Part 328 states that features excavated from uplands are not considered waters of the United States. For example, a drainage ditch excavated in the uplands, and/or located along a roadway, runway, or railroad that only carries water from upland areas, is not considered jurisdictional, even if it supports hydrophytic vegetation. Other common examples of non-jurisdictional areas excavated from uplands include stormwater or other treatment ponds, detention basins, retention ponds, sediment basins, artificial reflecting pools, and golf course ponds. Gravel pits excavated from uplands are not considered jurisdictional, so long as the areas in question have not been abandoned (i.e., the area is under some sort of management plan related to the gravel operation, including use as a water supply or water storage area). Wetlands that form on top of a landfill are not subject to Corps jurisdiction.

Some activities taking place in jurisdictional waters of the United States are exempt from regulation. Definitions of discharges not requiring permits (i.e., exempt activities) are found in 33 CFR 323.4. Many of the exempt activities listed in this section are related to agriculture, forestry, or mining. For example, the list includes normal farming, silviculture, and ranching activities that include plowing, seeding, cultivating, minor drainage, and harvesting for the production of food, fiber, and forest products. As long as these activities are part of an established farming, silviculture, or ranching operation they are exempt from regulation under

Section 404 of the CWA. It is the responsibility of the Corps to determine if the activities are part of an ongoing operation, and therefore, are exempt unless the Environmental Protection Agency (EPA) declares a special case under the 1989 MOA on jurisdiction in advance of the Corps determination. Another common example of an exempt activity would be the discharge of sediment, removed from a roadside ditch, into waters of the United States when that ditch was constructed through waters of the United States, prior to the enactment of Corps regulation. The ditch and adjacent waters would be considered a jurisdictional area, however, the discharge of sediments removed from the ditch needed to restore the contours to original design is exempt from regulation under Section 404 (33 CFR 323.4 (a)(3)). None of the exemptions listed in 33 CFR 323.4 allow for the conversion of waters of the United States to dry land (beyond any conversion that was authorized by the original project that is being maintained) through filling or drainage activities. Plowing a wetland and maintenance of existing drainage (ditches, drain tiles, etc.) may be exempt from regulation, but discharges associated with the installation of a new drainage system to convert additional wetlands to uplands require a Section 404 permit.

Pursuant to guidance provided on 11 April 1997, resulting from litigation on the Excavation Rule activities where incidental fallback of excavated materials is the only discharge are not regulated as a discharge of dredged material, pursuant to Section 404 of the CWA. However, activities that involve the discharge of the excavated material are regulated discharges where the material is sidecast, or otherwise discharged into waters of the United States. Examples include mechanized landclearing, and excavating a new drainage ditch where the material is sidecast into wetlands. Of course, navigational dredging continues to be regulated, pursuant to Section 10 of the RHA. The Corps regulates the discharge of dredged material into waters of the United States, pursuant to Section 404 and the transportation of dredged material for ocean dumping, pursuant to Section 103 of the Marine Protection, Research, and Sanctuaries Act.

The term drainage ditch is defined as a linear excavation or depression constructed for the purpose of conveying surface runoff or groundwater from one area to another. The term drainage ditch does not include drainage systems which also serve to hold and manage water flow (flood control systems). If a drainage ditch is constructed entirely in uplands, it is not a water of the United States unless it becomes tidal or otherwise extends the ordinary high water mark of existing Section 10 navigable waters. However, if a ditch is excavated in waters of the United States, including wetlands, it remains a waters of the United States, even it if is highly manipulated (RGL 87-7, dated 17 August 1987).

3. Wetland Delineations.

- *Prioritize "mom and pop" JDs*
- *Advise the public via PN*

Field wetland delineations are essential to timely and accurate processing and evaluation of permit applications in these areas. However, delineations are time and resource intensive and, in some districts, require an inordinate amount of time that the district could be devoting to other aspects of the process. In addition, Corps workload is generally comprised of "mom and pop" applications, as well as those from applicants who could hire consultants to perform the field wetland delineation for the district to verify. For these reasons, districts should advise (e.g., via public notice (PN)) their respective constituencies that they will prioritize their efforts as follows. Conduct verification of applicant prepared delineations on all applications and conduct field wetland delineations for "mom and pop type operations" in conjunction with permit applications first. The remaining wetland delineations in conjunction with permit applications will be conducted second, and other delineations (not in accordance with permit applications) as resources and time allow.

4. Forms of Permits.

- *Use GPs/LOPs whenever possible*
- *GPs can address ESA/NHPA issues*
- *Develop GPs for periodic emergencies*

Permits. The overall goal of the Regulatory Program is to provide for a timely permit decision that protects the aquatic environment and is fair, reasonable, and flexible for the applicant. The Corps should evaluate projects using the least extensive and time consuming review process, while still providing protection for the aquatic environment.

For projects that involve minimal net adverse effects on the aquatic environment, after consideration of compensatory mitigation, general permits (GP) or letter of permissions (LOPs) should be used rather than a standard permit (SP) review. GPs include nationwide permits (NWPs) and regional general permits (GPS). The use of GPs are encouraged, because GPs can be conditioned sufficiently to provide the same environmental protection as an SP (including conditions addressing endangered species and compensatory mitigation concerns), and are just as enforceable as an SP. Therefore, the district should not hesitate authorizing a project with GPs for fear that it will be unenforceable.

Projects which may cause more than minimal adverse effects to the aquatic environment, should receive an SP review. LOPs, that are subject to Section 10 of the RHA, should be used on a case-by-case basis to authorize activities where the work will have only minor individual or cumulative impacts on the environment or navigation, and where the work would encounter no appreciable opposition. LOPs in cases subject to Section 404 CWA (as well as Section 10 RHA), may be used to authorize projects that exceed the aforementioned thresholds, provided a PN has been issued establishing the suite of activities and geographic area where the LOP will be in effect. This process is essentially the same as establishing an RGP. Normally, coordination with applicable Federal and State agencies project by project is included in an LOP and adjacent property owners are notified. The short agency notification period, lack of need for a PN and the additional activity conditions associated with an LOP result in abbreviated documentation needed to authorize a project under an established LOP, which saves time over the preparation of an SP decision document.

Emergency Procedures. Division engineers are authorized to approve special procedures in emergency situations. Each division should develop emergency permit authorization procedures. An "emergency" is defined in the regulations (33 CFR 325.2(e)(4)) as a situation which would result in an unacceptable hazard to life, a significant loss of property, or an immediate unforeseen, and significant economic hardship if corrective action, requiring a permit, is not undertaken within a time period less than the normal time to process the application under standard procedures. In these situations, the district engineer explains the situation and the associated permit procedures to the division engineer then issues the permit after the division engineer concurs. This entire process may be verbal in extreme emergencies. Even in emergency situations, districts should make a reasonable effort to obtain comments from the involved Federal, State, local agencies, and the public. A decision document may be prepared after the fact and should include an environmental assessment (EA). In addition, districts should publish a notice regarding the special procedures and their rationale and prepare an EA and statement of findings (SOF) as soon as practicable after the emergency permit is issued. Districts should also maximize the use exemptions, as well as available regional, programmatic, and nationwide GPs in emergency situations. Some districts have developed GPs for emergency situations, which the district believes will periodically reoccur. This provides a more efficient, predictable permit mechanism to deal with the emergency when it reoccurs without reinventing the wheel, as well as an opportunity to efficiently coordinate with the involved agencies and the public. In any event, districts and divisions should establish procedures for the coordination of emergency permits whether or not GPs have been developed.

5. Discretionary Authority.

- *Do not use discretionary authority just because project controversial*

Discretionary authority is a tool (33 CFR 330.8) used to assert a more rigorous review of projects eligible for a GP (NWP or RGP) due to perceived adverse impacts to the aquatic environment. Generally speaking, evaluators should carefully consider the need for asserting discretionary authority over a project that would otherwise qualify for a GP. It is inappropriate to assert discretionary authority over a project, merely because it is controversial. A RGP or NWP can be issued quickly and provide maximum environmental protection, through effective permit conditioning. It is also critical that when a NWP is moved into an SP review, the administrative record supports the Corps action, including all notification time constraints associated with the NWP program. If a project meets the requirements of a NWP or other GPs, the Corps should carefully examine the facts and options available, commence the processing of the application in the manner that is the most efficient and the one that provides adequate protection of the aquatic environment. Of course, the SP process is appropriate for proposed projects that do not meet applicable GPs criteria.

6. Pre-application Meetings.

- *Corps arranges*
- *Not for minor impact projects*
- *Be candid with applicants*

Pre-application meetings, whether arranged by the Corps or requested by permit applicants, are encouraged to facilitate the review of potentially complex or controversial projects, or projects which could have significant impacts on the human environment (see 33 CFR Part 325.9(b)). Pre-application meetings can help streamline the permitting process by alerting the applicant to potentially time-consuming concerns that are likely to arise during the evaluation of their project. Examples include historic properties issues, endangered species impacts, dredging contaminated sediments, 404(b)(1) compliance statements, mitigation requirements, etc. Pre-application meetings are not recommended for projects that will result in only minor adverse impacts to the aquatic environment. Each Corps district is responsible for determining if a pre-application meeting is necessary (not the applicant or another agency) and if so, who will host/facilitate the meeting. Applicants usually appreciate a candid dialogue on their project, even if the discussion means substantial modification of the project may be necessary. It is vital to the success of the pre-application meeting to have knowledge of the project, prior to the meeting date and to be prepared to discuss ways to avoid/minimize project impacts. It should be remembered that the issuance of a PN should not be delayed to obtain information necessary to evaluate an application (33 CFR 325.1(d)(9)). An application is determined to be complete when sufficient information is received to issue a PN (33 CFR 325.1(d) and 325.3(a)). See 7. below.

7. Complete Application.

- *Refer to 33 CFR 325.1(d)*
- *15 days to request information*
- *15 days to publish a PN*

The information needed to deem a permit application complete is listed in 33 CFR 325.1(d). Upon initial receipt of an application, the Corps has 15 days to review it for completeness (33 CFR 325.2 (a)(1)). The Corps will contact the applicant within 15 days if additional information is necessary. The Corps should encourage applicants to provide a wetland delineation as part of their permit application. In some instances, submission of a wetland delineation is required for a permit application to be considered as complete (e.g., some NWPs). While additional information may eventually be needed to complete the evaluation of the project (e.g., an alternative analysis under the Guidelines), only that information identified at 325.1(d) is required to publish a PN. Once an application is deemed complete, the next step is to determine what form of permit review is appropriate for the proposed project and issue a PN (for SPs), within 15 days (33 CFR 325.2(a)(2)).

8. Project Purpose.

- *Basic project purpose used for water dependency*
- *Overall project purpose used for alternatives analysis*
- *Corps determines*

Defining the project purpose is critical to the evaluation of any project and in evaluating project compliance with the Section 404(b)(1) Guidelines (Guidelines). As stated in HQUSACE's guidance resulting from the Section 404(q) elevation of the Twisted Oaks Joint Venture and Old Cutler Bay permits, defining the purpose of a project involves two determinations, the basic project purpose, and the overall project purpose. The Twisted Oaks project proposed impounding a stream, by constructing a dam, to create a recreational amenity for a residential development proposed on adjacent uplands. HQUSACE agreed that, from a basic project purpose perspective, the dam was water dependent, but that the housing was not. Since the project included two elements, a project purpose excluding either was inappropriate. Therefore, it was concluded that the overall project purpose was a residential development with a water-related amenity. In Old Cutler Bay, the overall project purpose was properly determined to be "to construct a viable, upscale residential community with an associated regulation golf course in the south Dade County area." This overall project purpose recognizes that an essential part of the upscale residential project was to include a full size (regulation) golf course, and identified a reasonable geographic area for the alternatives analysis.

The basic purpose of the project must be known to determine if a given project is "water dependent." For example, the purpose of a residential development is to provide housing for people. Houses do not have to be located in a special aquatic site to fulfill the basic purpose of the project, i.e., providing shelter. Therefore, a residential development is not water dependent. If a project is not water dependent, alternatives, which do not involve impacts to waters of the United States are presumed to be available to the applicant (40 CFR 320.10(a)(3)). Examples of water dependent projects include, but are not limited to, dams, marinas, mooring facilities, and docks. The basic purpose of these projects is to provide access to the water. Although the basic purpose of a project may be water dependent, a vigorous evaluation of alternatives under National Environmental Policy Act (NEPA) and the Guidelines will often be necessary due to expected impacts to the aquatic environment (e.g., a marina that involves substantial impacts to or the loss of marsh or sea grass bed).

The overall project purpose is more specific to the applicant's project than the basic project purpose. The overall project purpose is used for evaluating practicable alternatives under the Section 404(b)(1) Guidelines. The overall project purpose must be specific enough to define the applicant's needs, but not so restrictive as to preclude all discussion of alternatives. Defining the overall project purpose is the responsibility of the Corps, however, the applicant's needs must be considered in the context of the desired geographic area of the development, and the type of project being proposed. Defining the overall purpose of a project is critical in its evaluation, and should be carefully considered. For example, a proposed road through wetlands or across a stream to provide access to an upland residential development would have an overall project purpose of "to construct road access to an upland development site." Based on this overall project purpose, the Corps would evaluate other potential access alternatives. However, the Corps would not consider alternatives in any way for the residential community or otherwise "regulate" the upland housing.

9. Preparing Public Notices.

- *Refer to 33 CFR 325.3(a)*
- *Include required information*
- *No EAs or other extraneous info*

The information that must be referenced in a PN is defined in 33 CFR 325.3(a). An alternatives analysis will only be included in a PN if offered by the applicant. Applicants are not required to provide a mitigation plan for PN purposes, but if an applicant includes a mitigation plan in the application it should be included in the PN. The PN should contain a concise description of the project, the overall project purpose, and its anticipated impacts on the aquatic environment. It should contain the minimum number of exhibits needed to adequately illustrate the plan, and should be as brief as possible to minimize printing and mailing costs. PNs will be printed on both sides of the paper.

The PN should also contain a preliminary assessment of project impacts on endangered species and historic properties, as well as an assessment on the need of an Environmental Impact Statement (EIS). The omission of a statement addressing the need for an EIS in the PN is by default a district preliminary determination that an EIS is not required for project review. If a statement is made in the PN that a given project "may effect" a federally listed species, the 90-day timeframe for consultation with the U.S. Fish and Wildlife Service (USFWS) and National Marine Fishery Service (NMFS)) under the Endangered Species Act (ESA) should begin immediately. All districts should establish agreements with the appropriate service (USFWS or NMFS) office to coordinate the preliminary review of proposed projects for the purposes of complying with the ESA.

Districts will not publish an EA, or other unnecessary materials, in a PN. Inclusion of an EA in a PN is not required by the regulations, and does not represent economic stewardship. Districts should not expend any significant effort toward the preparation of an EA until after the close of the PN comment period, when sufficient information has been gathered to address the public interest factors.

The PN comment period should be no more than 30 days, nor less than 15 days. The nature of the project and the geographic distribution of the notice are factors that should be used in determining how long to advertise the PN. In many cases, a PN of 15 to 21 days is sufficient time to allow the public to comment. To expedite the permit process, districts are encouraged to issue PNs for shorter periods of time where appropriate (e.g., emergency situations).

If comments generated by the PN cause the applicant to modify the plans for the project, consider whether the probable impacts to the aquatic environment resulting from the changes are substantially different from those described in the original PN before republicizing the project. If project impacts are similar to, or less than the original submittal, the Corps will proceed with a decision, without issuing another PN. To reduce cost, save time, and prevent possible confusion, avoid issuing an additional PN unless necessary.

To improve efficiency and reduce costs, districts may post PNs on the District Regulatory Home Page. Districts should coordinate via PN, prior to removing anyone from established mailing lists.

10. Internal Coordination.

- *Conduct routine, cost-effective coordination with other Corps elements*

PNs and pre-construction notications (PCNs) should routinely be coordinated with other district elements (e.g., Operations, Engineering, or Planning) to determine if the proposed action could affect a Federal project (e.g., Corps recreation areas, flood control, or navigation projects) located in the proximity of the proposal. Although the Corps office contacted may not have any concerns, it may be able to provide advice regarding other parties interested in the proposed work (e.g., marine trade groups, recreation associations). However, the Corps regulatory chief should also ensure that any added cost of this coordination is both legitimate and reasonable. Not all internal coordination should be funded from GRF funds. For example, other district elements should use project funds to determine if a proposed action (that is the subject of a permit application) could affect a Federal project.

11. Permit Evaluation/Public Hearings.

- *Corps runs the permit process*
- *Corps determines public hearing need*
- *Do not hold public hearings just because project controversial*
- *Use public meetings and workshops*

The Corps is the Decision-maker. Always remember that the Corps is in charge of the Regulatory Program and is responsible and accountable for all aspects of the decision as well as the quality and efficiency of its administration. This is particularly true for projects that generate considerable controversy and/or comments from other Federal, State, local environmental agencies and the public. The Corps Regulatory Program does not rely on reaching consensus, but relies on gathering sufficient information for the Corps to make its decision. The Corps determines the project purpose, the extent of the alternatives analysis, determination of which alternatives are practicable, which are less environmentally damaging, the amount and type of mitigation and all other aspects of the decision-making process (RGL 92-1). Once the appropriate information is gathered, the Corps must move in a timely manner to make a decision. The Corps decides what is relevant in evaluating projects. This responsibility must not be transferred to another agency or the public. Additional coordination after the close of the PN period should focus on the issues under consideration, be managed by the Corps and be concluded as soon as possible. The Corps must stay involved in the project, and be decisive when evaluating information from all sources.

Level of Review. Delegate signature authority to the lowest necessary level making sure there are always two levels of review for all SPs and GPs with PCNs. Districts may choose to have two levels of review for non-reporting GPs, depending on the specific circumstances, or for controversial determinations that no permit is required. However, as a general rule, project managers may sign letters confirming the applicability of non-reporting GPs, including NWPs (without additional review), as well as letters confirming that no permit is required.

Forwarding Comments. In the letter (normally one letter succinctly stating Corps position) from the Corps to the applicant forwarding comments received on the proposed project, the Corps will alert the applicant of substantial unresolved issues for which the Corps requires information. This is a critical part of the permit evaluation process. The most important issues needing resolution should be stressed in the subject letter. Formulate questions clearly and ask for detailed information directed at resolving specific issues raised during the comment period. The Corps will state its information needs, as well as the Corps position on comments forwarded to the applicant. The Corps is mandated to protect the public interest. However, the scope of project analysis limits the issues that are relative to the permit decision. Tell the applicant which issues must be addressed and which do not need to be addressed. Anticipated impacts should be measurable, not conjecture and be related to the project under consideration. In the comment letter, notify the applicant of relevant issues raised in the comment letters and those identified by the Corps independent of the PN process. This is also a good opportunity to reiterate program requirements. Also, inform the applicant of the affects of special conditions that may be warranted due to compliance with Section 401 certification, Coastal Zone Management Act (CZM), Section 106 of the National Historic Preservation Act (NHPA), or the ESA.

Evaluating the Applicant's Response. The Corps must determine the adequacy of the applicants' response. Applicants will only be asked by the Corps to respond to comments on concerns that the Corps has determined are relevant to the decision process. If the applicant's response does not adequately address the issues, the Corps must respond accordingly and in a timely manner. A phone call can generally suffice for one or two deficiencies, however, a letter reiterating the unresolved issues should follow for more controversial applications. The Corps should not tell applicants what to write, however, the Corps should be informative and advise applicants exactly what information is required and the questions that need to be answered. Document phone calls in the file. Emphasize that information was asked for previously. The Corps will not bargain with the applicant about the type of information needed, or the nature of the response. In most cases, applicants are cooperative. However, if necessary, advise the applicant that if the required information is not provided, the Corps will withdraw the permit application. Do not allow projects to become unmanageable by accepting a series of partial responses. The Corps must have sufficient information to make and substantiate a decision on the permit application. If the applicant asks for additional time to complete the response, the request should normally be granted. Should the applicant's response generate additional concerns or questions, the Corps may request additional information from the applicant, and re-coordinate the project with interested agencies.

Public Hearings. Public hearings are held at the discretion of the district commander only where a hearing would provide additional information not otherwise available which would enable a thorough evaluation of pertinent issues. Districts will receive numerous requests for public hearings, especially in connection with controversial projects with high public visibility. However, unless a public hearing would serve the aforementioned purpose, district commanders will deny the requests for public hearings. Districts should consider: (1) the extent to which the issues identified in conjunction with a request for a public hearing are consistent with the Corps need to make its Guidelines and public interest determinations, that is the extent to which the issues are within the Corps scope of analysis; (2) the extent to which the issues identified in conjunction with a request for a public hearing represent information not otherwise available to the Corps; and (3) whether the issues identified are already addressed by responses to the PN. Districts should also consider public meetings or workshops, which can be targeted towards a particular group of objectors and/or issues. These are more informal forums, much less expensive, that provide a higher interaction with a smaller segment of the concerned public.

12. Appropriate Level of Analysis Required for Evaluating Compliance with the Section 404(b)(1) Guidelines Alternatives Requirements (see RGL 93-2).

- *Corps determines appropriate level of Guidelines analysis*
- *Focus on environmental impacts, ensure rigor of alternatives analysis is commensurate with impacts to aquatic environment*

The amount of information needed to make such a determination and the level of scrutiny required by the Section 404(b)(1) Guidelines must be commensurate with the severity of the environmental impact (as determined by the functions of the aquatic resource and the nature of the proposed activity) and the scope/cost of the project. The Corps must focus its resources on evaluations that focus on the aquatic impacts, and provide value added for protection of the aquatic environment.

Although all requirements in 40 CFR 230.10 must be met, the compliance evaluation procedures will vary to reflect the seriousness of the potential for adverse impacts on the aquatic ecosystems. The Corps must keep in mind that it regulates the total aquatic environment, not just wetlands. Often other aquatic resources such as seagrass beds, submerged fresh water aquatic vegetation, and hard bottom areas are as or more ecologically valuable than wetlands.

It always makes sense to examine first whether potential alternatives would result in no identifiable or discernible difference in impact on the aquatic ecosystem. Those alternatives that do not, may be eliminated from the analysis since Section 230.10(a) of the Guidelines only prohibits discharges when a practicable alternative exists which would have less adverse impact on the environment (this includes consideration of impacts of the proposed project and alternatives on non-aquatic as well as aquatic resources).

By initially focusing the alternatives analysis on the question of impacts on the aquatic ecosystem, it may be possible to limit (or in some instances eliminate altogether) the number of alternatives that have to be evaluated for practicability (an inquiry that is difficult, time consuming, and costly for applicants).

The level of analysis required for determining which alternatives are practicable will vary depending on the type of project proposed. The determination of what constitutes an unreasonable expense should generally consider whether the projected cost is substantially greater than the costs normally associated with the particular type of project.

It is important to emphasize, however, that it is not a particular applicant's financial standing that is the primary consideration for determining practicability, but rather characteristics of the project and what constitutes a reasonable expense for these types of projects that are most relevant to practicability determinations.

The analysis of alternatives, pursuant to the Guidelines will also satisfy NEPA, which also requires the analysis of alternatives. Therefore, districts should not conduct or document separate alternatives analyses for NEPA and the Guidelines. The only fundamental difference between alternatives analyses for NEPA and the Guidelines is that under NEPA, alternatives outside of the applicant's control may be considered. If such an analysis is conducted, simply document the findings, under the appropriate title, within the combined alternatives discussion.

Other Federal agencies (e.g., Federal Highway Administration/State Departments of Transportation (DOT)) routinely prepare NEPA documentation, containing alternatives analyses, in conjunction with projects which require Corps permits. Districts should strive to communicate the Guidelines alternatives analysis requirements to the lead agency to enable that agency to conduct an analysis of alternatives to satisfy Guidelines requirements and avoid the need for the district to have to conduct a subsequent analysis. For example, State DOT prepare NEPA documents to analyze alternative corridor alignments for new highways. To the extent that any of these proposed alignments involve waters of the United States and require an SP, the State DOT NEPA document should incorporate Guidelines alternatives requirements into the analysis of corridor alternatives.

13. The Public Interest Determination.

- *Corps responsibility*
- *Refer to 33 CFR 320.4*
- *Provides environmental as well as public interest protection*
- *Require mitigation*
- *Analyze alternatives*

The public interest determination involves much more than an evaluation of impacts to wetlands. Once the project has satisfied the Guidelines, the project must also be evaluated to ensure that it is not contrary to the "public interest" (33 CFR Part 320.4). There are 20 public interest factors listed in 33 CFR 320.4. A project may have an adverse effect, a beneficial effect, a negligible effect or no effect on any or all of these factors. The Corps must evaluate the project in light of these factors, other relevant factors and the interests of the applicant to determine the overall balance of the project with respect to the public interest.

The following general criteria of the public interest review must be considered in the evaluation of every permit application (33 CFR 320.4(a)(2)):

a. The extent of the public and private need for the project.

b. Where unresolved conflicts exist as to the use of a resource, whether there are practicable alternative locations or methods that may be used to accomplish the objective of the proposed project.

c. The extent and permanence of the beneficial or detrimental effects the proposed work is likely to have on the private and public uses to which the project site is suited.

The decision whether to authorize or deny the permit application is determined by the outcome of this evaluation. The specific weight each factor is given is determined by its relevance to the particular proposal. It is important to remember that the Corps can perform an alternatives analysis, and must require compensatory mitigation, or other conditions to address environmental impacts for all permits, including Section 10 only permits. For each application, a permit will be granted unless the district engineer determines that it would be contrary to the public interest to do so (33 CFR 320.4(a)).

Level of Review. Delegate signature authority to the lowest necessary level making sure there are always two levels of review for all SPs and GPs with PCNs. Districts may choose to have two levels of review for non-reporting GPs depending on the specific circumstances, or for controversial determinations that no permit is required. However, as a general rule, project managers may sign letters confirming the applicability of non-reporting GPs, including NWPs (without additional review) as well as letters confirming that no permit is required.

14. Section 401 Certification and Coastal Zone Management.

- *Use provisional permits*
- *Establish dates and deadlines*
- *Use discretion in enforcement of 401/CZM conditions*

Provisional Permits. Obtaining Section 401 Water Quality Certification (WQC) and CZM consistency can result in substantial delays in issuing Section 404 and Section 10 permits. To avoid unreasonable delays in Corps permit processing, the following actions are recommended. In cases where the Corps has finished its evaluation of a permit proposal and the only action remaining is the issuance of the Section 401 certification or CZM consistency concurrence, the Corps should send a provisional permit to the applicant. Sending a provisional permit completes the Corps action on the proposal and notifies the applicant of the need to obtain a Section 401 certification or CZM concurrence from the certifying agency before the Section 404 permit is valid. The provisional permit also places the only remaining action with the certifying agency, properly focusing the applicant on the State. RGLs 93-1 and 92-04 discuss provisional permits and Section 401 coordination.

401/CZM Deadlines. In the interest of expediting the review of permit applications, districts are encouraged to establish deadlines for reaching certification or issuing/assuming waiver of certification with the appropriate 401 certifying agencies and consistency with CZM. The recommended deadlines are between 60 days and 120 days on 401 certifications and 60 days to 90 days for CZM. If an established deadline is passed without action by the Section 401 certifying agency the Corps will consider the 401 certification waived, and/or presume CZM consistency and proceed with the issuance of a permit. In addition, districts should establish an agreement with the certifying agencies stating that the issuance of a PN by the Corps constitutes a request for Section 401 certification and CZM consistency, and the PN date identifies the day from which the aforementioned deadlines should be established. The maximum amount of time allowed for a 401 certification agency to reach a decision is one year. The maximum amount of time for a coastal zone consistency certifying agency to reach a decision is six months.

401/CZM Conditions. For projects that require Section 401 authorization and/or CZM consistency concurrence, all special conditions of the 401 and CZM certification must be incorporated into the DA permit. The incorporation of Section 401 and CZM conditions is a necessary part of the Corps permit program. In all cases, the 401 and CZM conditions will be incorporated by reference to the 401/CZM certification and a copy of the 401/CZM certification must be attached to the permit. In addition, copies of the CZM consistency decision must be attached to the DA permit. Section 401 or CZM conditions are subject to discretionary enforcement by the Corps. In addition, Section 401 and CZM conditions may not be appealed, pursuant to the appeals process.

Withdrawal of 401/CZM Certification. If a State withdraws its WQC or CZM certification after the Corps permit is issued, the district does not have to automatically suspend or revoke the Corps permit. Rather, the district should review the circumstances which lead to the State's actions and, to the extent these circumstances bear upon the district's Guidelines compliance or public interest determinations, consider whether to leave the permit in place, modify, or rescind the permit. If the district decides to leave the permit in place, subsequent work by the permittee may be considered in violation of State law and subject to State enforcement procedures.

Applicability to Section 10 Activities. Section 401 certification of Section 10 only activities is not required unless the Section 10 permit involves a discharge into waters of the United States. Section 10 only projects normally do not result in a discharge, which is what triggers the need for 401 certification. For example, the installation of piers, docks, and similar structures do not require a Section 401 certification because there is normally no discharge associated with the project. If the state issues a 401 certification for a Section 10 only activity (no discharge) then those conditions do not become part of the DA permit unless the Corps adopts them, as Corps required conditions.

15. Endangered Species Act (ESA).

- *Corps determines scope of analysis*
- *Corps determines no effect/may effect and no jeopardy/jeopardy*
- *Corps decides whether to include RPAs*
- *Applicant must comply with take statement*

For the purpose of evaluating permit applications, the scope of analysis under the ESA is limited to boundaries of the permit area, plus any additional area outside Corps jurisdiction where there is sufficient Federal control and responsibility. The ESA scope of analysis is the same as that used for the consideration of historic properties under Section 106 of the NHPA and the NEPA. Within the properly defined scope of analysis for the ESA, the Corps must consider all direct and indirect impacts of the proposed discharge of dredged or fill material, structure or work on the federally listed species and its critical habitat. Indirect impacts of interrelated actions (part of the larger action and dependent upon the larger action for their justification) and interdependent actions (having no independent utility apart from the Federal action) must also be considered. This determination of indirect impacts to evaluate must include an evaluation of the causal relationship between the activity being authorized by the Corps, and the resultant physical effect of the activity on the listed species.

With respect to the conditioning of permits for applications that have undergone consultation as defined in the ESA, the Corps will decide what if any conditions are appropriate for inclusion into a Section 404/10 permit (if the Corps issues a permit all elements of the incidental take statement must be included by reference – see below). The ESA consultation may result in the drafting of a biological opinion (BO) by the USFWS/

NMFS. The BO may include a number of reasonable and prudent alternatives (RPAs), an incidental take statement, and conservation recommendations.

The incidental take statement contains limits, reasonable and prudent measures, terms and conditions, and instructions to handle or dispose of any taken species. The applicant must implement and comply with the requirements of the incidental take statement in order to legally take a listed species, which may result from the construction or operation of the permitted activity. The Corps issuance of a permit is not considered to be a take of listed species by the Corps, should such take occur as a result of the applicant's activities. The Corps permit will contain a condition indicating that the applicant must comply with the incidental take statement should the applicant take a listed species. Such permit condition will further indicate that the Services will be informed of and enforce any known violations of the incidental take statement.

Reasonable and prudent alternatives are items identified in the USFWS/NMFS BO that the USFWS/NMFS believes will ensure a determination of no jeopardy to a federally listed species or no degradation or adverse modification of critical habitat. The Corps considers all RPAs developed by the USFWS/NMFS, and decides which will be included in any Section 404/10 permit issued by the Corps. RPAs will be required by the Corps only to the extent that they are necessary for the Corps to make its determination that the authorized activity is not likely to jeopardize the continued existence of a listed species or result in the destruction or adverse modification of critical habitat. If the Corps decides to issue a permit without including the RPAs in the BO, the permit decision document must explain why the species will not be jeopardized.

Conservation recommendations will only be included as permit conditions, if the applicant so requests in accordance with 33 CFR 325.4.

16. Documentation - Environmental Assessment, Statement of Findings, 404(b)(1) Guidelines Compliance, EISs, and Standard Compliance Statements.

- *Focus on issues relevant to decision in documentation*
- *Focus on impacts within Corps scope of analysis to determine if EIS required*
- *Applicants pay for Regulatory Program EISs*

Documentation. The documentation of this process is variously referred to as a SOF/EA or a combined decision document which incorporates both the EA and SOF. A combined decision document should be used to consolidate discussion on project impacts and alternatives. The decision document will describe the proposed project, including any important on-site environmental features directly affected by the project, the regulation(s) under which the proposal is being reviewed (RHA, CWA, NEPA, ESA, etc.), a chronology of events since the application was received, including a summary of all comments received, and a final decision. The decision document must also include an EA that supports the decision (a findings of no significant impact (FONSI) or a findings of significant impact necessitating the preparation of an EIS), and if the project involves a discharge of dredged or fill material into waters of the United States, analysis of compliance with the 404(b)(1) Guidelines. The decision document should discuss all relevant public interest review factors, incorporate the findings of this review, include the 404(b)(1) Guidelines review and conclude with a public interest determination. The decision document must include the items listed above, even if no comments are received on the PN. However, brief one sentence discussions or checklists for public interest factors and the 404(b)(1) Guidelines should be used for routine SPs and LOPs.

Evaluating PN comments can be one of the more difficult and time-consuming tasks associated with completing a decision document. The issues raised in comment letters can vary widely, and may or may not be relevant to the Corps jurisdiction, or to the public interest factors. Nevertheless, all comments received need to be addressed in the decision document including the irrelevant ones and those that exceed the scope of the project under consideration. The farther an issue is removed from the area of jurisdiction or from the identified public interest factors (water quality, aquatic habitat, endangered species, navigable waters, historic resources, etc.), the less weight it will have in the decision making process. The level of importance each issue receives in the decision process, and therefore, the amount of attention in the decision document should be proportional to its association with the regulated activity. Comments on issues that are outside the Corps Regulatory Program scope can be addressed with one sentence stating that the issue is outside the Corps Regulatory Program.

It is often helpful to summarize the comments received into statements that can be categorized for evaluation, especially for projects where many comments are received (e.g., wetlands, air and water quality, noise, traffic, floodplain impacts). Some of the comments may be unrelated to the Corps regulatory mandate, such as alleged overcrowding of local public schools, project compliance with local land use plans and local ordinances. These issues can be grouped together, and briefly discussed in one paragraph of the decision document, with emphasis that they are beyond the scope of Corps review. Lastly, the comments that are essential to the public interest review (e.g., Corps jurisdiction or regulations, and other Federal involvement) should be identified and examined in the level of detail necessary to reach a decision. Not all issues are equally important; focus on the major issues that form the basis or the "equation" for the district's final permit decision. Remember that it is the district's responsibility to establish a point in the evaluation process when there is sufficient information on the major issues to proceed with a decision.

Consider comments from other Federal and State agencies carefully, keeping in mind that these comments presumably reflect the mandated interests of that agency and should be focused on the agencies expertise (e.g., fish and wildlife values from the USFWS). However, the other agencies concerns may not be relevant to the action under review. For example, under its mandate a State's natural resource agency may be interested in preserving upland habitat for game species or songbirds within a subdivision development. However, those upland concerns may not outweigh impacts to the aquatic environment that are germane to the Corps conclusions regarding the public interest and the Guidelines. To foster good Government and to inform the applicant, the Corps can transmit the natural resource agency's concerns, to the applicant. However, no response is required from the applicant for such items unless there is a strong link to the Corps regulatory responsibility. In addition, the Corps should refrain from placing special conditions on a permit that are beyond reasonable Federal control (e.g., posting signs for hours or operations of a boat ramp, upland lighting requirements), unless requested to do so by the applicant (33 CFR 325.4).

Carefully consider all requests for a time extension of the PN comment period, whether the request comes from an agency or from a private individual. The requesting party must justify why the PN should be extended. If the stated reasons for requesting the extension are not reasonable and substantive, the request should be denied. For the Corps to consider such a request, it must be received prior to the expiration of the PN.

EISs. The Corps determines if an EIS is required on any particular IP. EISs should only be prepared when they are legally required; that is when the district concludes that its permit decision will significantly affect the quality of the human environment after consideration of any mitigation the Corps would require. Districts will receive numerous requests to prepare EISs, especially in connection with controversial projects with high public visibility. In some instances, despite a relatively minor permit action, districts may find themselves pressed to prepare EISs because the Corps represents the only Federal agency with permit authority and, therefore, the only authority to require that such a document be prepared (e.g., a GP for a minor road crossing in conjunction with a large project for which non-related upland endangered species and/or historical property impacts are projected). It is inappropriate for districts to prepare EISs under these circumstances, i.e., the Corps is the only agency with jurisdiction or "the only game in town." The district must carefully evaluate the proper scope of analysis under 33 CFR 230.7, 1988 NEPA regulation (33 CFR 325, Appendix B) and not expand that scope of analysis just because others request such action. Unless, the district's permit decision, with the proper scope of analysis, would result in impacts significantly affecting the quality of the human environment, districts will not prepare an EIS. Districts should apply paragraph 7. of Appendix B, 33 CFR 325 of the Corps regulations and determine: (1) the scope of analysis (that is the activities and geographic range associated with project impacts); and (2) whether those impacts within the Corps scope of analysis constitute a significant impact on the quality of the human environment. If an EIS is required, the applicant must furnish the necessary information for the Regulatory Branch's review. The CEQ's 40 questions fully supports the Corps approach of preparing mitigated "FONSI." The determination of significance of impacts is done AFTER considering all mitigation. Where a limited scope of analysis is involved, the Corps should describe briefly the basis of the scope of analysis in the decision document.

Regarding the financing of EIS preparation, refer to the HQUSACE memorandum of 17 December 1997, from the Director of Civil Works to Major Subordinate Commanders and District Commanders, subject: Guidance on EIS Preparation, Corps Regulatory Program. The substance of that memorandum is repeated below.

"1. Appendix B, 33 CFR Part 325," provides policy guidance on preparation of NEPA documents for the Corps Regulatory Program. This regulation provides that the district engineer may prepare an EIS or may obtain information to prepare an EIS, either with his/her own staff or by choosing a contractor, either at the expense of the Corps or the expense of the applicant, who reports directly to the district engineer (see paragraph (3), 8b, 8c, and 8f). Due to budgetary constraints, preparing a project specific EIS at the expense of the Corps can no longer be funded.

2. Effective immediately, any Corps district preparing an EIS on a permit action will use a "third party contractor" as the primary method to prepare all or part of a project specific EIS or to obtain required information (40 CFR 1500-1508). "Third party contract" refers to the preparation of an EIS by a contractor paid by the applicant but who is selected and supervised directly by the district engineer (Corps Regulatory Branch). (See 40 CFR 1506.5(c) and Council on Environmental Quality's (CEQ) Forty Most Asked Questions Concerning CEQ's NEPA Regulations, #16 and #17). Contractors election by the Corps for a Regulatory Program EIS will be as follows: The Corps will select from the applicant's list the first contractor that is fully acceptable to the Corps, using the applicant's

order of preference. The procedures outlined in 40 CFR 1500-1508 and CEQ's forty questions must be followed. Furthermore, the Corps is responsible for final acceptance of the draft and final EIS.

3. Appendix B, 33 CFR Part 325, provides that the district engineer may require the applicant and/or his/her consultant to furnish information required for an EIS. The applicant and/or his/her consultant will then provide the information for the Corps use in preparing an EIS. This is an option which may be utilized in preparing a project specific EIS; however, to manage Corps resources more efficiently and equitably, this approach will be utilized by a district in preparing a project specific EIS only when for some reason the third party contracting cannot be used. If this method is used, the applicant is responsible for providing required information and data to the Corps. The Corps is responsible for review and acceptance of required information, data, or drafts and must be especially vigilant in identifying and eliminating any bias that could exist in a draft NEPA document prepared by a contractor selected and supervised by an applicant. The district engineer (Corps Regulatory Branch) has the final determination for EISs prepared by the applicant and his consultant of whether the data provided is adequate and accurate. The Corps will carefully review the applicant's drafts to ensure they are technically adequate and not biased.

4. Of course, a programmatic EIS will still have a substantial portion of the effort conducted and funded by the Corps. However, even for programmatic EISs, the Corps can, and should, identify applicant groups, States, and/or local Governments to cost share in the effort. Whenever an agency prepares a programmatic EIS, the requirements of 40 CFR 1506. 1(c) present potential legal and practical problems for processing any Corps permit related to the programmatic EIS (especially if the permit would require a project specific EIS). For that reason and due to budget implications, any decision to do a programmatic EIS will be reviewed and approved by CECW-OR before a commitment is entered into for any programmatic EIS.

5. Due to Regulatory Program budget limits, all Regulatory Program EIS's must be managed in the Regulatory Branch and primarily reviewed in the Regulatory Branch. The Regulatory Branch will only contract out work to other Corps elements, other Federal agencies, or private consultants, when additional expertise beyond that available in the Regulatory Branch is necessary or where it makes good business sense for the Regulatory Program.

Statement of Findings Standard Compliance Statements. The following statements must be included in all SOFs.

a. Section 176 (c) of the Clean Air Act General Conformity Rule Review: A statement must be included in the SOF that the proposed project has been analyzed for conformity applicability, pursuant to regulations implementing Section 176(c) of the Clean Air Act and that it has been determined that the activities proposed under this permit will not exceed de minimis levels of direct emissions of a criteria pollutant or its precursors and are exempted by 40 CFR Part 93.153. Statements must also be included that any later indirect emissions are generally not within the Corps continuing program responsibility, that these emissions generally cannot be practicably controlled by the Corps, and, for these reasons, a conformity determination is not required for a permit.

b. Public Hearing Request: A statement must be included in the SOF as to whether a public hearing was requested and by whom (e.g., a Senator, 200 requests). This statement should be followed by the district engineer's decision on whether to hold a public hearing and the summary rationale. It is sufficient to state, "I have reviewed and evaluated the requests for a public hearing." There is sufficient information available to evaluate the proposed project; a public hearing would not result in information that is not already available, therefore, the requests for a public hearing were denied," in the SOF. However, there may be a need to elaborate on this in the EA.

c. Compliance with 404(b)(1) Guidelines: If applicable, a statement must be included that the project has been evaluated for compliance with the Section 404(b)(1) Guidelines. This should be followed with statements and summary information which conclude whether the project represents the least environmentally damaging, practicable alternative, whether the project complies with all applicable State and Federal criteria, whether the project will result in significant degradation of the aquatic environment, and whether appropriate and practicable mitigation has been required to offset the permitted loss of aquatic functions. A summary statement that the project complies or does not comply with the Guidelines should also be included.

d. Public Interest Determination: A statement must be included noting whether the project is or is not contrary to the public interest and contain summary information on each relevant public interest factor.

e. Finding of No Significant Impact: A statement that the project will/will not result in significant impacts to the quality of the human environment and, therefore, the preparation of an EIS is/is not required.

Special Studies. The Corps is often approached to contribute Regulatory Program funds or to seek additional Federal funding in support of special studies. Districts should check with divisions first, then CECW-OR before committing Regulatory Program funds.

17. Conditioning Permits.

- *Easily enforceable*
- *Must relate to environmental protection or the public interest*
- *Must be justified in documentation*
- *Include 401/CZM conditions*

Special conditions placed on a Corps permit should be limited to those necessary to comply with the Federal law, while affording the appropriate and practicable environmental protection, including offsetting aquatic impacts with compensatory mitigation (see 33 CFR 325.4 – read carefully). There must be sufficient justification in the administrative record to warrant the conditions and the conditions should be easily enforced. All Corps imposed conditions should be substantially related to the impacts (e.g., Corps authorizes impacts to the aquatic environment, thus protection of uplands, except as buffers to open waters, are not appropriate unless the applicant specifically requests such conditions). Under the Corps Administrative Appeal Process (subject to final rule making), once a standard permit is issued, its terms and conditions may be appealed to the division engineer. In the appeal process, any condition that is not sufficiently justified in the administrative record is subject to removal or modification. Section 401 and CZM conditions may not be appealed through the Corps administrative appeals process.

Permit conditions should be developed so they are easily enforced, and relate to issues raised in the public interest review process (e.g., the aquatic environment, the ESA, navigation, cultural resources). Conditions to State 401 WQCs must be included as conditions of all Section 404 permits. CZM consistency conditions must also be included as conditions of Corps permits. Both 401 and CZM conditions should normally be in the form of one condition referencing the enclosed State conditions. Special conditions should not be used for the purpose of complying with State laws or local ordinances. An example of a well structured condition for a mitigation plan may require the date of the drawings, a monitoring and planting plan (with success rates), deed restrictions, and other narratives that describe the mitigation area. In many situations, more than one individual works on a project, particularly when compliance and/or enforcement are involved. Therefore, it is essential that permit conditions should be developed with a focus towards ease in interpretation. The number of special conditions should also be held to the minimum necessary to protect the aquatic environment, and to ensure compliance with Federal law. Given the high workload requirements associated with mitigation requirements (e.g., multiple inspections, review of annual reports, agency coordination), enforcement of all conditions may not be possible. Therefore, the importance of writing clear, easily enforceable conditions cannot be overstated. While all Section 401 conditions must be attached to the Section 404 permit, the Corps should only seek to ensure compliance of those conditions that most closely relate to our jurisdiction. Use enforcement discretion for 401 and CZM conditions, dependent on resource limitations.

It some cases it may be appropriate to require the recording of deed restrictions, land covenants, or conservation easements as a condition of a DA permit. The deed restrictions may involve wetlands or uplands, when required to sufficiently protect/mitigate the aquatic resources associated with the permitted action. The need for covenants should also be justified in the administrative record. Special justification should be referenced where upland buffers are incorporated into the decision. The inclusion of buffer areas, although not usually associated with the requirements of the CWA, can add numerous benefits to the value of an aquatic environment. The record should identify specific functions that the upland buffer is performing, such as water quality benefits, aquatic species habitat support (e.g., shading), or watershed protection and should not be based solely upon habitat values to wildlife.

18. Compensatory Mitigation.

- *Corps determines mitigation*
- *Replace lost functions*
- *Require permittee reporting*
- *Compliance inspections essential*
- *Provide clear, enforceable permit conditions*
- *Use best professional judgement*

Mitigation is a critical part of the Corps Regulatory Program. The following is a focus on the mitigation policy and the source of that policy, under which the Corps should be operating the Regulatory Program, as well as some basic concepts that districts should consider in formulating compensatory mitigation.

Mitigation Policy.

a. Corps 1986 Consolidated Rule (33 CFR 320.4(r)): The mitigation policy in the Corps consolidated rule applies to all permit actions (unless superceded by the Mitigation MOA with regard to Section 404 actions), including Section 10 permits. The mitigation policy in the consolidated rule: (1) reiterates the CEQ types of mitigation, avoiding, minimizing, rectifying, reducing, or compensating for resource losses; (2) provides that compensatory mitigation may be on site or off site (no preference); (3) stipulates that losses will be avoided to the extent practicable; (4) considers project modifications as a form of mitigation; (5) pertains to legal requirements (ESA, 106, 404(b)(1) Guidelines, etc.); (6) addresses public interest impacts; (7) provides for the Corps to accept any additional mitigation the applicant requests to be added; and (8) stipulates that all mitigation should be directly related to the proposed work, should have a link to the aquatic environment (this includes upland buffers to flowing or other open waters) and appropriate to the degree and scope of the proposed impacts to waters of the United States.

b. CEQ Mitigation Policy (CEQ 1978 Regulation (Definitions Section at 1508.20) and 40 Questions: As previously indicated, the CEQ regulations contain the same definition of mitigation as the Corps consolidated rule. The CEQ 40 questions on its regulation state that mitigation can be used to offset impacts to the level of non-significance, thus our NEPA documents are essentially all EAs, with a FONSI, based on mitigation. This is one of the required compliance statements in all Corps SOFs documents.

c. The DA – EPA 1990 MOA on Mitigation: The Mitigation MOA, which applies only to standard individual permits (IPs) under Section 404: (1) establishes the mitigation sequence, avoid, minimize, and then compensate to extent practicable, for remaining unavoidable losses; (2) reiterates that significant degradation (40 CFR 230.10(c)) can be offset by compensatory mitigation to the level of non-significance; and (3) reiterates the requirement that the Corps should require mitigation to ensure no significant degradation of the waters of the United States occurs.

d. 1995 Mitigation Banking Guidance (Federal Register site): The mitigation banking guidance provides detailed guidance on the formulation and management of mitigation banks. This guidance establishes criteria regarding both on site versus off site and in kind versus out of kind mitigation and concludes that the Corps should require the mitigation that is best for the aquatic environment (no a priori preferences). It is the district's call on what is best ecologically, normally on a watershed basis. In other words, the Corps should keep mitigation in some "district defined" watershed to offset the impacts (increasingly the Corps will probably use United States Geological Survey watershed catalogue units). Sometimes it is not possible to keep mitigation in the watershed where the impact occurs, nor does it make ecological sense to do so. In such cases, the Corps may require mitigation outside of the watershed. This approach (what is best for the aquatic environment) should guide all of the district's mitigation decisions, not just for mitigation banks.

e. NWP Regulation (33 CFR 330): The NWP regulations: (1) stipulate that the Corps will require compensatory mitigation to the extent necessary to ensure no more than minimal adverse environmental effects, both individually and cumulatively; (2) contain Condition 13, which has explicit requirements for mitigation to the extent that adverse effects on the aquatic environment are minimal; (3) contain Condition 20 (current 404 only condition 4), which requires onsite minimization and avoidance to the extent practicable; and (4) reiterates the Guidelines policy that offsite alternatives analysis does not apply to any GP, including NWP's.

Mitigation Concepts. In addition to the above, the following concepts should be considered in determining mitigation requirements for projects:

a. The amount of mitigation required should be commensurate with the anticipated impacts of the project. The goal of mitigation is to replace aquatic resource functions and other impacts.

b. The aforementioned 1990 EPA/DA MOA establishes a preferred sequence of on-site/in-kind mitigation to off-site/out-of-kind approaches, however, districts should not consider this a hard and fast policy. Corps field experience has shown ecological value in pursuing practicable and successful mitigation within a broader geographical context. This approach, combined with innovations such as mitigation banks and in-lieu-fee programs, provides proportionately higher ecological gains where the aquatic functions are most needed. The Corps depends on regulators reviewing relevant agency and public comments and applying their best professional judgement in requiring appropriate and practicable mitigation for unavoidable, authorized aquatic resource impacts. The bottom line test for mitigation should be what is best for the overall aquatic environment.

c. Districts may choose to use a timely and efficient aquatic resource assessment methodology. It is useful, but not required, that the assessment method be agreed upon by the coordinating Federal and State agencies. It is particularly important to work with State and local Regulatory agencies. The use of an agreed upon method will help ensure a degree of uniformity and fairness concerning the assessment of functions as well as mitigation requirements, including a quantified replacement of functions.

d. The Corps decides what is appropriate and practicable mitigation, not the coordinating agencies. The Corps should consult with other agencies in determining appropriate mitigation, but the Corps makes the final decision. Different agencies operate under different mandates and may have conflicting recommendations. It is up to the Corps to review comments received and reach a defendable and reasonable conclusion.

e. The use of conservation easements or deed restrictions can be an important part of a mitigation plan. The use of these tools is encouraged where feasible. Each district should develop a standard procedure and format to facilitate the use of deed restrictions and easements.

f. Districts will inspect a relatively high percentage of compensatory mitigation, to ensure compliance with permit conditions. This includes SPs and GPs. This is important because many of the Corps permit decisions require (and presume the success of) compensatory mitigation to offset project impacts. To minimize field visits and the associated expenditure of resources, SPs and GPs with compensatory mitigation requirements should require applicants to provide periodic monitoring reports and certify that the mitigation is in accordance with permit conditions. Districts will review all monitoring reports.

g. Districts will inspect all mitigation banks to ensure compliance with the banking agreement. Mitigation banking is an important part of the Regulatory Program because it represents an efficient and effective way to offset authorized aquatic resource impacts.

19. Duration of Permits.

- *Permit duration is the Corps discretion*

The timeframe that an IP is valid is at the discretion of the district engineer. The time limit commonly used for IPs involving construction is 3 to 5 years. Use of 5-year construction periods reduces the need to extend permits. However, the regulations (33 CFR 325.6) allow for flexibility on this issue. The regulations state that the duration of a permit should be reasonable, and commensurate with the nature of the work. For example, Disney World in Orlando, Florida was issued a permit with a duration of 20 years in order to accommodate the construction of this huge project in manageable phases. There are two exceptions. The first is for permits issued for the transport and disposal of dredged material in ocean waters, which can be valid for no more than 3 years (33 CFR 325.6(c)). The second is for maintenance dredging permits, which can be valid for a maximum 10 years from date of issuance of the permit originally authorizing the dredging (33 CFR 325.6(e)).

20. Permit Modifications and Time Extensions.

- *Extend/modify permits to the extent practicable*
- *Requests for extensions must precede expiration date*

Project managers are encouraged to use time extensions and permit modifications to the extent practicable to increase efficiency. Time extensions on existing permits will normally be granted where the project has not changed, and the regulation and policy framework are substantively the same as existed for the original decision. While time extension requests are generally reviewed favorably, it is imperative that written requests for the extension are received prior to the expiration of the permit in question (preferably within 30 days prior to their expiration). As long as the request for extension is received prior to expiration of the permit, the Corps decision may be made after the original expiration date.

When reviewing a request for a permit modification, it is important to remember that changes to the site plan are not the deciding factor in determining the need for a new permit review. The deciding factor is the probable impacts to the aquatic environment or other aspects of the public interest, not the configuration of the project, or the preferred method for carrying out the work. If issues to Corps jurisdiction are substantial (e.g., substantial increases in the fill footprint are proposed) a modification will normally be subject to a PN. If impacts to the aquatic environment are less than the original proposal, a modification to the original permit should be executed quickly without a PN. Agency coordination may not be necessary for non-controversial projects, unless the District believes the net overall changes/impacts to the aquatic environment are greater. Modifications for controversial projects (e.g., those that involve an endangered species or significant historic resource) should normally involve agency coordination. Not all agencies need to be notified of proposed modifications. For example, a modification that would only affect historic resources might be of little interest to the EPA and the USFWS. Notification in such a case would be appropriately restricted to the State Historic Preservation Officer or to parties who commented on historic resource issues during the PN comment period. A brief supplemental decision document should be prepared for permit modifications since these are final permit decisions.

21. Enforcement/Compliance.

- *Maintain a viable enforcement program*
- *Prioritize workload*
- *Require permittees' compliance reports*
- *No surveillance as a discrete activity*

It is important to maintain a viable enforcement program. Enforcement of unauthorized activities as well as activities that are not in compliance with permit requirements is a necessary component to the Regulatory Program. Districts will prioritize inspections and subsequent actions based on compensatory mitigation, local hot spots, endangered/threatened species, cultural resources, navigation concerns, or other controversial issues which the district considers important (e.g., Corps enforcement of Section 401 or CZM conditions is discretionary). There are national and district-specific performance measures currently in development that may serve to prioritize compliance inspections. Districts should require permittees (recipients of SPs as well as GPs) to periodically supply reports with pictures on compensatory mitigation projects to assist in the district's review of compliance. Districts will require all permittees (SPs, LOPs, RGPs, NWPs) to submit a self certification statement of compliance as provided in the NWP program.

Every attempt should be made to resolve violations of permit conditions and unpermitted activities by restoration or other non-litigation means, wherever possible. Districts should not be expending funds on surveillance as a discrete activity. Rather, surveillance should be performed in conjunction with other field activities such as jurisdictional delineations or other field activities performed in conjunction with permit or enforcement actions.

22. File Maintenance.

- *Organize files chronologically*
- *Purge drafts and unnecessary documents*

The administrative record for projects should be organized chronologically once the Corps has made a decision on authorization of the project. All non-essential items not relative to the decision process (e.g., personal notes, phone messages, old plans, draft documents, and duplicate documents) it should be discarded. If keeping outdated plans or project descriptions is necessary, the authorized plans and descriptions should be identified with a stamp that says "Permitted Plans" for easy identification. Files should be purged of unnecessary data/forms/notes to save on storage and to aid in future inquiries that might occur such as a request for modification or a compliance investigation.

23. Reporting.

- *Report honestly*
- *All districts are under-funded to meet workload challenges*

Follow the updated QPDS guidelines contained in Appendix III as well as supplemental guidance. When using IPs and NWPs or combinations of NWPs, to authorize a project, report all permits that were issued and not just one corresponding to one project. For example, for a single and complete project involving both a NWP 13 for several hundred feet of bank protection with a NWP 14 minor road crossing, report both the NWP 13 and 14. When using the same NWP to authorize portions of linear projects, report all NWPs and not just one corresponding to one overall project.

PART II

Regulatory Program Priorities

Part II divides the Regulatory Program into six distinct segments and lists the work that should be prioritized within each segment as well as the lower priority work for each segment. It is recognized that all districts are performing lower priority work within these segments and it is not intended that Part II of the SOP dissuade districts from doing so. However, all districts should perform the high priority work within each segment before expending resources on the lower priority work. In addition, the degree to which districts perform lower priority work, along with the associated resource implications, will be considered when responding to districts' requests for additional resources. Part II is a work in progress that we believe will become more comprehensive as revisions are made.

It should be noted that four of the segments have associated ranges of percentages which indicate the percent range of the districts' annual budget that should be expended on that segment. The Public Outreach and Watershed Approaches segments have a lower range of 0% which indicates that these are not mandatory segments of Regulatory Program work. However, these are important segments because their execution can make districts' regulatory efforts more efficient. The other two segments, Mitigation and Staff Support/Administration do not have percentages because they must be accomplished on an as needed basis.

As previously stated, Part II is a work in progress that we believe will become more comprehensive as revisions are made. For example, the lists within each of the segments contain policy as well work types and are somewhat redundant to Part I. We intend to work with the MSCs, as well as the Workload Indicator Task Force, to refine the lists so as to identify priority work types in each of the segments for funding purposes.

WATERSHED APPROACHES (0%-20%)

ABOVE THE LINE

1. Watershed Plans (SAMPs) should result in a more expedient permit process (RGPs, PGPs, no permit by permit alternatives analysis).
2. Focus priority on watersheds with a high volume of regulatory activity.
3. Coordinate Corps programs to minimize conflict and identify potential opportunities.
4. Coordinate with other agencies to minimize conflict and identify potential opportunities.

BELOW THE LINE

5. Be innovative and flexible.
6. Use for cumulative impact assessment.
7. Work/develop partnerships with state/federal/local agencies. Encourage partner funding of watershed plans/efforts.
8. Use USGS mapping and stream classification.
9. Use and encourage GIS analysis
10. Use HGM

PERMIT EVALUATION (60%-80%)

ABOVE THE LINE

1. Resource permit evaluation for timely decisions.
2. Maximize use of lowest form of authorizations (RGP, NWP).
3. Ensure decision documents are concise and minimal length necessary.
4. Ensure scope of analysis is properly defined.
5. Define overall project purpose and ensure alternatives analysis is commensurate with project impacts.

BELOW THE LINE

6. Field wetland delineations for non-"Mom and Pop".
7. Extensive negotiation with other agencies to reach consensus.
8. Project specific EIS, where "significance" of impact is questionable.
9. Expend GRF funds on special studies and external reviews.
10. Multiple site visits/meetings of extensive pre-applications.

PUBLIC OUTREACH (0%-10%)

ABOVE THE LINE
1. Provide public information about the Regulatory Program on website, including Public Notices.
2. Keep website updated/current.
3. Encourage use of Public Meetings in lieu of Public Hearings.
4. Use national survey to get feedback from the regulated public.
5. Use P.A.O. more effectively to deal with media on regulatory matters.

BELOW THE LINE
6. Conduct regular meetings with applicant and public interest groups.
7. Respond to all requests for speakers.
8. Develop/update district's regulatory brochures.
9. Develop extensive information on issues that are related to the regulatory program (e.g., plant guides brochures on wetlands).

ENFORCEMENT (10%-25%)

ABOVE THE LINE
1. Assure permit conditions are enforceable.
2. Implement self-reporting and certification for compliance.
3. Prioritize compliance actions, on high quality environmental loss and navigation.
4. Strive for environmental results in all enforcement actions, including extra mitigation for the loss prior to resolution.
5. Prioritize on-going violations, and consider interim measures.

BELOW THE LINE
6. Low Priority for compliance on 401, CZM, and RPM conditions.
7. Low priority for either unauthorized or compliance effort where low environmental loss is involved.
8. Ensure that litigation and criminal actions are only initiated where clearly appropriate.

STAFF SUPPORT/PROGRAM ADMINISTRATION

ABOVE THE LINE
1. Provide effective training – for staff both Prospect and within district.
2. Optimize use of Field Offices.
3. Provide adequate travel budget for training and field visits.
4. Report all QPDS data accurately and timely.
5. Budget: Schedule 100% / Obligate 100% / Expend 96%.

_____

BELOW THE LINE
6. Additional data entry (beyond that required for QPDS) requirements for PMs.
7. More than two reviewers for NPR, NWPs, RGPs, LOPs, Denials w/o Prejudice.
8. More than three reviewers for SPs, EISs, Denials w/Prejudice, MBIs, and SAMPs.

MITIGATION

ABOVE THE LINE
1. Mitigation should be reasonable, practicable, commensurate with impacts, and what's best for the aquatic environment.
2. Require and review monitoring reports on mitigation banks and other substantial mitigation, including in-lieu fee approaches to assure success.
3. Encourage development/use of mitigation banks; and encourage in-lieu fee approaches, where responsible third parties are available, e.g., State and local agencies, conservation organizations.
4. *Corps* determines appropriate type and level of mitigation and whether mitigation is required.
5. Delegate MBI signature to district, and simplify MBIs.

_____

BELOW THE LINE
6. Compliance inspections for all mitigation.
7. Looking for mitigation options for non-mom and pop applicants.
8. Multiple site visits to a mitigation site.

PART III

Quarterly Permit Data System (QPDS)
Reporting And Definitions

QPDS
Quarterly Regulatory Data

A-1. On a quarterly basis, each district regulatory office will provide the information described herein. Use of the Quarterly Permit Data System (QPDS) is mandatory at both District and Division levels and is a critical tool in the Corps management of the Regulatory Program at all levels. Accuracy of the data, consistent with these definitions is of paramount importance in the assessment of district and division workloads. The same action should not be entered more than once in QPDS. For example, if a meeting is held at a project site, it should not be entered as both a site visit and a public information meeting in QPDS. Districts will enter data in QPDS and forward a copy of the quarterly data file to the division office within 10 calendar days of the end of each quarter. Divisions will prepare a consolidated data file and forward copies of the consolidated and each district data file to the Commander, USACE (ATTN: CECW-OR), within 15 calendar days of the end of each quarter.

A-2. Data elements required for input into QPDS, are defined in Section A-3, paragraphs, 1 through 5. Data categories are:

1. Evaluation Workload.
2. Evaluation Days.
3. Other Workload Items.
4. Staffing.
5. Enforcement Workload.

A-3. Data Element Definitions.

1. Evaluation Workload. Paragraphs A-3: 1(a) through 1(h) only apply to individual permit actions (i.e. standard permits, letters of permission (LOP), and denials) completed during the quarter and should not be used to account for general permits [regional general permits (RGPs), programmatic general permits (PGPs), state program general permits (SPGs), or nationwide permits (NWPs)]. All permit applications will be date stamped immediately upon receipt and logged into the district database [RAMS field = Date Received (table=dates, column=rcvd); RAMS II field = Date Activity Received"]. All applications will be reviewed to

determine the type of permit action and to determine completeness within 15 days of receipt. This review includes a conscious decision by a qualified regulator as to the anticipated type of final action. If an application is incomplete, requests for additional information must be made within the same 15 days. Applications should *not* be routinely logged in as an individual permit, later withdrawn and reentered as another type of permit.

(a) "Carryover". QPDS generates by authority (Section 10, 404, 10/404, or 10/103), the number of pending individual permit applications at the beginning of the quarter. These figures will agree with the number reported for "Pending", 1(h) on the previous quarterly report. If errors were made in the previous quarter, corrections will be made in the current quarter using the "Withdrawn", A-3, 1(c) data element.

(b) "Received". Enter by authority, the number of applications received for evaluation as an individual permit. An application need not be complete to be reported here. Applications that start out as an individual permit (i.e., entered in received category) but, because of project modifications, etc., are found to qualify for a general permit, will be entered in category A-3, 1(c) as withdrawn and reentered in category A-3: 1(i) or 1(j) as a general permit. However, this category will not be used to initially log in actions that clearly should be reviewed under a general permit (RGPs and NWPs).

(c) "Withdrawn". Enter by authority, the number of individual permit applications withdrawn by the applicant or the Corps. This applies to both applications that were considered complete and those that were not. Withdrawal will occur only when: 1) requested by the applicant, 2) there is a conversion from an individual permit to a general permit (NWP, RGP, SPGP or PGP) or no permit required, or 3) there is a clear indication from the applicant that he/she does not intend to take further action. If the applicant responds within 45 days of a written request for information that he/she intends to provide the additional information, the application will not be withdrawn [(33 CFR 325.2(d)(5)]. Any application entered in this category that is resubmitted or reactivated will be considered a new application and entered as "Received". Applications entered in this category are not considered when calculating evaluation times. This category is also to be used to correct errors from the previous quarter (This includes errors in totals due to applications coming in at the last minute, etc.). It is not to be used to report applications that are suspended and then reinstated without significant modifications.

(d) "Standard". Enter, by authority, the number of standard permits issued (i.e., required a public notice). If portions of a project are also authorized by NWP, report each type of NWP used. See paragraph A-3, 1(j).

(e) "LOP". Enter by authority, the number of LOP issued.

(f) "Denials W/P". Enter by authority the number of standard permits denied with prejudice. Denial "with prejudice" occurs when the permit is denied on its merits and the applicant has met all other necessary prerequisites.

(g) "Denials WO/P". Enter by authority the number of standard permits denied without prejudice. Denial "without prejudice" occurs when the permit is denied because the applicant lacks a necessary prerequisite (denial of 401 certification or lack of consistency with the Coastal Zone Management Act).

(h) "Pending". QPDS generates by authority, the number of individual permits in process at the end of the quarter. This figure is calculated using the data from A-3: 1(a) through 1(g). To check your input use the formula: h = [(a + b) -(c + d + e + f + g)].

(i) "Regional". Enter by authority, the number of RGP actions completed, by letter of verification, and the number of actions authorized by PGPs during the quarter. This is not an estimate and includes only those cases where formal contact, via letter, from the Corps has been established with the individual proposing work or the agency administering the PGP. Telephone contacts are not to be included. The issuance of the initial RGP or PGP is not entered.

(j) "NWP". Enter by authority, the number of NWP actions completed, by letter of verification, during the quarter. This is not an estimate and includes only those NWP actions that result in written contact from the Corps with the individual proposing the work. Telephone contacts are not to be included. Per regulation, each type of NWP can only be used once in authorizing a single and complete project. However, two or more different NWPs can be used to authorize portions of a single and complete project. If portions of a project are authorized by more than one type of NWP, report each different type of NWP. Generally each stream crossing (or crossings of other waters of the U.S.) on linear projects (such as pipelines and highways) is considered as a single and complete project (33 CFR 330.2(j)). If NWP conditions are met in such cases, NWPs can be used and reported for each crossing of waters of the U.S. associated with the linear project. Therefore, if one or more NWPs can be used to authorize a linear project, then each crossing of waters of the U.S. should be reported separately in QPDS.

2. Evaluation Days. Evaluation time (in calendar days) begins from the date an application is accepted as complete [33 CFR 325.1 (d)(9)] until the date the permit, denial, or general permit verification letter is mailed to the applicant. For the purpose of calculating evaluation time on standard permits and LOP, the clock stops only when the unvalidated permit is mailed to the applicant for signature [33 CFR 325.2(a)(7)]. No other

action can stop the clock. An applicant's failure to sign or mail back the permit does not affect evaluation time. Withdrawn permits are not included in this calculation. For general permits (RGP, PGP, or NWP), the term "application" includes any request to perform work approved by a general permit. Evaluation time for general permits is calculated from the date the application is complete to the date of the letter of verification. A general permit application is considered completed when all the information required by the predischarge notification (PCN) has been received. If a PCN is not required, then an application is complete when adequate information has been received for the Corps to evaluate the activity.

(a) "Standard". Enter by authority, the total number of evaluation days for the standard permits entered in A-3, 1(d).

(b) "LOP". Enter by authority, the total number of evaluation days for the LOPs entered in A-3, 1(e).

(c) "Denied W/P". Enter by authority, the total number of evaluation days for the permits denied with prejudice entered in A-3, 1(f).

(d) "Denied WO/P". Enter by authority, the total number of evaluation days for the permits denied without prejudice entered in A-3, 1(g).

(e) "Regional". Enter by authority, the total number of evaluation days for the RGP verifications entered in A-3, 1(i).

(f) "NWP". Enter by authority, the total number of evaluation days for the NWP verifications entered in A-3, 1(j).

(g) "Standard & Denial". Enter by range of processing time (0-60, 61-120, over 120) the number of standard permits issued and denied [A-3: 1(d) + 1(f) + 1(g)].

(h) "LOP". Enter by range of processing time, the number of LOPs issued [A-3, 1(e)].

(i) "Regional". Enter by range of processing time, the number of RGP verifications [A-3, 1(i)].

(j) "Nationwide". Enter by range of processing time, the number of NWP verifications [A-3, 1(j)].

(k) Primary Cause of Delay. Enter the primary cause of delay for all standard permits with an evaluation time of greater than 120 days. In cases involving more than one reason for delay, districts should make a determination and enter only the reason that was the most significant, chronologically, in contributing to the delay.

(1) Applicant ("Appl"). Cases where the applicant has delayed evaluation by not providing timely replies to information requests, or has requested a delay.

(2) 401 Certification ("401 Cert"). Cases where the 401 certification process has delayed a permit decision.

(3) Coastal Zone Management ("CZM"). Cases where the CZM certification process has delayed a permit decision.

(4) Historic Properties ("Hist Prop"). Cases where compliance with the National Historic Preservation Act (See Appendix C to 33 CFR 325) has delayed a permit decision.

(5) Section 404(q) Memorandum of Agreement (MOA) informal delays ("404Q/Inf"). Cases resolved through the informal process at the District or Division level.

(6) Section 404(q) formal delays ("404Q/Formal"). Cases requiring formal notification or referral (404(q) MOA Part IV).

(7) Internal Delays ("Internal"). Cases where the cause for delay is a result of something under the control of the Corps but not included in the other categories. A supplemental report explaining the reasons for these delays is required. This category will also include delays related to the Endangered Species Act, until the QPDS report form is modified.

(8) "Total". QPDS generates the total number of permits decisions delayed. This number should equal the total number of standard & denied permits that took over 120 days to evaluate.

3. Other Workload Items.

(a) "No Permit Required". Enter the total number of evaluations that result in a determination, by written verification, that there is no Corps jurisdiction or that result in a finding that no permit is required. Telephone contacts are not to be included.

(b)"Applications Modified". Enter the total number of standard permits issued [A-3, 1(d)] that were substantially modified in order to reduce environmental, or other public interest factor impacts. Modifications can include changes in design, orientation, size and/or location that were made as a result of objections, the 404(b)(1) process, the Section 404 Mitigation MOA, preapplication meetings, etc. This would not include those applications where only minor changes were made that involved minimal effort. Compensatory and/or offsetting mitigation, as required by the sequencing process, is not considered to be a "modification". Multiple modifications to the same application are entered only once.

(c) "Permit Modified". Enter the total number of previously issued individual permits, which were modified, at the request of a permittee, during the quarter. Modifications, which occur during permit evaluation, are entered in paragraph A-3, 3(b). Multiple modifications of the same permit, at different times, should be entered and reported separately [Reference 33 CFR 325.7(b)].

(d) "Site Visits". Enter the total number of all site visits made during the quarter in connection with permit evaluations or jurisdictional determinations (JDs). Multiple visits to a single site are entered separately. Each site visit must be documented by field notes, memorandums for the record, photographs, etc. This category does not include verified JDs/ wetland delineations, preapplication meetings held on site, enforcement or compliance inspections which are entered in paragraph A-3: 3(g), 3(f), and 3(k).

(e) "EIS Pending". Enter the number of regulatory EIS's that are currently being developed with the district as the lead agency. Report each quarter until the final EIS is issued.

(f) "EIS Commented On". Enter the number of EIS's, including cooperating agency EISs, where the Regulatory office provided written comments to other agencies or offices. This includes any EIS's, not just ones requiring regulatory action.

(g) "Jurisdiction Determinations (Office)". Enter the number of JDs made in the office, without recourse to a site visit. This includes verifying wetland delineations, ordinary high water determinations made in association with permit applications and enforcement actions. Determinations must be followed by written documentation. Only enter an application/project in this category once. If both office and field determinations are performed on a project, only the field JD will be reported in paragraph A-3, 3(h).

(h) "Jurisdiction Determinations (Field)". Enter the number of jurisdiction determinations made through a field visit. This includes verification of wetland delineations, ordinary high water determinations made in association with permit applications and enforcement actions. Determinations must be followed by written documentation. Only enter an application/project in this category once. If both office and field determinations are performed on a project, only the field JD will be reported.

(i) "Public Hearings". Enter the number of public hearings, as described in 33 CFR 327 that were held during the quarter.

(j) "Public Information Meetings". Enter the number of public meetings held to publicize the Regulatory Program. This includes participation in workshops, seminars, and meetings on specific permit actions held in lieu of a public hearing. This category does not include routine meetings with applicants, such as preapplication meetings.

(k) "Pre-application Consultations". Enter the number of pre-application consultations. This process may or may not involve other agencies. Telephone inquiries from potential applicants or telephone calls with other agencies are not included. It is possible to have more than one pre-application consultation on a proposed project or enforcement action.

4. Staffing. This section will be used to obtain information on regulatory full time equivalents (FTEs). FTE levels are to be reported quarterly. The numbers reported should be consistent with FORCON numbers.

(a) "FTE Allocated". Enter the number of FTEs allocated by the District Engineer, or his designee, for the fiscal year. This number will include district and field regulatory staff only, and not counsel, planning, or other district elements. This number may or may not change from quarter to quarter, depending on senior management (above regulatory element level) and FORCON adjustments.

(b) "FTE Expended". Enter the number of FTEs actually expended by the regulatory office (district and field regulatory staff only). This number will change from quarter to quarter and will reflect actual staffing levels through the current quarter; it is cumulative. The annual (5th quarter) report will be the same as the fourth quarter.

5. Enforcement / Compliance Workload.

(a) "Carryover". QPDS generates by authority (Section 10, 404, 10/404, 10/103), the number of unresolved, unauthorized activities (UA) files at the beginning of the quarter. These figures must agree with the numbers entered for "Pending," A-3, 5(j), on the previous quarterly report. If errors were made in the previous quarter, corrections will be made in the current quarter using the "Other Resolution" data element.

(b) "Reported". Enter by authority, the total number of UAs reported or detected.

(c) After-the-Fact ("ATF"). Enter by authority, the number of UAs resolved by acceptance of an ATF permit. The UA file should not be closed before a decision is reached on the ATF application.

(d) Voluntary Restoration ("REST"). Enter by authority, the number of UA files closed after obtaining voluntary restoration, without litigation.

(e) Litigation ("LIT"). Enter by authority, the number of UA cases closed after (1) obtaining a Federal court decision (e.g. consent decree, pre-trial settlement, court order, etc.), and (2) the completion of all required work. Since additional enforcement work will likely be required, do not close the file at the time it is sent to the Department of Justice (DOJ) for consideration of legal action.

(f) Penalty ("PEN"). Enter by authority, the number of UA cases closed after obtaining a penalty pursuant to Section 309 of the Clean Water Act.

(g)Administrative Closure ("ADMIN"). Enter by authority, those UA cases closed after DOJ, U.S. Environmental Protection Agency (USEPA), and/or Corps Office of Counsel (OC) declines to take action. It is not necessary for the regulatory element to inquire on each and every case. Cases may be closed where it is known that DOJ, USEPA, and/or OC have not pursued this type of case in the past [33 CFR 326.5(e)].

(h) Other Resolution ("OTHER"). Enter by authority, the total number of activities that do not result in an enforcement case but require an initial investigation, such as no jurisdiction, work covered by an existing authorization, or discretionary closure by the Corps. This data element is also used to correct any errors made in the previous quarterly report.

(i) "Total". QPDS generates by authority, the number of UA cases resolved at the end of the quarter. This figure is calculated using the data from A-3: 5(a) through 5(h). To check your input use the formula: $i = (c + d + e + f + g + h)$.

(j) "Pending". QPDS generates by authority, the number of UA cases unresolved at the end of the quarter. This figure is calculated using the data from A-3: 5(a) through 5(i). To check your input use the formula: $j = [(a + b) - i]$. Pending files include those submitted to DOJ, USEPA, or OC for legal action that have not been closed.

(k) "UA Site Visits". Enter the total number of on-site visits made in conjunction with a reported or on-going enforcement action. This includes all site visits made in connection with enforcement actions. Multiple visits to a single site are entered separately. Documentation of each visit must be made in writing. This data element does not include verified JDs/wetland delineations, which should be entered in paragraph A-3, 3(g), or 3(h).

(l) "UAs Detected by COE". Enter the number of UAs detected by the Corps. This number will constitute part of the total in data element A-3, 5(b). The number is not limited to those detected by Corps regulatory personnel, but includes all actions detected by any Corps personnel.

(m) "Carryover". QPDS generates the total number of noncompliance (NC) cases unresolved by using data from the previous quarter. If errors were made in the previous quarter corrections will be made using the "Minor Resolution" data element of the current quarter.

(n) Compliance Inspections ("Comp Insp"). Enter the total number of compliance inspections conducted during the quarter. This includes only the inspection of permitted activities. Multiple visits to the same site are entered separately. Documentation of each visit must be made in writing.

(o) Noncompliance ("Noncom"). Enter the total number of inspections that resulted in a determination of NC with the terms and/or conditions of a permit.

(p) "MOD". Enter the total number of NC cases resolved through modification of the permit.

(q) Voluntary Restoration ("Vol Rest"). Enter the total number of NC cases resolved by voluntary restoration, without litigation.

(r) Litigation ("LIT"). Enter the total number NC cases resolved by legal action (e.g. court order, consent decree, fine, etc.) at the DOJ level.

(s) "Penalty". Enter the total number of NC cases resolved through an administrative civil penalty.

(t) Minor Resolution ("Minor"). Enter the total number of NC cases closed based on a determination that neither the OC or DOJ will not pursue the case and the minor nature of the activity does not justify further administrative action. Cases may be closed where it is known that OC or DOJ has in the past, not taken this type of case. It is not necessary for the regulatory element to inquire on each and every case. This data element is also used to correct any errors made in the previous quarterly report and as a "catch all" for NC cases that are not resolved by one of the methods in paragraphs A-3: 5(p) through 5(s).

(u) "Pending". QPDS generates the number of NC cases unresolved at the end of the quarter. This figure is calculated using the data with the formula: $u = [(m + o) - (p + q + r + s + t)]$. Pending files include those submitted to OC and/or DOJ for legal action, until the case is closed.

6. Attached for reference are representations of the data entry screens as they appear when using the QPDS program.

Appendix H

Selected Attributes of
40 Common Wetland
Functional Assessment Procedures

TABLE H-1 Selected Attributes of 40 Common Wetland Functional Assessment Procedures

Assessment Procedure[1]	Scale[2] Non-tidal	Habitat Type[3]	Functions[4]	Social Values, Red Flags
AREM	SS	TW	HA	+
Coastal method	SS, WS	NT	HY,WQ,HA	+
CT method	SS	TW	HY,WQ,HA	+
Descriptive approach	SS	NT,TW	HY,WQ,HA	+
EPW	SS	NT,TW	HY,WQ,HA	+
HAT	SS	NT,TW	HA	+
HEP	SS	NT,TW,UP	HA	
HGM approach	SS	NT,TW	HY,WQ,HA	+
Hollands-Magee method	SS	TW	HY,WQ,HA	+
IBI	SS,LS	NT,TW,UP	5	
Interim HGM	SS	NT,TW	HY,WQ,HA	+
IVA	SS	NT,TW	HY,WQ,HA	+
Larson method	SS	NT	HY,HA	+
MDE method	SS	NT	HA	+
ME Tidal method	SS,WS	TW	HA	+
MNRAM	SS	NT	HY,WQ,HA	+
MT form	SS	NT	HY,WQ,HA	+
NBM	SS,WS	TW	HY,WQ,HA	
NC-CREWS	LS	NT,TW	HY,WQ,HA	
NC guidance	SS	NT	HY,WQ,HA	+
NEFWIBP	SS	NT	WQ(?),HA	+
NH method	SS	NT	HY,WQ,HA	+
NJ watershed method	LS	NT	HY,WQ,HA	
OFWAM	SS	NT	HY,WQ,HA	+
PAM HEP	SS	NT,TW,UP	HA	
PFC	SS	NT,TW,UP	HY,WQ,HA	
RA	SS	NT,TW	HY,WQ,HA	+
Rapid Assessment Procedure	WS	NT	HY,WQ,HA	+
Synoptic Approach	LS	NT,TW,UP	HY,WQ,HA	+
VIMS Method	SS	NT	HY,WQ,HA	+
WAFAM	SS	NT,TW	HY,WQ,HA	
WCHE	SS,LS	NT	HA	
WET	SS	NT,TW	HY,WQ,HA	+
WEThings	SS	NT,TW	HA	
WHAMS	SS	NT,TW,UP	HA	
WHAP	SS	NT,TW,UP	HA	
WI RAM	SS	NT	HY,WQ,HA	+
WQI	SS	NT	HY,WQ,HA	
WRAP	SS	NT,TW	HY,WQ,HA	
WVA	SS	NT,TW	HA	

[1]AREM = Avian Richness Evaluation Method (Adamus 1994); Coastal Method (Cook et al. 1993); CT Method = Connecticut Method (Ammann et al. 1996); Descriptive Method = Wetland Functions and Values: A Descriptive Method (USACE 1995); EPW = Evaluation for Planned Wetlands (Bartoldus et al. 1994); HAT = Habitat Assessment Technique (Cable

Weighting Applied	Scaled by Area	Functional Categories Assessed[6]	Method Description[7]	Factor Measured[8]	Number States Where Used[9]
+	+	AV,HS,SC	M	I	3
		F	M	Q	3
	+	FV	M	Q	3
+		F,V	L	Q	5
	+	F	M	E	5
	+	HQ	M	BS	3
	+	HS	M	V	50
	+	F	M	V	6
		F	M	E	8
		BC	M	M	5
	+	F	M	V	9
+	+	F,V	M		2
+	+	RF	C	C,V	11
+	+	F	M	I	1
+(?)		EI,F,V	M	Q	1
		F	L	Q	1
+	+	F,V	M	FM	1
		EH,TR	L	Q	4
		SF,F,ES	M	V	1
+		F,V	FC	F	2
		BI	M	I,M	6
	+	FV	M	Q	7
+		EI,PI	M	I	1
+(?)		F	AC	Q	2
+	+	HS	M	V	3
		PFC	CL	E	15
		F,V	BPJ		10
	+	F	M	V	15
+	+	F,V,FL,RP	M	I	15
		F	L	F	2
	+	F	M	V	1
	+	PS,TS,NR	M	V	1
+		F,V	L	Q	50
		HP	M	Q	7
	+	HS	M	V	1
+	+	HQ	FC	E	1
		FV	L	Q	5(?)
+		WQ	EM	V	1
		F	RI	V	1
		HS	M	V	1

et al. 1989); HEP = Habitat Evaluation Procedure (USFWS 1980); HGM Approach = Hydrogeomorphic Approach (Smith et al. 1995); Hollands-Magee Method = A Method for Assessing the Functions of Wetlands (Hollands and Magee 1985); IBI = Index of Biological Integrity (Karr 1981); Interim HGM (NRCS 1997); IVA = Indicator Value Assessment (Hruby et al. 1995); Larson Method = Methods for Assessment of Freshwater Wetlands (Golet 1976); MDE Method = Method for the Assessment of Wetland Function (Fugro East, Inc. 1995); ME Tidal Method = Maine Citizens Tidal Marsh Guide (Bryan et al. 1997); MNRAM = Minnesota Routine Assessment Method (MBWSR 1998); MT Form = Montana Wetland Field Evaluation Form (Berglund 1996); NBM = Narragansett Bay Method (Lipsky 1996); NC-CREWS = North Carolina Coastal Regional Evaluation of Wetland Significance (Sutter and Wuenscher 1996); NC Guidance = Guidance for Rating the Values of Wetlands in North Carolina (NCDENR 1995); NEFWIBP = New England Freshwater Wetlands Invertebrate Biomonitoring Protocol (Hicks 1997); NH Method = New Hampshire Method (Ammann and Stone 1991); NJ Watershed Method = Watershed-Based Wetland Assessment Methodology (Zampella et al. 1994); OFWAM = Oregon Freshwater Wetland Assessment Methodology (Roth et al. 1996); PAM HEP = Pennsylvania Modified 1980 Habitat Evaluation Procedure (Palmer et al. 1985); PFC = Process for Assessing Proper Functioning Condition (Pritchard et al. 1993); RA = Wetland Assessment: A Regulatory Assessment Method (Kusler 1997); Rapid Assessment Procedure = Rapid Assessment Procedure for Assessing Wetland Functional Capacity (Magee 1998); Synoptic Approach = Synoptic Approach for Wetlands Cumulative Effects Analysis (Leibowitz et al. 1992); VIMS Method = Technique for the Functional Assessment of Virginia Coastal Plain Nontidal Wetlands (Bradshaw 1991); WAFAM = Washington State Wetland Functional Assessment Method (Hruby et al. 1998); WCHE = Wildlife Community Habitat Evaluation (Schroeder 1996); WET = Wetland Evaluation Technique (Adamus et al. 1987); WEThings (Whitlock et al. 1994); WHAMS = Wildlife Habitat Assessment and Management System (Palmer et al. 1993); WHAP = Wildlife Habitat Appraisal Procedure (Frye 1995); WI RAM = Wisconsin Rapid Assessment Methodology (WDNR 1992); WQI = Wetland Quality Index (Lodge et al. 1995); WRAP = Wetland Rapid Assessment Procedure (Miller and Gunsalus 1997); WVA = Wetland Value Assessment Methodology (LDNR 1994).

[2]SS = site specific, which assesses functions at all parts of wetlands; WS = wetland system, designed to look at entire wetland system as compared to site-specific projects; LS = landscape, wherein functions of wetlands are assessed within the mosaic of wetlands and uplands at the landscape scale.

[3]NT = nontidal wetlands, TW = tidal wetlands, UP = uplands.

[4]HY = hydrological, WQ = water quality, HA = habitat, O = other.

[5]IBI measures biological condition using samples of living organisms. Biological condition is expressed as the divergence from biological integrity. Most other procedures use measures of habitat structure to assess habitat suitability (Bartoldus 1999).

[6]AR = avian richness, BC = biological condition, BI = biological integrity, EH = ecological health, EI = ecological integrity, ES = ecological significances, F = function(s), FL = functional loss, FV = functional value, HP = habitat potential, HQ = habitat quality, HS = habitat suitability, NR = native richness, PI = potential impact, PFC = proper functioning condition, PS = plot suitability, RF = resource factor (function), RP = replacement potential, SC = species composition, SF = subfunction, TR = tidal restrictions, TS = tract suitability, V = value, WQ = wetland quality.

[7]AC = assessment criteria, BPJ = best professional judgment, C = classification, CL = checklist, EM = evaluation matrix, FC = flow chart or evaluation key, L = list of considerations or questions, M = assessment, functional or habitat relationship model, RI = rating index.

[8]BS = bird species, C = assessment criteria, E = element or component, F = factor, FM = field measurement, I = indicator, M = metric, Q = evaluation question or considerations, V = variable or parameter.

[9]Including adapted or considered applicable.

[10]RA has not been used in any state but reflects the overall approach applied in many.

SOURCE: Adapted Bartoldus (1999).

References

Adamus, P.R. 1994. User's Manual Avian Richness Evaluation Method (AREM) for Lowland Wetlands of the Colorado Plateau. EPA/600/R93/240. U.S. Environmental Protection Agency, Environmental Research Laboratory, Corvallis, OR.

Adamus, P.R., E.J. Clairain, R.D. Smith, and R.E. Young. 1987. Wetland Evaluation Technique (WET), Vol. II: Methodology. NTIS ADA 189968. Vicksburg, MS: Department of the Army, Waterways Experiment Station.

Ammann, A.P., and A.L. Stone. 1991. Method for the Comparative Evaluation of Nontidal Wetlands in New Hampshire. NHDES-WRD-1991-3. Concord, NH: New Hampshire Department of Environmental Services.

Ammann, A.P., R.W. Frazen, and J.L. Johnson. 1996. Method for the Evaluation of Inland Wetlands in Connecticut. DEP Bulletin 9. Connecticut Department of Environmental Protection, Hartford, CT.

Bartoldus, C.C. 1999. A Comprehensive Review of Wetland Assessment Procedures: A Guide for Wetland Practitioners. St. Michaels, MD: Environmental Concern, Inc. 196 pp.

Bartoldus, C.C., E.W. Garbisch, and M.L. Kraus. 1994. Evaluation for Planned Wetlands (EPW): A Procedure for Assessing Wetland Functions and a Guide to Functional Design. St. Michaels, MD: Environmental Concern, Inc.

Berglund, J. 1996. MDT Montana Wetland Field Evaluation Form and Instructions. Prepared for the Montana Department of Transportation, Helena. (July 1, 1996; Appendix A revised 9/23/1997). 19 pp+ appen.

Bradshaw, J.G. 1991. A Technique for the Functional Assessment of Nontidal Wetlands in the Coastal Plain of Virginia. Special Report 315 in Applied Marine Science Ocean Engineering. Gloucester Point, VA: Virginia Institute Marine Science, College of William and Mary.

Bryan, R.R., M. Dionne, R. Cook, J. Jones, and A. Goodspeed. 1997. Maine Citizens Guide to Evaluating, Restoring, and Managing Tidal Marshes. Brunswick, ME: Maine Audubon Society.

Cable, T.T., B. Brack, Jr., and V.R. Holmes. 1989. Simplified method for wetland habitat assessment. Environ. Manage. 13(2):207-213.

Cook, R.A., A.J.L. Stone, and A.P. Ammann. 1993. Method for the Evaluation and Inventory of Vegetated Tidal Marshes in New Hampshire: (Coastal Method). Concord, NH: Audubon Society of New Hampshire.

Frye, R.G. 1995. Wildlife Habitat Appraisal Procedure (WHAP). PWD RP N7100-145 (2/95) Texas Parks and Wildlife Department, Austin, TX. [Online]. Available: http://www.tpwd.state.tx.us/conserve/whap/body.html [May 20, 2001].

Fugro East, Inc. 1995. A Method for the Assessment of Wetland Function. Prepared for Maryland Department of the Environment, Baltimore, by Fugro East, Inc., Northborough, MA.

Golet, F.C. 1976. Wildlife wetland evaluation model. Pp. 13-34 in Models for Assessment of Freshwater Wetland, J.S. Larson, ed. Pub. 32. Water Resource Research Center, University of Massachusetts, Amherst.

Hicks, A.L. 1997. New England Freshwater Wetlands Invertebrate Biomonitoring Protocol (NEFWIBP). The Environmental Institute, University of Massachusetts at Amherst.

Hollands, G.G., and D.W. Magee. 1985. A method for assessing the functions of wetlands. Pp. 108-118 in Proceedings of the National Wetland Assessment Symposium, J. Kusler and P. Riexinger, eds. Association of Wetlands Managers, Berne, N.Y.

Hruby, T., W.E. Cesanek, and K.E. Miller. 1995. Estimating relative wetland values for regional planning. Wetlands 15(2):93-107.

Hruby, T., T. Granger, K. Brunner, S. Cooke, K. Dublanica, R. Gersib, L. Reinelt, K. Richter, D. Sheldon, A. Wald, and F. Weinmann. 1998. Methods for Assessing Wetland Functions. Vol. I. Riverine and Depressional Wetlands in the Lowlands of Western Washington. Washington State Department of Ecology Pub. 98-106. Olympia, WA: Department of Ecology.

Karr, J.R. 1981. Assessment of biotic integrity using fish communities. Fisheries 6(6):21-27.

Kusler, J. 1997. Wetland Assessment: A Regulatory Assessment (RA). Association of State Wetland Managers, Institute of Wetland Science and Public Policy, Berne, NY.

LDNR (Louisiana Department of Natural Resources). 1994. Habitat Assessment Models for Fresh Swamp and Bottomland Hardwoods within the Louisiana Coastal Zone. Louisiana Department of Natural Resources, Baton Rouge, LA.

Leibowitz, S.G., B. Abbruzzese, P.R. Adamus, L.E. Hughes, and J.T. Irish. 1992. A Synoptic Approach to Cumulative Impact Assessment: A Proposed Methodology. EPA/600/R-92/167. U.S. Environmental Protection Agency, Environmental Research Laboratory, Corvallis, OR.

Lipsky, A. 1996. Narragansett Bay Method: A Manual for Salt Marsh Evaluation. Save the Bay, Providence, RI.

Lodge, T.E., H.O. Hillestad, S.W. Carney, and R.B. Darling. 1995. Wetland Quality Index (WQI): A Method for Determining Compensatory Mitigation Requirements for Ecologically Impacted Wetlands. Proceedings of American Society of Civil Engineers South Florida Section Annual Meeting, Sept. 22-23, 1995, Miami, FL.

Magee, D.W. 1998. A Rapid Procedure for Assessing Wetland Functional Capacity based on Hydrogeomorphic (HGM) Classification. Berne, NY: Association of State Wetland Managers.

Miller, R.E., and B.E. Gunsalus. 1997. Wetland Rapid Assessment Procedure (WRAP). Tech. Pub. REG-001. West Palm Beach, FL: South Florida Water Management District, National Resource Management Division.

MBWSR (Minnesota Board of Water and Soil Resources). 1998. Minnesota Routine Assessment Method for Evaluating Wetland Functions (MinRAM). Draft Version 2.0. Minnesota Board of Water and Soil Resources, St. Paul, MN.

NCDENR (North Carolina Department of Environment and Natural Resources). 1995. Guidance for Rating the Values of Wetlands in North Carolina. North Carolina Department of Environment and Natural Resources, Raleigh, NC.

NRCS (Natural Resource Conservation Service). 1997. Interim Guidance for Making Wetland Minimal Effect/Mitigation Decisions for USDA Programs in Alaska. August 1997. [Online]. Available: http://159.189.24.10/wlistates/alaska.htm. [May 10, 2001].

Palmer, J.H., M.T. Chezik, R.D. Heaslip, G.A. Rogalsky, D.J. Putman, R.W. McCoy, and J.A. Arway. 1985. Pennsylvania Modified 1980 Habitat Evaluation Procedure Instruction Manual. U.S. Fish and Wildlife Service, State College, PA.

Palmer, J.H., R.H. Muir, and T.M. Sabolcik. 1993. Wildlife Habitat Assessment and Management System: Habitat Evaluation Procedure Technology for Wildlife Management Planning. Pennsylvania Game Commission, Bureau of Land Management, Harrisburg, PA.

Pritchard, D., H. Barrett, J. Cagney, R. Clark, J. Fogg, K. Gebhart, P. L. Hansen, B. Mitchell, and D. Tippy. 1993. Riparian Area Management: Process for Assessing Proper Functioning Condition. TR 1737-9. BLM/SC/ST-93/003+1737+REV95+REV98, U.S. Dept. of Interior, Bureau of Land Management, Service Center, CO.

Roth, E., R. Olson, P. Snow, and R. Sumner. 1996. Oregon Freshwater Wetland Assessment Methodology, 2nd Ed., Oregon Division of State Lands, Salem, OR.

Schroeder, R.L. 1996. Wildlife Community Habitat Evaluation: A Model for Deciduous Palustrine Forested Wetlands in Maryland. Tech. Report WRP-DE-14. Vicksburg, MS: U.S. Army Engineers Waterways Experiment Station.

Smith, R.D., A. Ammann, C. Bartoldus, and M.M. Brinson. 1995. An Approach for Assessing Wetland Functions Using Hydrogeomorphic Classification, Reference Wetlands, and Functional Indices. Tech. Report WRP-DE-9. Vicksburg, MS: U.S. Army Engineers Waterways Experiment Station.

Sutter, L.A., and J.R. Wuenscher. 1996. NC-CREWS: A Wetland Functional Assessment Procedure for the North Carolina Coastal Area. Draft. Division of Coastal Management, North Carolina Department of Environment and Natural Resources, Raleigh, NC. 61 pp + appen.

USACE (U.S. Army Corps of Engineers). 1995. The Highway Methodology Workbook Supplement. Wetland Functions and Values: A Descriptive Approach. NENEP-360-1-30a. Waltham, MS: U.S. Army Corps of Engineers, New England Division.

USFWS (U.S. Fish and Wildlife Service). 1980. Habitat Evaluation Procedure (HEP) Manual (102ESM). Fort Collins, CO: U.S. Fish and Wildlife Service.

WDNR (Wisconsin Department of Natural Resources). 1992. Rapid Assessment Methodology for Evaluating Wetland Functional Values. Wisconsin Department of Natural Resources.

Whitlock, A.L., N.M. Jarman, and J.S. Larson. 1994. WEThings: Wetland Habitat Indicators for Non-game Species. Vol. II. TEI Pub. 94-2. Amherst, MA: The Environmental Institute, University of Massachusetts.

Zampella, R.A., R.G. Lathrop, J.A. Bognar, L.J. Craig, and K.J. Laidig. 1994. A Watershed-based Wetland Assessment Method for the New Jersey Pinelands. Pinelands Commission, New Lisbon, NJ.

Appendix I

Functions, Factors, and Values Considered in Section 404 Permit Reviews

In 33 CFR 320.4, there is a series of 17 factors that must be reviewed prior to making a final determination on the project. Paragraphs (a) and (b) of this section are the public interest review and the review of the project's effect on wetlands. Both list a series of functions that must be considered in each case, where appropriate.

Factors considered in the "Public Interest" review include[1]

conservation,
aesthetics,
wetlands,
fish and wildlife values,
floodplain values,
navigation,
recreation,
water quality,
safety,
mineral needs,

economics,
general environmental concerns,
historic properties,
flood hazards,
land use,
shore erosion and accretion,
water supply and conservation,
energy needs,
food and fiber production,
considerations of property ownership,
and the needs and welfare of the people.

---

[1]Factors listed in 33 CFR 320.4(a)(1).

Functions, factors and values considered in the "Effect on Wetlands" review include[2]

significant natural biological functions, including food chain production, general habitat and nesting, spawning, rearing and resting sites for aquatic or land species;

study of the aquatic environment or as sanctuaries or refuges;

natural drainage characteristics;

sedimentation patterns;

salinity distribution;

flushing characteristics, current patterns, or other environmental characteristics;

shielding other areas from wave action, erosion, or storm damage;

storage areas for storm and flood waters;

ground water discharge areas that maintain minimum baseflows important to aquatic resources;

prime natural recharge areas;

significant water purification functions; and

wetlands which are unique in nature or scarce in quantity to the region or local area.

---

[2]33 CFR § 320.4 (b)(2) Public Interest Review

Appendix J

Biographical Sketches of Committee Members

JOY ZEDLER (*Chair*) is a professor of botany and Aldo Leopold Chair of Restoration Ecology at the University of Wisconsin. She earned her Ph.D. in botany from the University of Wisconsin. Previously she was a professor of biology at San Diego State University and director of the Pacific Estuarine Research Laboratory. Dr. Zedler's research interests include wetland ecology, structure and functioning of restored ecosystems, use of scientific information in management and restoration, and use of mesocosms in experimental studies. She has served as a member of the National Research Council's Committee on Characterization of Wetlands, the Committee on Restoration of Aquatic Ecosystems, and the Water Science and Technology Board. She is a member of the Nature Conservancy Governing Board and the Environmental Defense Fund Board of Trustees.

LEONARD SHABMAN (*Vice Chair*) is a professor in the Department of Agricultural and Applied Economics at Virginia Polytechnic Institute and State University and director of the Virginia Water Resources Research Center. He earned his Ph.D. in resource and environmental economics from Cornell University. His research interests include water supply, water quality, and flood hazard management; fishery management; and the role of economists in public policy formulation. Dr. Shabman was a member of the National Research Council's Committee on Watershed Management; the Committee on U.S. Geological Survey Water Resources Research; the Committee on Flood Control Alternatives in the American

294

River Basin; and the Committee on Restoration of Aquatic Ecosystems: Science, Technology, and Public Policy.

VICTORIA ALVAREZ is senior environmental planner and liaison to the U.S. Army Corps of Engineers, San Francisco District for the California Department of Transportation (CalTrans). She earned a B.A. (botany) and M.A. (biology) from California State University. Her professional experience includes research, regulation, consulting, and management involving plant biotechnology, endangered species identification, vegetative mapping, botanical studies, Clean Water Act regulation, compensatory habitat mitigation and restoration, and mitigation monitoring. She represented the California Department of Transportation at the National Research Council's Transportation Research Board's Environmental Research Needs in Transportation Conference in November 1996, where she participated on the Wetlands Subcommittee.

ROBERT O. EVANS is associate professor and extension leader in the Department of Biological and Agricultural Engineering at North Carolina State University, where he earned his Ph.D. in 1991. His research interests include wetland restoration of prior converted cropland; instream wetland construction, restoration, and enhancement; establishment of riparian vegetative buffers; and restoration and enhancement of riparian wetlands. He has 15 years of experience applying, testing, and validating hydrological simulation models for a variety of field applications, including wetland delineation and restoration. Dr. Evans currently serves on the North Carolina Senate Select Committee on Coastal River Water Quality and Fish Kills and is technical adviser to the Water Quality Section, North Carolina Division of Water Quality, North Carolina Wetlands Restoration Program, North Carolina Conservation Enhancement and Reserve Program, and North Carolina Irrigation Society. He is also chair of the American Society of Civil Engineers (ASCE) Water Resources Engineering Division's Water Quality Committee and the ASCE Task Committee establishing "Guidelines for Restoration of Prior Converted Cropland."

ROYAL C. GARDNER is a professor of law and director of Graduate and International Programs at Stetson University College of Law in St. Petersburg, Florida. Previously he served in the Army General Counsel's office as the Department of the Army's principal wetland attorney, advising the Assistant Secretary of the Army (civil works) on legal and policy issues related to administration of the Clean Water Act Section 404 program by the Corps of Engineers.

J. WHITFIELD GIBBONS is a professor of ecology and a senior research ecologist at the Savannah River Ecology Laboratory in Aiken, South Carolina, and a research associate in the Department of Vertebrate Zoology at the National Museum of Natural History, Smithsonian Institution. Dr. Gibbons earned his Ph.D. (zoology) at Michigan State University and M.S. and B.S. (biology) degrees at the University of Alabama. His research interests focus on the population dynamics and ecology of aquatic and semiaquatic vertebrates and has involved detailed population studies of fish, amphibians, and reptiles, particularly turtles. An emphasis of his research has been on the application of basic research to environmental impact and conservation issues, particularly regarding natural and degraded wetlands.

JAMES WENDELL GILLIAM is William Neal Reynolds Professor of Soil Science at North Carolina State University. He earned his Ph.D. at Mississippi State University. His research interests include the effects of agriculture, forestry, and various best management practices on the quality of water draining from watersheds; water management to increase yields and minimize nonpoint-source pollution; and the effectiveness of riparian wetlands and other vegetative filters for improving the quality of agricultural drainage waters. He was the 1999 recipient of the Soil Science Society of America's Applied Research Award. In 1996 he received the U.S. Department of Agriculture Award for Superior Service for Research on Riparian Buffers. Dr. Gilliam also served as a member of the National Research Council's Committee on Long-Range Soil and Water Conservation.

CAROL A. JOHNSTON is a senior research associate with the Natural Resources Research Institute at the University of Minnesota in Duluth; currently she is on assignment as program director for the Ecosystem Studies Program at the National Science Foundation. She earned her Ph.D. in soil science at the University of Wisconsin. Her research interests include wetland soils, biogeochemistry, and mapping; effects of land/water interactions on surface water quantity and quality; spatial and temporal variability of wetland processes; and geographic information systems. Dr. Johnston was a member of the National Research Council's Water Science and Technology Board for 6 years, and served as its vice-chair from 1997 to 2000. She previously served on the National Research Council's Committee on Watershed Management and Committee on Wetlands Characterization.

WILLIAM J. MITSCH is a professor of natural resources and environmental science at Ohio State University and director of the Olentangy

River Wetland Research Park. He earned his Ph.D. in environmental engineering sciences (systems ecology) from the University of Florida. His research interests include wetland ecology and management, wetland restoration and creation, wetland biogeochemistry, ecological economics of wetlands and other ecosystems, ecological engineering, ecosystem ecology and modeling, wetlands and global climate change, wetland vegetation dynamics, and primary productivity in aquatic systems. Dr. Mitsch is editor-in-chief of *Ecological Engineering.* He served as a member of the National Research Council's Committee on Characterization of Wetlands.

KAREN PRESTEGAARD is an associate professor of geology at the University of Maryland, College Park. She earned her Ph.D. (geology) from the University of California, Berkeley. Her research interests include mechanisms of streamflow generation and their variations with watershed scale, geology, and land use; effects of floods on rivers; relationships among channel morphology, channel hydraulics, and aquatic ecology; and hydrology of coastal and riparian wetlands. Dr. Prestegaard served on the National Research Council's Committee on Yucca Mountain Peer Review: Surface Characteristics.

ANN M. REDMOND is the Regional Manager of Ecological and Water Resources for the Tallahassee office of WilsonMiller, Inc. She was formerly Vice President of Development for Wetlandsbank, Inc., where she was responsible for Wetlandsbank's expansion into new markets, as well as its regulatory and legislative activities. Ms. Redmond was with the Florida Department of Environmental Protection (DEP) for 12 years as the agency's expert on mitigation and mitigation banking, leading the development and implementation of rules and legislation in these areas. She has worked on the assessment of watershed-level cumulative impacts due to development impacts, watershed and ecosystem management initiatives, the conceptual framework for restoration planning, development of functional assessment methods, and sustainability of mitigation projects. Prior to her tenure at DEP, she worked with a regional water management agency and as an environmental consultant. She earned her Bachelor and Master of Science degrees in Biological Sciences from Florida State University, with emphases on botany and ecology.

CHARLES SIMENSTAD is a research associate professor at the School of Aquatic and Fishery Sciences, University of Washington, Seattle, where he earned B.S. and M.S. degrees in fisheries. He is also coordinator of the University of Washington's Wetland Ecosystem Team. His research interests include estuarine and nearshore marine ecosystem structure and food web dynamics; estuarine ecology of juvenile salmonid fishes; landscape structure of estuarine fish habitats; effects of anthropogenic change to estuarine habitat support of fish and shellfish; coastal wetland restoration ecology; and functional assessment of restored, created, and enhanced wetlands. He serves on the editorial board of *Estuaries* as associate editor for habitat restoration and wetlands.

R. EUGENE TURNER is a professor in the Department of Oceanography and Coastal Sciences and director of the Coastal Ecology Institute at Louisiana State University. He earned a B.A. degree from Monmouth College (Illinois), an M.S. degree from Drake University, and a Ph.D. from the University of Georgia. He is chairman of INTECOL Wetlands Working Group and treasurer of INTECOL and is active in scientific matters concerning aspects of coastal environmental management, including the low-oxygen zone off the Mississippi River (the DEAD ZONE). He serves on the Book Board of the American Geophysical Union (AGU), is managing editor of AGU's *Coastal and Estuarine Science Series*, and is editor-in-chief of *Wetlands Ecology and Management*. He served as a member of the National Research Council's Committee on the Role of Technology in Marine Habitat Protection and Enhancement.

Glossary

33 CFR 331—Regulations of the USACE that govern how the Corps implements its regulatory program.

40 CFR 230—Regulations promulgated by the EPA that the Corps applies when determining whether to issue a Section 404 permit.

Abiotic—Refers to nonliving components of an ecosystem.

ADID—Advance identification.

Advance identification (ADID)—Process of the CWA in which studies are undertaken by EPA in cooperation with the Corps and in consultation with states and tribes to collect information on the location and functions of wetlands in specified areas.

Avoidance—Design of a facility so as to have no impact on wetlands; first and most desirable of the sequencing steps in wetland mitigation.

California Coastal Act—1976 act that regulates coastal development, with specific guidelines for wetland habitat impacts.

CEQ—Council on Environmental Quality.

Clean Water Act (CWA)—Primary federal law that establishes a dredge-and-fill permit program to protect the waters of the United States, including many wetlands. When originally enacted, the statute was formally known as the Federal Water Pollution Control Act (FWPCA). In 1977 amendments to the FWPCA, Congress adopted the statute's more popular name, the Clean Water Act.

Compensation wetlands planning—Process using a watershed-based approach that involves advance planning to protect and restore wetland resources within the watersheds.

Compensatory mitigation—Final step in the mitigation sequencing process to offset the loss of wetland or other aquatic resources if adverse impacts remain after avoidance and minimization.

Connectivity—The extent of terrestrial habitat between wetlands that is suitable for overland migration by animals.

Constructed wetland—A created wetland that has been developed on a former upland environment with poorly drained soils for the primary purpose of contaminant or pollution removal from wastewater or runoff; also referred to as a treatment wetland.

Council on Environmental Quality (CEQ)—A federal agency that has promulgated NEPA regulations, which are binding on other federal agencies.

Cropped wetlands (CW)—Agricultural lands with hydric soils.

CWA—Clean Water Act.

CZMA—Coastal Zone Management Act.

EPA—U.S. Environmental Protection Agency.

Ephemeral wetland—A wetland in which water is present for short intervals between periods in which the wetland has no water.

ESA—Endangered Species Act.

Exotics—Nonnative species of plants or animals that have been introduced to a region.

Fish and Wildlife Coordination Act—Federal law that requires federal agencies to consult with the USFWS about habitat loss.

Food Security Act—Federal law that contains the Swampbuster provision.

FWPA—Freshwater Wetlands Protection Act.

FWPCA—Federal Water Pollution Control Act.

GAO—General Accounting Office.

General permit—Regional, nationwide, and programmatic permits issued by the Corps pursuant to Section 404(e) for activities that should result in only minimal individual and cumulative impacts.

GIS—Geographic information system.

HGM—Hydrogeomorphic.

Hydric soils—Soils that remain wet at least part of the year and that support hydrophytic vegetation.

Hydrogeomorphic (HGM)—Referring to a wetland's location in the landscape, the source of water for the wetland, and the hydrodynamics of the system.

Hydrological conditions—Characteristics of a habitat in terms of the structure and dynamics of the water.

Hydroperiod—A measure of the aquatic character of a wetland based on the timing and duration of water in the wetland during designated intervals.

Hydrophytic vegetation—Plants associated with wetland soils.

Individual permit—Standard permit or letter of permission issued by the Corps pursuant to Section 404 of the CWA.

In-kind compensation—Creation, restoration, enhancement, or preservation of wetlands similar to those being impacted.

In-lieu fee—Defined by the Corps and EPA as a payment "to a natural resource management entity for implementation of either specific or general wetland or other aquatic resource development projects" for projects that "do not typically provide compensatory mitigation in advance of project impacts."

Invasive species—Species of nonnative plants or animals that have become established in a region and are detrimental to one or more indigenous species.

Management-oriented planning—Process in a watershed-based approach involving in-lieu fees in the planning of wetlands to solve existing management problems.

MBRT—Mitigation Banking Review Team.

Metapopulation—Suite of populations of a species among which genetic exchange occurs in the landscape.

Minimization—Second most desirable of the sequencing steps in wetland mitigation, in which an activity that cannot avoid some impact on wetlands is designed in a manner to have minimal impact.

Mitigation—Avoiding, minimizing, rectifying, reducing, or compensating for resource losses.

Mitigation banking—Defined by the Corps and EPA as restoring, creating, enhancing, or preserving wetlands and other aquatic resources for purposes of providing compensatory mitigation in advance of authorized impacts to similar resources at another site.

Mitigation performance standards—Measurable, legally enforceable special conditions contained in a Section 404 permit that will lead to successful completion of the overall mitigation goals and objectives.

MOA—Memorandum of Agreement.

Mycorrhizae—Soil fungi that enable and enhance nutrient uptake by plant roots.

National Environmental Policy Act (NEPA)—Federal law that requires federal agencies to consider the environmental impacts of their proposed actions.

National Pollutant Discharge Elimination System (NPDES)—A federal permit program under the CWA pertaining to discharges of pollutants other than dredge-and-fill material.

NEPA—National Environmental Policy Act.

NMFS—National Marine Fisheries Service.

NPDES—National Pollutant Discharge Elimination System.

NRCS—Natural Resources Conservation Service, formerly the Soil Conservation Service.

NWP—Nationwide permit.

Out-of-kind compensation—Restoration, creation, enhancement, or preservation of wetlands that provide different functions than those of wetlands being adversely affected by a project.

PCN—Preconstruction notification.

PN—Public notice.

Prior converted cropland (PC)—A delineation of agricultural land with hydric soil that was planted to a crop at least once between 1983 and 1985 and was previously drained at an intensity consistent with the local NRCS standards.

Programmatic general permit (PGP)—Under a PGP issued by the Corps, a state or local government assumes primary responsibility for the issue of dredge-and-fill permits.

Protection-oriented planning—Process in a watershed-based approach involving the identification of high-quality wetlands that will be used for protection.

Rivers and Harbors Acts (RHAs)—Federal laws enacted in the 1890s that regulate work conducted in traditionally navigable waters.

SAMP—Special Area Management Plan.

SCWRP—Southern California Wetland Recovery Project.

Section 10 of the Rivers and Harbors Act of 1899—One of the earliest regulations governing the placement of dredge-and-fill material in waters of the United States.

Section 404—Provision of the CWA that governs the discharge of dredge-and-fill material into waters of the United States, including many wetlands.

Section 404 permit of the CWA—Permit issued by the Corps, or a state with an EPA-approved program, that authorizes the discharge of dredged or fill material into waters of the United States, including many wetlands.

Section 404(b)(1) of the CWA—Provision of the CWA that authorizes the EPA to promulgate guidelines for the Corps to apply when making Section 404 permit decisions.

Sequencing—Permit review process for mitigation that involves the consideration of mitigation in three steps: (1) seeking avoidance of the impact, (2) minimizing any unavoidable impacts, and (3) compensating for any remaining impacts.

Soil macrofauna—Animals living in salt marsh sediments on which wading birds feed by probing into the bottom of shallow water habitat.

SOPs—Standard Operating Procedures.

Source-sink—Concept in which populations forming a metapopulation contribute disproportionately to the numbers of individuals comprising various populations.

Southern California Wetland Recovery Project (SCWRP)—A partnership of public agencies that work to acquire, restore, and enhance coastal wetland and watersheds between Point Conception and the U.S.-Mexico border.

Swampbuster—Provision of the FSA under which landowners who planted an agricultural commodity crop in wetlands that had not been cropped prior to December 23, 1985, may lose USDA benefits.

USACE—U.S. Army Corps of Engineers.

USDA—U.S. Department of Agriculture.

USFWS—U.S. Fish and Wildlife Service.

USGS—U.S. Geological Survey.

Watershed—Land area that drains into a stream, river, or other body of water.

Wetland—An ecosystem that depends on constant or recurrent shallow inundation or saturation at or near the surface of the substrate.

Wetland creation—The conversion of a persistent upland or open water area into a wetland by human activity at a site where a wetland did not previously exist.

Wetland enhancement—An increase in one or more functions of an existing wetland by human modification.

Wetland preservation—Protection of an already existing and well-functioning wetland; wetland preservation has been used in some instances as mitigation for the loss of wetlands elsewhere.

Wetland restoration—Return of a wetland from a disturbed or altered condition by human activity to a previously existing condition.

Wetland Restoration Fund (WRF)—Funds for initiating the planning of a wetland restoration program in North Carolina.

WRP—National Agricultural Wetland Reserve Program.

Index

A

Abiotic components, 299
Activity, completing compensatory mitigation before permitting, 7, 139, 167
Advance identification (ADID), 146, 299
Agency technical capacity, 158–160
Agricultural uses, wetland losses due to, 57–58
Alnus, 32
Amphibians, major component of wetland biodiversity, 40
Analyses
  of 404(b)(1) guidelines, for processing Department of the Army permit applications, 253–254
  of compliance for 17 mitigation projects with field investigation in Western Washington, 120
  scope of, in processing Department of the Army permit applications, 241–242
  of soil, plant, and animal communities for mitigation sites compared with reference sites, 211–216

Animal communities present in a wetland, 39–40
  analyses of, for mitigation sites compared with reference sites, 211–216
Animal dispersal corridors, in watersheds, wetlands as, 51–53
Application, in processing Department of the Army permit, 247
Approaches
  floristic, 129–130
  to the nationwide permit process, 77
  third-party compensation, 9, 93, 168
  watershed, 3–5, 45, 59, 140–149, 273
Area basis compliance, for mitigation that was attempted based on field inspection or monitoring reports, 119
Army Corps of Engineers. *See* U.S. Army Corps of Engineers
Assessment of function, over broad range of performance conditions, 136
Atlantic coast, coastal wetlands on, 41
Authority, discretionary, in processing Department of the Army permit applications, 246
Avoidance, 299

B

Baltimore District's Guidelines, 227–228
Basin Wetlands and Riparian Restoration
    Plan, 210
Biological dynamics, evaluating in terms of
    regional reference models, 5, 45
Biological opinion (BO), 257–258
BO. *See* Biological opinion
Board on Environmental Studies and
    Toxicology, 20
Bogs, 26–27
Bush, George W., 156

C

California Coastal Act, 299
California Department of Fish and Game,
    South Coast Region, guidelines for
    wetland mitigation, 217–218
California Department of Transportation,
    203, 205
Carolina bays, 52
Case studies, 199–210
    Coyote Creek mitigation site, 201–208
    Everglades National Park, 34, 199–201
    North Carolina Wetland Restoration
        Program, 208–210
CEQ. *See* Council on Environmental
    Quality
Channelization, 58
Chicago District Mitigation Guidelines,
    229–230
Cladium jamaicence, 30
Clean Water Act (CWA), 1–4, 6, 60, 299. *See
    also* Section 404 permits
    objective of, 11–13, 15, 53, 240
Clinton Administration Wetland Plan, 145
Coastal wetlands, on the Atlantic and Gulf
    of Mexico coasts, 41
Coastal zone management, in processing
    Department of the Army permit
    applications, 255–257
Coastal Zone Management Act (CZMA),
    252, 255–256, 264–265
Code of Federal Regulations, 65
Commanders, memorandum for, 234–238
Commercial mitigation banks, 86
Committee on Mitigating Wetland Losses,
    2, 12–13, 20–21

Community structure, setting goals
    addressing, 7, 45
Comparative studies, of mitigation and
    natural wetlands, 189–198
Compensation
    in-kind, 301
    out-of-kind, 302
    permittee-responsible, 8, 167
    self-sustaining, 53–57
    third-party, 9, 93, 168
Compensation wetland planning, 146–147,
    300
Compensatory mitigation, 300
    in California, parameters measured in,
        107
    completing before permitting activity,
        7, 139, 167
    defined, 14
    designing and constructing individual
        sites on watershed scale, 7, 139, 167
    establishing long-term stewardship for,
        8, 168
    guidelines for implementing, 9, 93
    initiation of, 150
    performance standards from selected
        Section 404 permits requiring, 222
    in processing Department of the Army
        permit applications, 265–268
Compensatory mitigation mechanisms
    under Section 404, 82–93
    legal responsibility for the mitigation,
        86–88
    location of the compensatory mitigation
        action, 83–86
    MBRT process, 91
    recommendation, 93
    relationship of mitigation actions to
        permitted activities, 88–91
    stewardship requirements, 91–92
    a taxonomy, 92
Compliance, 94–122
    based on area, for mitigation that was
        attempted based on field inspection
        or monitoring reports, 119
    based on permit number, for when the
        mitigation plan was fully
        implemented, 118
    improving monitoring of, 8, 168
    inspection and enforcement of, 156–157
    with mitigation design standards, 97–
        101

in mitigation planning, 95–97
with mitigation ratios, 108–110
monitoring duration of, 112–113
monitoring of, 110–112
in processing Department of the Army
    permit applications, 270
in project implementation, 101–103
recommendations, 122
record of, 113–121
with requirements for permittee-
    responsible compensation, 8, 167
Compliance with permit conditions, 103–108
    design standards and detailed
        performance standards, 104–108
Compliance workload terms, 282–284
Conditioning permits, in processing
    Department of the Army permit
    applications, 264–265
Conduit for groundwater, soil in wetlands
    as, 32
Connectivity, 300
Conservation. *See* Wetland conservation
Constructed salt marshes
    at a mitigation site in North Carolina, 42
    in natural sites in Paradise Creek,
        Southern California, 115
    in San Diego Bay, long-term data for, 43
Constructed wetlands, 300
    defined, 13
Contaminants, from soil in wetlands, 32
Continuous parametric scale, creating, 136
Corps of Engineers. *See* U.S. Army Corps
    of Engineers
Corridors. *See* Animal dispersal corridors
Council on Environmental Quality (CEQ),
    62, 266, 300
Cowardin system. *See* Wetland
    classification system
Coyote Creek mitigation site, 201–208
    duration of monitoring, 205–207
    long-term success criterion, 205
    monitoring and site development, 205–
        207
    monitoring parameters, 204
    short-term success criterion, 204–205
    site installation and postinstallation site
        review, 205
Craven County, N.C., comparison between
    observed and DRAINMOD
    simulated water-table depths for a
    wetland restoration site in, 55

Creation of wetlands, 22–27. *See also*
    Constructed wetlands
    wetland types that are difficult to
        create, 24–27
    wetland types that have been created,
        22–24
Creation of wetlands that are ecologically
    self-sustaining, 123–128
    adopting a dynamic landscape
        perspective, 124–125
    attending to subsurface conditions, 127
    avoiding overengineered structures in
        the wetland's design, 126
    choosing wetland restoration over
        creation, 125–126
    conducting early monitoring, 128
    considering complications in degraded
        or disturbed sites, 128
    considering the hydrogeomorphic and
        ecological landscape and climate,
        123–124
    incorporating appropriate planting
        elevation, depth, soil type, and
        seasonal timing, 126–127
    providing appropriately heterogeneous
        topography, 127
    restoring or developing naturally
        variable hydrological conditions, 125
Credits, wetland, 67
Cropped wetlands (CW), 300
CWA. *See* Clean Water Act
CWA Section 404 program. *See* Section 404
    permits
CZMA. *See* Coastal Zone Management Act

D

DA. *See* Department of the Army
Data
    long-term, for salt marshes constructed
        in San Diego Bay, 43
    quality assurance measures for entry of,
        3, 122
Degraded sites, 44
    considering complications in, 128
Denials, 278
Denitrification, 27
Department of the Army (DA) permit
    applications, policies and
    procedures for processing, 240–271

Depth factors, incorporating as appropriate, 126–127
Design reference manual, developing to help projects achieve permit requirements, 8, 168
Design standards, 104–108
Discretionary authority, in processing Department of the Army permit applications, 246
Dispersal corridors. *See* Animal dispersal corridors
District commands, memorandum for, 234–238
Disturbed sites, considering complications in, 128
Documentations, EA/SOF/guidelines compliance, in processing Department of the Army permit applications, 258–264
DRAINMOD (hydrological model) simulated water-table depths, compared with observed water-table depths, 55
Duration
of inundation or saturation, 29
of monitoring, 112–113
of monitoring, at the Coyote Creek mitigation site, 205–207
of permits, in processing Department of the Army permit applications, 268
Dynamic landscape perspective, adopting, 124–125

E

EA. *See* Environmental assessment
Early monitoring, conducting, 128
Echinochloa crusgalli, 230
Ecological functionality
percentage of permits meeting various tests of, 117
of small, isolated wetlands, 52
Ecological parameters
landscape and climate, 123–124
in paired replacement and reference wetlands, 116
Ecoregional perspectives
in setting wetland project priorities, encouraging states to use, 9, 167
on where a wetland occurs, 38

Effect on Wetlands review, factors considered in, 293
EIS. *See* Environmental Impact Statements
Emergency procedures, 245
Endangered Species Act (ESA)
general mitigation requirements of, 62, 101, 249
in processing Department of the Army permit applications, 257–258
Enforcement
in processing Department of the Army permit applications, 270
in regulatory program priorities, 274
Enforcement workload terms, 282–284
Enhancement. *See* Wetland enhancement
Environmental assessment (EA), 245, 249
Environmental Impact Statements (EIS), 261–263
Environmental Protection Agency. *See* U.S. Environmental Protection Agency
EPA. *See* U.S. Environmental Protection Agency
Ephemeral wetlands, 300
Equivalency, functional, for constructed salt marshes in relationship to natural sites in Paradise Creek, Southern California, 115
ESA. *See* Endangered Species Act
Evaluation days, 278–280
Evaluation workload terms, 276–278
Everglades National Park, case study at, 34, 199–201
Excavation Rule, 243
Exotics, 300
Expectations for the permittee, 149–154
initiation of compensatory mitigation projects, 150
monitoring for performance, 151–152
permit compliance conditions for permittee-responsible mitigation, 153–154
permit conditions, 150–151
transfer of long-term responsibility, 152–153
Expectations for the regulatory agency, 154–160
agency technical capacity, 158–160
compliance inspection and enforcement, 156–157
long-term stewardship and management, 157–158

recognizing temporal lag, 155
recognizing watershed needs, 154–155

F

Federal actions, regarding wetland permit and mitigation requirements, 61
Federal Register, 69–70, 74, 76
Federal Water Pollution Control Act (FWPCA), 64
Federal wetland program, expanding state wetland programs to fill gaps in, 9, 168
FEDR. *See* Florida Department of Environmental Regulation
Fens, 25–26
Field inspection, area basis compliance for mitigation that was attempted based on, 119
File maintenance, in processing Department of the Army permit applications, 270–271
Financial assurances, ensuring for long-term site sustainability, 7, 139, 167
Findings, 1–10
    advantages of third-party compensation approaches, 9, 93, 168
    advantages of watershed approach, 3–5, 45, 59
    goal of no-net-loss-of-wetlands, 2–3, 122, 168
    inadequate support for regulatory decision making, 8–9, 167–168
    problems with Section 404 permits, 45, 137, 139, 167–168
Findings of no significant impact (FONSI), 259, 261
Fish and Wildlife Coordination Act, 63–64, 74, 300
    general mitigation requirements of, 61
Fish and Wildlife Service. *See* U.S. Fish and Wildlife Service (USFWS)
Flood-control practices, wetland losses due to, 58
Florida Department of Environmental Regulation (FEDR), 125–126
Floristic Quality Assessment, 130
FONSI. *See* Findings of no significant impact
Food Security Act (FSA), 1, 17, 300
    general mitigation requirements of, 62

Forested wetlands, 23
Forms of permits, in processing Department of the Army permit applications, 244–245
40 CFR 230, 299
Frequency of monitoring for permits that required mitigation, 111
Freshwater emergent marshes, 22–23
Function. *See* Wetland functions
Functional assessment, 132–136
    of all recognized functions, 136
    creating a continuous, parametric scale, 136
    defined, 14
    of function over broad range of performance conditions, 136
    including reliable indicators of important wetland processes, 136
    integrating over space and time, 136
    selected attributes of 40 common procedures, 285–291
Functional equivalency, for constructed salt marshes, in relationship to natural sites, 115
Functionality, percentage of permits meeting various tests of, 117
Funds, for staff professional development, committing, 8, 168
FWPCA. *See* Federal Water Pollution Control Act
FWS. *See* U.S. Fish and Wildlife Service

G

General Accounting Office (GAO), 162
General permits (GP), 76, 244, 300
Geographical information system (GIS) data, 48, 52
Germination medium, soil in wetlands as, 32
GIS. *See* Geographical information system data
Global positioning system (GPS) technology, 48
Goals
    addressing both community structure and wetland functions setting, 7, 45
    of no-net-loss-of-wetlands, 2–3, 122–137, 168
GP. *See* General permits

GPS. *See* Global positioning system technology
Groundwater withdrawals, wetland losses due to, 58
Gulf of Mexico (GOM) coast, coastal wetlands on, 41

H

Habitat evaluation procedures (HEPs), 131
Habitat support
    for fauna, 31
    for mycorrhizae and symbiotic bacteria, 32
    for soil macrofauna, 32
Headquarters, Operations, Construction, and Readiness Division, 19, 83
HEPs. *See* Habitat evaluation procedures
Herbaceous wetlands, 22–23
Heterogeneous topography, providing, 127
HGM. *See* Hydrogeomorphic Method
Hole-in-the-Donut, 34, 199–201
Hydric soils, 300
Hydrogeomorphic Method (HGM), 114–115, 131–136, 159–160, 301
    landscape and climate, 123–124
Hydrological continuum, 29
Hydrological function of wetlands, 28–29, 35–36, 301
    restoring or developing naturally variable, 125
Hydrological variability, incorporating into wetland mitigation design and evaluation, 5, 45, 135
Hydrology, effect of wetland function and position in the watershed on, 48–49
Hydroperiods, 301
Hydrophytic vegetation, 301

I

Impact sites
    area permitted, as a result of permits issued by the U.S. Army Corps of Engineers, 19
    evaluating with same tools as mitigation sites, 7, 137
In-kind compensation, 301
In-lieu fees, 87, 301

Inadequate support, for regulatory decision making, 8–9, 167–168
Indicators of important wetland processes, reliability of, 136
Individual permits, 301
Installation review, at the Coyote Creek mitigation site, 205
Institutional reforms for enhancing compensatory mitigation, 138–168
    expectations for the permittee, 149–154
    expectations for the regulatory agency, 154–160
    guidelines for, 9, 168
    improvements in permittee-responsible mitigation, 149–154
    introduction, 138–140
    recommendations, 166–168
    support for increased state responsibilities, 165–166
    third-party mitigation, 160–164
    watershed-based approach to compensatory mitigation, 140–149
Interagency Wetland Plan, 146
Internal coordination, in processing Department of the Army permit applications, 250
Inundation, duration and timing of, 29
Invasion of *Schinus terebinthifolius*, conceptual model of factors facilitating the, 200
Invasive species, 301
Iva frutescens, 228

J

Juncus roemerianus, 228
Jurisdictional issues, 53
    in processing Department of the Army permit applications, 242–243

L

Legal assurances, ensuring for long-term site sustainability, 7, 139, 167
Legal compliance, defined, 15
Legal responsibility for the mitigation, under Section 404, 86–88
Letter of permissions (LOPs), 244–245
Local watershed plans (LWPs), 208–210

Location of the compensatory mitigation action, under Section 404, 83–86
Long-term data, for salt marshes constructed in San Diego Bay, 43
Long-term effects of wetland creation, enhancement, and restoration, research into, 9, 168
Long-term responsibility, transfer of, 152–153
Long-term site sustainability, ensuring legal and financial assurances for, 7, 139, 167
Long-term stewardship
  establishing for compensatory mitigation sites, 8, 168
  and management, 157–158
LOPs. *See* Letter of permissions
Los Angeles District's Proposed Guidelines for Riparian Habitat, 229
Losses, by cause and acres lost, 18
Losses of wetland area and functions. *See also* No-net-loss-of-wetlands goal
  due to agricultural uses, 57–58
  due to flood-control practices, 58
  due to groundwater withdrawals, 58
  due to urbanization, 57
  tracking, 3, 122
Ludwigia peploides, 206
LWPs. *See* Local watershed plans
Lythrum salicaria (purple loosestrife), 30, 230

M

Major subordinate, commands for, 234–238
Management-oriented wetland planning, 145–146, 301
Manual. *See* Design reference manual
Marine Protection, Research, and Sanctuaries Act, 240, 243
Massachusetts, ecological parameters in paired replacement and reference wetlands in, 116
MBRT. *See* Mitigation Banking Review Team
Measurable performance standards, in permits, writing, 7, 122
Measured parameters, in compensatory wetland mitigation projects in California, 107

Memorandum for commanders, major subordinate commands, and district commands, April 8, 1999, 101, 234–238
Memorandum of Agreement (MOA), 12, 65, 71, 90–92, 108, 141, 241
Memorandum of Understanding (MOU), 63, 160, 162–163
Metapopulations, 52, 301
Method for Assessment of Wetland Function (MDE method), 133
Minimization, 301
Minnesota Routine Assessment Method for Evaluating Wetland Functions, 133
Mitigation, 301
  approaching on a watershed scale, 4, 59
  California Department of Fish and Game, South Coast Region guidelines for, 217–218
  design standards for, 97–101
  federal actions regarding, 61
  incorporating hydrological variability into design and evaluation, 5, 45
  initiating, 102
  permittee-responsible, 149–154
  proposed, 95, 97
  in regulatory program priorities, 275
  relationship to permitted activities under Section 404, 88–91
  required as a result of permits issued by the U.S. Army Corps of Engineers, 19
  sequencing, 66
  specified but not carried out, 101–103
Mitigation banking, 301
Mitigation Banking Review Team (MBRT) process under Section 404, 68–70, 82, 88, 91, 151, 160–164
Mitigation compliance, 94–122
  with design standards, 97–101
  keeping record of, 113–121
  with permit conditions, 103–108
  in planning, 95–97
  and project implementation, 101–103
  recommendations, 122
Mitigation Memorandum of Agreement (MOA), 12, 65, 71, 125
Mitigation monitoring, 110–112
  at the Coyote Creek mitigation site, 204–207
  of duration, 112–113

Mitigation performance standards, 301
Mitigation permits, with special conditions, 101
Mitigation planning, 95–97
  area to be lost and proposed mitigation, 95, 97
Mitigation projects, monitoring of, 110–112
Mitigation ratios, 108–110
  and achievement rates for different wetland types in southern California, 109–110
Mitigation requirements, 61–63
  defined, 15
  Endangered Species Act, 62
  Fish and Wildlife Coordination Act, 61
  Food Security Act, 62
  National Environmental Policy Act of 1969, 62
Mitigation site performance
  ecoregion in which a wetland occurs, 38
  factors that contribute to, 35–45
  hydrological regime, 35–36
  kinds of animals present, 39–40
  kinds of plants present, 38–39
  time factors, 40–44
  wetland place in the landscape, 37–38
  wetland size, 36–37
Mitigation sites
  compared with reference sites, analyses of soil, plant, and animal communities for, 211–216
  evaluating with same tools as impact sites, 7, 137
Mitigation wetlands, making self-sustaining, 4–5, 45
MOA. *See* Memorandum of agreement
Monitoring
  conducting early, 128
  duration of, 112–113, 205–207
  frequency of, for permits that required mitigation, 111
  for performance, 151–152
Monitoring reports, attempting area-based compliance with mitigation based on, 119
MOU. *See* Memorandum of Understanding
Mycorrhizae, 302
  habitat for, 32

N

National Environmental Policy Act (NEPA), 248, 254, 257, 302
  general mitigation requirements of, 62
National Historic Preservation Act (NHPA), 252, 257
National Marine Fisheries Service (NMFS), 62, 249
National Mining Association v. *U.S. Army Corps of Engineers*, 72
National Oceanic and Atmospheric Administration (NOAA), 50, 68, 70
National Pollutant Discharge Elimination System (NPDES), 302
National Research Council (NRC)
  Committee on Mitigating Wetland Losses, 2, 12–13, 20–21
  defining wetland hydrology, 35
Nationwide permits (NWPs), 70, 76–79, 267
  approach to process for, 77
  listing of current, 78
Natural recruitment, seeding versus, 39
Natural Resources Conservation Service (NRCS), 50, 68, 131
Natural Resources Defense Council v. *Callaway*, 165
Naturally variable hydrological conditions, 125
Nature Conservancy, 87
NCWRP. *See* North Carolina Wetland Restoration Program
NEPA. *See* National Environmental Policy Act
New England District's Guidelines, 226–227
NHPA. *See* National Historic Preservation Act
Nitrate reduction, 50
NMFS. *See* National Marine Fisheries Service
No-net-loss-of-wetlands goal, 2–3, 122–137, 168, 217
  establishing watershed organizations for tracking, monitoring, and managing wetlands, 3, 168
  expanding and improving quality assurance measures for data entry, 3, 122
  floristic approach, 129–130

habitat evaluation procedures and the hydrogeomorphic approach, 131–132

HGM as a functional assessment procedure, 132–136

operational guidelines for creating or restoring wetlands that are ecologically self-sustaining, 123–128

recommendations, 136–137

and the Section 404 program, 16–20

technical approaches toward achieving, 123–137

tracking wetland area and functions lost and regained, 3, 122

wetland functional assessment, 128–129

NOAA. *See* National Oceanic and Atmospheric Administration

Nonriverine systems, 28

Norfolk District's Guidelines, 227

North Carolina, created salt marsh constructed as a mitigation site in, 42

North Carolina Wetland Restoration Program (NCWRP), 147, 208–210

NPDES. *See* National Pollutant Discharge Elimination System.

NRCS. *See* Natural Resources Conservation Service

NWPs. *See* Nationwide permits

O

Observed water-table depths, compared with DRAINMOD simulated water-table depths, 55

Ohio, permit conditions and compliance for replacement wetlands investigated in, 114

Oligotrophic conditions, 26

Operational guidelines for creating or restoring wetlands that are ecologically self-sustaining, 123–128

adopting a dynamic landscape perspective, 124–125

attending to subsurface conditions, 127

avoiding overengineered structures in the wetland's design, 126

choosing wetland restoration over creation, 125–126

conducting early monitoring, 128

considering complications in degraded or disturbed sites, 128

considering the hydrogeomorphic and ecological landscape and climate, 123–124

incorporating appropriate planting elevation, depth, soil type, and seasonal timing, 126–127

providing appropriately heterogeneous topography, 127

restoring or developing naturally variable hydrological conditions, 125

Organizations. *See* Watershed organizations

Out-of-kind compensation, 302

Outcomes of wetland restoration and creation, 22–45

factors that contribute to the performance of mitigation sites, 35–45

five wetland functions, 27–34

possibility of restoring or creating wetland structure, 22–27

recommendations, 45

replaceability of wetland functions, 27

Overengineered structures, avoiding in the wetland's design, 126

P

Paired replacement and reference wetlands, ecological parameters in, 116

Palustrine nonriverine systems, 28

Paradise Creek, Southern California, functional equivalency for constructed salt marshes in relationship to natural sites in, 115

Parameters, measured in compensatory wetland mitigation projects in California, 107

Parametric scale, creating a continuous, 136

PC. *See* Prior converted cropland

PCNs. *See* Preconstruction notifications

PDNs. *See* Predischarge notifications

Peltandra virginica, 228

Percent loss, by cause and acres lost, 18

Percent plant cover, on created or restored coastal wetlands on the Atlantic and Gulf of Mexico coasts, 41

Performance conditions, assessing function over broad range of, 136
Performance of mitigation sites
    ecoregion in which a wetland occurs, 38
    factors that contribute to, 35–45
    hydrological regime, 35–36
    kinds of animals present, 39–40
    kinds of plants present, 38–39
    time factors, 40–44
    wetland place in the landscape, 37–38
    wetland size, 36–37
Performance standards
    an approach to developing, 219–233
    defined, 15
    detailed, 104–108
    in permits, writing measurable, 7, 122
    from selected Section 404 permits requiring compensatory mitigation, 222
    for wetland creation and restoration in Section 404 permits, 219–233
Permit applications. See Processing Department of the Army permit applications
Permit conditions, 150–151
    and compliance for replacement wetlands investigated in Ohio, 114
    compliance with, 103–108
    for permittee-responsible mitigation, 153–154
Permit evaluation
    in processing Department of the Army permit applications, 250–253
    in regulatory program priorities, 273
Permit modifications, in processing Department of the Army permit applications, 269
Permit number, compliance based on, 118
Permit process, approach to the nationwide, 77
Permit-specific mitigation, 88
Permits, percentage meeting their requirements and percentage meeting various tests of ecological functionality or viability, 117
Permits issued by the U.S. Army Corps of Engineers regulatory program, results of, 19
Permitted activities (timing), relationship to mitigation actions under Section 404, 88–91

Permittee compensation, using watershed perspective in establishing, 7, 167
Permittee expectations, 149–154
    initiation of compensatory mitigation projects, 150
    monitoring for performance, 151–152
    permit compliance conditions for permittee-responsible mitigation, 153–154
    permit conditions, 150–151
    transfer of long-term responsibility, 152–153
Permittee-responsible compensation, enforcing clear compliance requirements for, 8, 167
Permittee-responsible mitigation, improvements in, 149–154
PGP. See Programmatic general permits
Phalaris arundinacea (reed canary grass), 30, 230
Phragmites australis/communis (giant reed grass), 30, 50, 230
Planning and measuring tools for wetland function, broadening, 7, 45
Plant communities
    analyses of, for mitigation sites compared with reference sites, 211–216
    present in a wetland, 38–39
Plant cover, on created or restored coastal wetlands on the Atlantic and Gulf of Mexico coasts, 41
Planting. See Wetland planting
Planting elevation, incorporating as appropriate, 126–127
PNs. See Public notices
Poa compressa, 230
P. pratensis, 230
Pogogyne abramsii (mesa mint), 25
Policies and procedures for processing Department of the Army applications, 240–271
    appropriate level of analysis, 404(b)(1) guidelines for, 253–254
    compensatory mitigation, 265–268
    complete application, 247
    conditioning permits, 264–265
    discretionary authority, 246
    documentations, EA/SOF/guidelines compliance, 258–264
    duration of permits, 268

Endangered Species Act, 257–258
enforcement/compliance, 270
file maintenance, 270–271
forms of permits, 244–245
internal coordination, 250
jurisdiction, 242–243
permit evaluation/public hearings, 250–253
permit modifications and time extensions, 269
pre-application meetings, 246–247
preparing public notices, 249–250
project purpose, 247–249
public interest determination, 254–255
reporting, 271
scope of analysis, 241–242
Section 401 certification and coastal zone management, 255–257
wetland delineations, 244
Postinstallation review, at the Coyote Creek mitigation site, 205
Prairie potholes, 49, 58
Pre-application meetings, in processing Department of the Army permit applications, 246–247
Preconstruction notifications (PCNs), 76–77
Predischarge notifications (PDNs), 279
Prior converted cropland (PC), 302
Problems with Section 404 permits, 6–8, 45, 137, 139, 167–168
assessing wetland function using scientific procedures, 7, 136–137
broadening wetland function planning and measuring tools, 7, 45
completing compensatory mitigation before permitting activity, 7, 139, 167
designing and constructing individual compensatory mitigation sites on watershed scale, 7, 139, 167
enforce clear compliance requirements for permittee-responsible compensation, 8, 167
ensuring legal and financial assurances for long-term site sustainability, 7, 139, 167
establishing long-term stewardship for compensatory mitigation sites, 8, 168

evaluating impact sites with same tools as mitigation sites, 7, 137
improving compliance monitoring, 8, 168
setting goals addressing both community structure and wetland functions, 7, 45
using a watershed perspective in establishing permittee compensation, 7, 167
writing measurable performance standards in permits, 7, 122
Procedures. *See* Policies and procedures for processing Department of the Army permit applications
Processing of Department of the Army permit applications, 240–271
appropriate level of analysis, 404(b)(1) guidelines, 253–254
compensatory mitigation, 265–268
complete application, 247
conditioning permits, 264–265
discretionary authority, 246
documentations, EA/SOF/guidelines compliance, 258–264
duration of permits, 268
Endangered Species Act, 257–258
enforcement/compliance, 270
file maintenance, 270–271
forms of permits, 244–245
internal coordination, 250
jurisdiction, 242–243
permit evaluation/public hearings, 250–253
permit modifications and time extensions, 269
pre-application meetings, 246–247
preparing public notices, 249–250
project purpose, 247–249
public interest determination, 254–255
reporting, 271
scope of analysis, 241–242
Section 401 certification and coastal zone management, 255–257
wetland delineations, 244
Professional development, commiting funds for, 8, 168
Program administration, in regulatory program priorities, 275
Programmatic general permits (PGP), 276, 302

Project design standard, defined, 15
Project implementation, 101–103
  mitigation permit with special
    conditions, 101
  mitigation specified but not carried out,
    101–103
Project purpose, in processing Department
  of the Army permit applications,
  247–249
Proposed mitigation, area to be lost and,
  95, 97
Protection-oriented wetland planning, 146,
  302
Public hearings, in processing Department
  of the Army permit applications,
  250–253
Public interest determination, in processing
  Department of the Army permit
  applications, 254–255
Public Interest review, factors considered
  in, 292
Public notices (PNs), in processing
  Department of the Army permit
  applications, preparing, 244,
  249–250
Public outreach, in regulatory program
  priorities, 274

Q

Quality assurance measures for data entry,
  expanding and improving, 3, 122
Quarterly Permit Data System (QPDS)
  definitions, 271, 276–284
  enforcement/compliance workload,
    282–284
  evaluation days, 278–280
  evaluation workload terms, 276–278
  staffing, 282
  workload items, 280–282

R

RAMS. *See* Regulatory Analysis and
  Management System database
Ranking of compliance for sites in San
  Francisco Bay that were issued
  Section 404 permits, 120
Rapid Assessment Procedure, 133

Recommendations. *See also* Findings
  for compensatory mitigation
    mechanisms under Section 404, 93
  for institutional reforms for enhancing
    compensatory mitigation, 166–168
  for mitigation compliance, 122
  for outcomes of wetland restoration and
    creation, 45
  for technical approaches toward
    achieving no-net-loss-of-wetlands
    goal, 136–137
  for watershed setting, 59
Record keeping, 121
Reference models, evaluating biological
  dynamics in terms of regional, 5, 45
Reference sites. *See also* Design reference
  manual
  compared with mitigation sites,
    analyses of soil, plant, and animal
    communities for, 211–216
Reference wetlands, paired, ecological
  parameters in, 116
Regained wetland area and functions,
  tracking, 3, 122
Regional general permits (RGP), 246, 276
Regional reference models, evaluating
  biological dynamics in terms of, 5,
  45
Regulatory agency expectations, 154–160
  agency technical capacity, 158–160
  compliance inspection and
    enforcement, 156–157
  long-term stewardship and
    management, 157–158
  recognizing temporal lag, 155
  recognizing watershed needs, 154–155
Regulatory Analysis and Management
  System (RAMS) database, 3, 121–122
Regulatory decision making
  commiting funds for staff professional
    development, 8, 168
  developing design reference manual to
    ensure projects are likely to achieve
    permit requirements, 8, 168
  encouraging states to use ecoregional
    perspectives in setting wetland
    project priorities, 9, 167
  inadequate support for, 8–9, 167–168
  researching long-term effects of
    wetland creation, enhancement, and
    restoration, 9, 168

Regulatory Guidance Letters (RGLs), 67,
    241
Regulatory program priorities, 272–275. *See
    also* Standard Operating Procedures
    (SOPs) for the Army Corps of
    Engineers regulatory program
    enforcement, 274
    mitigation, 275
    permit evaluation, 273
    public outreach, 274
    staff support/program administration,
        275
    watershed approaches, 273
Replacement wetlands
    paired, ecological parameters in,
        116
    permit conditions and compliance for,
        114
Reporting, in processing Department of the
    Army permit applications, 271
Required mitigation, as restoration,
    creation, and enhancement for
    permits issued under permitting
    programs, 96
Research, into long-term effects of wetland
    creation, enhancement, and
    restoration, 9, 168
Restoration. *See* Wetland restoration
Review of Corps permits issued
    nationwide, 98
Revised Quarterly Permit Data System
    (QPDS) definitions, 276–284
    enforcement/compliance workload,
        282–284
    evaluation days, 278–280
    evaluation workload terms, 276–278
    staffing, 282
    workload items, 280–282
RGLs. *See* Regulatory Guidance Letters
RGP. *See* Regional general permits
RHA. *See* Rivers and Harbors Act of 1899
Riparian wetlands, 48
    giving special attention and protection
        to, 5, 59
Rivers and Harbors Act of 1899 (RHA)
    Section 10 of, 63, 240, 243, 302
Robertson v. *Methow Valley Citizens Council*,
    62
Rooting medium, soil in wetlands as, 32

S

Salix, 32
S. interior, 230
Salt marshes, 23. *See also* Constructed salt
    marshes
San Diego Bay, salt marshes constructed in,
    43
Saturation, duration and timing of, 29
Schinus terebinthifolius (Brazilian pepper)
    deterring invasion by, 34
    factors facilitating invasion of, 199–200
Scientific procedures, assessing wetland
    function using, 7, 136–137
Scirpus spp., 50
S. robustus, 228
SCWRP. *See* Southern California Wetland
    Recovery Project
Sea-level rise, and wetlands placement, 56
Seagrasses, 23
Seasonal timing, incorporating as
    appropriate, 126–127
Seattle District's Guidelines for Freshwater
    Wetlands, 228–229
Section 401 certification, in processing
    Department of the Army permit
    applications, 255–257
Section 10 of the Rivers and Harbors Act of
    1899, 302
Section 404 permits, 2, 10, 15, 73–79, 303
    assessing wetland function using
        scientific procedures, 7, 136–137
    broadening wetland function planning
        and measuring tools, 7, 45
    compensatory mitigation mechanisms
        under, 82–93
    completing compensatory mitigation
        before permitting activity, 7, 139, 167
    designing and constructing individual
        compensatory mitigation sites on
        watershed scale, 7, 139, 167
    enforce clear compliance requirements
        for permittee-responsible
        compensation, 8, 167
    ensuring legal and financial assurances
        for long-term site sustainability, 7,
        139, 167
    establishing long-term stewardship for
        compensatory mitigation sites, 8,
        168

evaluating impact sites with same tools
    as mitigation sites, 7, 137
factors considered in the Effect on
    Wetlands reviews, 293
factors considered in the Public Interest
    review, 292
general permits, 76
implementing, 12
improving compliance monitoring, 8,
    168
performance standards for wetland
    creation and restoration in, 219–233
problems with, 6–8, 45, 137, 139, 167–
    168
processing flow chart for, 75
setting goals addressing both
    community structure and wetland
    functions, 7, 45
standard permits, 73–74
using a watershed perspective in
    establishing permittee
    compensation, 7, 167
writing measurable performance
    standards in permits, 7, 122
Section 404(b)(1) of the CWA, 65, 140, 303
    level of analysis guidelines, in
        processing Department of the Army
        permit applications, 253–254
Sedge meadows, 23
Seed banks, soil in wetlands as, 32
Seeding, versus natural recruitment, 39
Self-design, wetland planting aiding, 39
Self-sustaining compensation projects,
    watershed position and, 53–57
Self-sustaining mitigation wetlands, 4–5, 45
Sequencing, 66, 303
Shrub swamps, 23
Sierra Club v. Alexander, 61
Simulated water-table depths, by
    DRAINMOD, compared with
    observed water-table depths, 55
Site review, installation and
    postinstallation, at the Coyote Creek
    mitigation site, 205
SOF. See Statement of findings
Soil communities, analyses of, for
    mitigation sites compared with
    reference sites, 211–216
Soil Conservation Service. See Natural
    Resources Conservation Service

Soil functions in wetlands, 31–34
    conduit for groundwater, 32
    germination medium, 32
    habitat for mycorrhizae and symbiotic
        bacteria, 32
    habitat for soil macrofauna, 32
    rooting medium, 32
    seed bank, 32
    source of contaminants, 32
    source of water and nutrients for plants,
        32
    water-quality functions, 32
Soil macrofauna, 303
    habitat for, 32
Soil type, incorporating as appropriate,
    126–127
Solid Waste Agency of Northern Cook County
    v. U.S. Army Corps of Engineers, 12,
    71
SOPs. See Standard operating procedures
Source-sink, 303
Southern California Wetland Recovery
    Project (SCWRP), 146–147, 303
SP. See Standard permits
Spartina alterniflora (smooth cordgrass), 23,
    126, 228
S. foliosa, 24
S. patens, 228
Special conditions, mitigation permit with,
    101
Sphagnum moss, 26
St. Paul District's Guidelines, 226
Stable-water ponds, 106
Staff support
    commiting funds for professional
        development, 8, 168
    in regulatory program priorities, 275
Staffing terms, 282
Standard Operating Procedures (SOPs) for
    the Army Corps of Engineers
    regulatory program, 101–103, 148,
    151, 156, 239–284
    policies and procedures for processing
        Department of the Army permit
        applications, 240–271
    regulatory program priorities, 272–275
    revised Quarterly Permit Data System
        definitions, 276–284
Standard permits (SP), 73–74, 244–245
State responsibilities, support for
    increased, 165–166

State Revolving Loan Fund, 164
State wetland programs
    encouraging use of ecoregional
        perspectives in setting project
        priorities, 9, 167
    expanding to fill gaps in federal
        wetland program, 9, 168
Statement of findings (SOF), 245, 258
Stewardship requirements, under Section
    404, 91–92
Stream order, 47
Subordinates, commands for, 234–238
Subsurface conditions, attending to, 127
Success criteria at the Coyote Creek
    mitigation site
    long-term, 205
    short-term, 204–205
Support
    for increased state responsibilities, 165–
        166
    for regulatory decision making, 8–9,
        167–168
    of vegetation by wetlands, 30–31
Swampbuster program, 17, 303
Symbiotic bacteria, habitat for, 32

T

Taxonomy
    of compensatory mitigation
        mechanisms, 84
    under Section 404, 92
Technical approaches toward achieving no-
    net-loss-of-wetlands goal, 123–137
    floristic approach, 129–130
    habitat evaluation procedures and the
        hydrogeomorphic approach, 131–
        132
    HGM as a functional assessment
        procedure, 132–136
    operational guidelines for creating or
        restoring wetlands that are
        ecologically self-sustaining, 123–128
    recommendations, 136–137
    wetland functional assessment, 128–129
Temporal lag, recognizing, 155
Terminology, 13–16
    compensatory mitigation projects, 14
    constructed wetlands, 13
    functional assessment methods, 14

legal compliance, 15
mitigation requirements, 15
performance standard, 15
project design standard, 15
treatment wetlands, 13
watersheds, 15
wetland creation, 13
wetland enhancement, 13
wetland functions, 14
wetland preservation, 13–14
wetland restoration, 13
wetland structure, 14–15
wetland types, 14–15
wetlands, 13
Thalassia testudinum, 24
Third-party compensation approaches
    advantages of, 9, 93, 168
    expanding state wetland programs to
        fill gaps in federal wetland program,
        9, 168
    guidelines for implementing
        compensatory mitigation, 9, 93
    guidelines for modifying institutional
        systems for, 9, 168
Third-party mitigation, 160–164
33 CM 331, 299
Time factors
    extensions in processing Department of
        the Army permit applications, 269
    in wetland restoration and creation, 40–
        44
Timing
    incorporating appropriate seasonal,
        126–127
    of inundation or saturation, 29
    toward equivalency for soil, plant, and
        animal components in wetland
        restoration projects compared with
        that of natural reference wetlands,
        42
Topography, providing appropriately
    heterogeneous, 127
Topography-based flow models, 49
Tracking wetland area and functions
    lost and regained, 3, 122
Transportation Equity Act, 69
Treatment wetlands, defined, 13
Typha spp. (cattails), 30, 50, 230
T. augustifolia, 228
T. domingensis, 30

U

Unified Federal Policy for Ensuring a
    Watershed Approach to Federal
    Land and Resource Management,
    140
United States v. *Riverside Bayview Homes,
    Inc.,* 11
United States v. *Wilson,* 165
Urbanization, wetland losses due to, 57
U.S. Army Corps of Engineers (USACE), 1,
    7–9, 60–77, 80–81, 125
    enforcement chart for inspection and
        noncompliance, 81
    Headquarters, Operations,
        Construction, and Readiness
        Division, 19, 83
    policies and procedures for processing
        Department of the Army permit
        applications, 240–271
    Regulatory Analysis and Management
        System (RAMS) database, 3
    regulatory program priorities, 272–275
    results of permits issued by, 19
    revised Quarterly Permit Data System
        (QPDS) definitions, 276–284
    Standard Operating Procedures (SOPs)
        for the regulatory program, 239–284
    Wetland Delineation Manual, 29, 227
U.S. Department of Agriculture (USDA)
    Natural Resources Conservation
        Service, 50
    Swampbuster program, 17, 303
U.S. Environmental Protection Agency
    (EPA), 2, 9, 65–72, 83, 108, 125, 243
    Interagency Wetland Plan, 146
U.S. Fish and Wildlife Service (USFWS), 3,
    16–18, 20, 62, 249, 257–258
    wetland classification system of, 14, 133
U.S. Geological Survey (USGS), 48
U.S. Supreme Court, 11, 71

V

Values considered in Section 404 permit
    reviews, 292–293
Vernal pools, 25
Viability, percentage of permits meeting
    various tests of, 117

W

Washington State Department of
    Transportation, 230–231
Water and nutrients for plants, from soil in
    wetlands, 32
Water quality
    effect of wetland function and position
        in the watershed on, 49–51
    function of soil in wetlands, 32
    improving in wetlands, 29–30
Water Quality Certification (WQC), 255–
    256
Water Science and Technology Board, 20
Water-table depths, comparison between
    observed and DRAINMOD
    simulated, 55
Water-table position and duration of root
    zone saturation for wetland site that
    satisfies the jurisdictional hydrology
    criteria, 105
Watershed approach, 15, 46–59, 303
    advantages of, 3–5, 45, 59
    approaching wetland conservation and
        mitigation on a watershed scale, 4,
        59
    avoiding wetlands that are difficult or
        impossible to restore, 4, 45
    basing wetland restoration and creation
        on broad range of sites, 5, 45
    evaluating biological dynamics in terms
        of regional reference models, 5, 45
    giving special attention and protection
        to riparian wetlands, 5, 59
    implications of, 141–144
    incorporating hydrological variability
        into wetland mitigation design and
        evaluation, 5, 45
    making all mitigation wetlands self-
        sustaining, 4–5, 45
    recommendations, 59
    in regulatory program priorities, 273
    watershed organization and landscape
        function, 46–47
    watershed template for wetland
        restoration and conservation, 58–59
    wetland function and position in the
        watershed, 47–57
Watershed approach to compensatory
    mitigation, 140–149
    compensation wetland planning, 146–147

implications of the watershed approach,
141–144
management-oriented wetland
planning, 145–146
protection-oriented wetland planning,
146
Watershed needs, recognizing, 154–155
Watershed organizations, for tracking,
monitoring, and managing
wetlands, 3, 168
Watershed-scale perspective, 57–58
designing and constructing individual
compensatory mitigation sites on, 7,
139, 167
on losses due to agricultural uses, 57–58
on losses due to groundwater
withdrawals, 58
on losses due to urbanization, 57
using in establishing permittee
compensation, 7, 167
on wetland losses due to flood-control
practices, 58
Wegener Ring, 40
Western Washington, analyses of
compliance for 17 mitigation
projects with field investigation in,
120
Wet meadows, 23
Wet prairies, 23
Wetland area, anticipated gain as a result
of permits issued by the U.S. Army
Corps of Engineers, 19
Wetland biodiversity, amphibians a major
component of, 40
Wetland classification system, of the U.S.
Fish and Wildlife Service, 14, 133
Wetland conservation, approaching on a
watershed scale, 4, 59
Wetland creation, 303
defined, 13
research into long-term effects of, 9, 168
Wetland credits, 67
Wetland Delineation Manual, 29, 227
Wetland enhancement, 303
defined, 13
research into long-term effects of, 9, 168
Wetland functions, 12, 27–34
assessing using scientific procedures, 7,
136–137
broadening planning and measuring
tools, 7, 45

considered in Section 404 permit
reviews, 292–293
defined, 14
effect on hydrology, 48–49
effect on water quality, 49–51
groundwater recharge, 12, 27–29
habitat support for fauna, 31
hydrological function, 28–29
and position in the watershed, 47–57
provision of a unique environment, 12,
27
replaceability of, 27
setting goals addressing, 7, 45
shoreline stabilization, 12
soil functions, 31–34
support of vegetation, 30–31
water-quality improvement, 12, 27, 29–
30
water retention, 12
watershed position and self-sustaining
compensation projects, 53–57
wetlands as animal dispersal corridors
in watersheds, 51–53
Wetland hydrology, NRC definition of, 35
Wetland mitigation. *See* Mitigation
Wetland permits, 60–81
Clean Water Act and the goal of no-net-
loss-of-wetlands, 70–73
data on implementation, compliance,
ecological success, and monitoring
frequency, 121
evolution of compensatory mitigation
requirements in the Section 404
program, 60
federal actions regarding, 61
general Corps mitigation requirements,
63–64
general mitigation requirements, 61–63
in-lieu fees, 69–70
inspection and enforcement, 80
mitigation banking, 67–69
Section 404 mitigation requirements,
64–67
Section 404 permit process, 73–79
Wetland planning
management-oriented, 145–146
protection-oriented, 146
Wetland planting, aiding self-design, 39
Wetland preservation, 304
defined, 13–14

Wetland processes, reliable indicators of, 136
Wetland programs. *See* Federal wetland program; State wetland programs
Wetland restoration, 22–27, 304. *See also* Mitigation
  basing on broad range of sites, 5, 45
  choosing over creation, 125–126
  defined, 13
  factors that contribute to the performance of mitigation sites, 35–45
  outcomes of, 22–45
  possibility of restoring or creating wetland structure, 22–27
  recommendations, 45
  replaceability of wetland functions, 27
  research into long-term effects of, 9, 168
  site in Craven County, N.C., 55
  types that are difficult to restore, 24–27
  types that have been restored, 22–24
Wetland Restoration Fund (WRF), 209, 304
Wetland types
  bogs, 26–27
  defined, 14–15
  fens, 25–26
  forested wetlands, 23
  freshwater emergent marshes, 22–23
  herbaceous wetlands, 22–23
  salt marshes, 23
  seagrasses, 23
  sedge meadows, 23
  shrub swamps, 23
  that are difficult to restore or create, 24–27
  that have been restored and created, 22–24
  used in processing Department of the Army permit applications, 244
  vernal pools, 25
  wet meadows, 23
  wet prairies, 23
Wetlands, 303. *See also* Coastal wetlands; Constructed wetlands; Creation of wetlands; Cropped wetlands; Ephemeral wetlands; Forested wetlands; Herbaceous wetlands; Mitigation wetlands; No-net-loss-of-wetlands goal; Reference wetlands; Regained wetland area and functions; Replacement wetlands; Riparian wetlands; Treatment wetlands
  defined, 13
  giving special attention and protection to riparian, 5, 59
  losses due to agricultural uses, 57–58
  losses due to flood-control practices, 58
  losses due to groundwater withdrawals, 58
  losses due to urbanization, 57
  losses of, 17
  paired replacement and reference, ecological parameters in, 116
  place in the landscape, 37–38
  size of, 36–37
  structure of, 14–15
  that are difficult or impossible to restore, avoiding, 4, 45
Wetlands placement, sea-level rise and, 56
Wetlands restoration fund (WRF), 147
Workload terms, 280–282
  compliance, 282–284
  enforcement, 282–284
  evaluation, 276–278
WQC. *See* Water Quality Certification
WRF. *See* Wetland Restoration Fund; Wetlands restoration fund
Writing measurable performance standards, in permits, 7, 122

Y

Year-to-year variation
  in the longest period that wetland hydrological criteria are satisfied, 107
  in water-table depth and duration of root zone saturation, 106

Z

Zabel v. Tabb, 62
Zostera marina (eelgrass), 23